MICROBIOLOGICAL SYNTHESES

MICROBIOLOGICAL SYNTHESES

Recent Advances

Edited by S. Torrey

NOYES DATA CORPORATION
Park Ridge, New Jersey, U.S.A.
1983

Library of Congress Catalog Card Number: 83-2457
ISBN: 0-8155-0937-5
ISSN: 0198-6880
Printed in the United States

Published in the United States of America by
Noyes Data Corporation
Mill Road, Park Ridge, New Jersey 07656

10 9 8 7 6 5 4 3 2 1

Library of Congress Cataloging in Publication Data

Main entry under title:

Microbiological syntheses.

(Biotechnology review ; no. 1) (Chemical technology review ; no. 219)
Includes index.
1. Microbiological synthesis--Patents. 2. Antibiotics--Patents. 3. Microbial metabolites--Patents. I. Torrey, S. II. Series. III. Series: Chemical technology review ; no. 219. [DNLM: 1. Genetic intervention--Patents. 2. Microbiology--Patents. QW 4 M6253]
QR148.M53 1983 660'.62 83-2457
ISBN 0-8155-0937-5

FOREWORD

The detailed, descriptive information in this book is based on U.S. patents, issued between December 1980 and November 1982, that deal with microbiological synthesis and genetic engineering. This is the first volume in the Noyes biotechnology review series.

The book is a data-based publication, providing information retrieved and made available from the U.S. patent literature. It thus serves a double purpose in that it supplies detailed technical information and can be used as a guide to the patent literature in this field. By indicating all the information that is significant, and eliminating legal jargon and juristic phraseology, this book presents an advanced commercially oriented review of recent developments in the field of microbiological synthesis and genetic engineering.

The U.S. patent literature is the largest and most comprehensive collection of technical information in the world. There is more practical, commercial, timely process information assembled here than is available from any other source. The technical information obtained from a patent is extremely reliable and comprehensive; sufficient information must be included to avoid rejection for "insufficient disclosure." These patents include practically all of those issued on the subject in the United States during the period under review; there has been no bias in the selection of patents for inclusion.

The patent literature covers a substantial amount of information not available in the journal literature. The patent literature is a prime source of basic commercially useful information. This information is overlooked by those who rely primarily on the periodical journal literature. It is realized that there is a lag between a patent application on a new process development and the granting of a patent, but it is felt that this may roughly parallel or even anticipate the lag in putting that development into commercial practice.

Many of these patents are being utilized commercially. Whether used or not, they offer opportunities for technological transfer. Also, a major purpose of this book is to describe the number of technical possibilities available, which may open up profitable areas of research and development. The information contained in this book will allow you to establish a sound background before launching into research in this field.

Advanced composition and production methods developed by Noyes Data are employed to bring these durably bound books to you in a minimum of time. Special techniques are used to close the gap between "manuscript" and "completed book." Industrial technology is progressing so rapidly that time-honored, conventional typesetting, binding and shipping methods are no longer suitable. We have bypassed the delays in the conventional book publishing cycle and provide the user with an effective and convenient means of reviewing up-to-date information in depth.

The table of contents is organized in such a way as to serve as a subject index. Other indexes by company, inventor and patent number help in providing easy access to the information contained in this book.

16 Reasons Why the U.S. Patent Office Literature Is Important to You

1. The U.S. patent literature is the largest and most comprehensive collection of technical information in the world. There is more practical commercial process information assembled here than is available from any other source. Most important technological advances are described in the patent literature.

2. The technical information obtained from the patent literature is extremely comprehensive; sufficient information must be included to avoid rejection for "insufficient disclosure."

3. The patent literature is a prime source of basic commercially utilizable information. This information is overlooked by those who rely primarily on the periodical journal literature.

4. An important feature of the patent literature is that it can serve to avoid duplication of research and development.

5. Patents, unlike periodical literature, are bound by definition to contain new information, data and ideas.

6. It can serve as a source of new ideas in a different but related field, and may be outside the patent protection offered the original invention.

7. Since claims are narrowly defined, much valuable information is included that may be outside the legal protection afforded by the claims.

8. Patents discuss the difficulties associated with previous research, development or production techniques, and offer a specific method of overcoming problems. This gives clues to current process information that has not been published in periodicals or books.

9. Can aid in process design by providing a selection of alternate techniques. A powerful research and engineering tool.

10. Obtain licenses–many U.S. chemical patents have not been developed commercially.

11. Patents provide an excellent starting point for the next investigator.

12. Frequently, innovations derived from research are first disclosed in the patent literature, prior to coverage in the periodical literature.

13. Patents offer a most valuable method of keeping abreast of latest technologies, serving an individual's own "current awareness" program.

14. Identifying potential new competitors.

15. It is a creative source of ideas for those with imagination.

16. Scrutiny of the patent literature has important profit-making potential.

CONTENTS AND SUBJECT INDEX

INTRODUCTION

This book is a survey of over three hundred patents granted since January 1, 1981 and having to do with synthesis of various substances by microorganisms or by the enzymes formed by microorganisms. Included also is the subject of genetic engineering, which must certainly be considered today's most exciting area of scientific research.

Most syntheses by living organisms may be looked upon as fermentation—chemical changes induced in various substrates by microorganisms such as bacteria, yeasts, molds or fungi, or the enzymes which such microorganisms produce. Some fermentation processes have been known and used since the beginning of man's history. Fermentation is essential in the manufacture of alcoholic beverages and the preparation of most breads. It has been used for the recovery of copper from the drainage water of mines since 1000 B.C.

The bulk of the processes reviewed in this book deal with the formation of new antibiotics. Since of all the antibiotics synthesized since the discovery of penicillin very few are now in use, it may seem surprising that in the past two years so many patents have been granted for totally new or chemically modified antibiotics, but two things must be recognized in this field that make it necessary for the search for new antibiotics to continue. First of all, no antibiotic known has antibacterial activity for all clinically important pathogenic bacteria, particularly those of Gram-negative type. Secondly, the problem of acquired resistance of bacteria to existing antibiotics has become very serious. It may also be mentioned that certain antibiotics have been shown in in-vivo animal studies to exhibit potent antitumor and antileukemic characteristics. Other antibiotics have antifungal, anthelmintic and/or antimildew characteristics.

This book has been divided into ten chapters. The first four concern products made primarily by the fermentation of various strains of microorganisms. The products include antibiotics, other pharmaceuticals and certain organic compounds—primarily acids. A few products, such as interferon in the third chapter, are produced by cell culture rather than fermentation.

The fifth, sixth and seventh chapters deal primarily with enzymatically catalyzed reactions—for the preparation of drugs and organic chemicals (fifth chapter), carbohydrates and artificial sweeteners (sixth chapter) and ethanol and other fuels (seventh chapter).

It will be obvious that the patents could have been arranged other ways and that many of the explained processes fall into more than one of the listed categories. For example, most of the processes for fermentation of starches could be carried to completion for the production of ethyl alcohol.

The polysaccharides described in the eighth chapter are produced either by microorganisms or by enzyme fermentations. They are of a particular type, being gums or thickeners used to control the flow properties of aqueous systems. They are of particular interest at this time because of their use in oil well drilling and fracturing, although they have utility also as food and paint additives and in textile printing.

The ninth chapter contains descriptions of the treatment of waste materials by various microorganisms and a collection of miscellaneous processes which do not fit comfortably in any other chapter.

The last chapter is the longest and most complicated. It deals with the most discussed and least understood field of research since the splitting of the atom and is considered by some people to have almost as much potential for the benefit and the ruin of the human race. It was felt that this chapter must contain the words "genetic engineering," despite the fact that to some persons the words conjure up visions of the creation of Frankensteins or Supermen, since those who are involved in this work use the words as the overall term to describe their research.

Twenty-two of the thirty patents deal with specific procedures devised for use in recombinant DNA technology. The others concern specific products which can be made using such techniques. The first patent in the chapter is the only one in the book which was granted before January 1, 1981 and has been included because of its importance as an historical first.

The study of the molecular structure of genes and the knowledge which has been gained by these studies have already produced new tools for the diagnosis and treatment of the diseases of plants and animals, as will be shown by some of the patented processes herein.

Another of the patents in this chapter involves production of genetically engineered microorganisms which demonstrate remarkable improvement in their degradation of complex hydrocarbons. It is hoped that this type of microorganism may provide a solution to the terrible ecological problem of oil spills.

It may well be that, when the next edition of this volume is published, the major thrust of the research in microbial biosynthesis and the larger number of patents granted will be in this field now covered by only one chapter.

ANTIBIOTICS PRODUCED BY MICROORGANISMS

In most of the patents in this chapter, the morphological characteristics and physiological properties of the microorganism strain used for the fermentation are described in detail. The description may include their appearance, particularly their color characteristics; their growth characteristics on various media; their utilization of various carbon sources, etc. In many cases, especially when a new strain of microorganism is described, the source of the new strain is given and the steps taken to classify the genus and species of the strain are thoroughly delineated.

In most cases, the processes and procedures for isolation and purification of the compound made by the fermentation are also carefully described. The chemical and physical characteristics of the product antibiotic are also given in detail. Such information may include the empirical formula, elemental analysis, optical rotation measurement, solubilities, antibacterial activity both in vitro and in vivo, etc. In many cases the patent contains tables or graphs giving the infrared and ultraviolet absorption spectra of the product.

When the antibiotics are capable of forming acid addition salts, the caveat that such salts must be nontoxic and pharmaceutically acceptable is understood.

Because of the space limitations of this book, the above data are not usually included in the abstract given for each patent. If any process is of particular interest to the reader the patent is, of course, obtainable to provide the details.

Reference is also made in these abstracts to "conventional" or "normal" fermentation procedures. Such procedures consist of the cultivation of the microorganism under aerobic conditions, for example in a stationary surface culture or, preferably, in a submerged culture that is supplied with oxygen, normally in the form of atmospheric oxygen, by shaking and/or stirring in shaking flasks or fermenters of known construction. Suitable temperatures are between about 20° and 35°C, preferably 28°C. The cultures are usually carried out at a pH of from 6.0 to 8.0 which normally needs no adjustment during fermentation. The time for completion of the fermentation varies, but seems usually to be between two and seven days.

The nutrient medium used for the cultivation of the microorganism must contain assimilable carbon and nitrogen sources and essential mineral salts. As sources of carbon starch, disaccharides such as lactose and saccharose, monosaccharides like glucose, and suitable carbohydrate-containing industrial raw materials, such as molasses may be used. Examples of nitrogen sources are amino acids, peptides and proteins and their decomposition products, ammonium salts and nitrates, and suitable industrial nitrogen-containing raw materials, such as meat extracts, yeast autolysate and extract, casein hydrolysate, soy protein, cereal seed fractions, corn steep liquor, etc. Apart from ammonium salts and nitrates, the nutrient medium may contain as inorganic salts chlorides, carbonates, sulfates, and especially phosphates of alkali metals and alkaline earth metals, and also trace elements, such as magnesium, iron, zinc and manganese.

To avoid repetition, the abstracts do not usually specify the carbon and nitrogen sources used in the culture, unless a preferred source is used for the microorganism in question. Quite detailed and typical examples of the procedure used to cultivate and ferment the microorganisms to produce antibiotics are given in the first section of this chapter under the two headings, "Production of Rachelmycin" and "Antibiotic SM-173B."

In a further means of saving space and avoiding repetition, names and locations of the most commonly used repositories for microorganisms and enzymes, together with the abbreviations used for their accession numbers, are listed below rather than each time they are used in the book. Repositories which are used only once or twice are fully identified in the patent.

Accession Number Prefix	Repository
ATCC	American Type Culture Collection, Rockville, Maryland
NRRL	Northern Regional Research Laboratory, U.S. Department of Agriculture, Peoria, Illinois
FERM	Fermentation Research Institute, Agency of Industrial Science and Technology, Chiba City, Japan
DSM	Deutsche Sammlung von Mikroorganismen, Goettingen, West Germany
NCIB	National Collection of Industrial Bacteria, Torry Research Station, Aberdeen, Scotland
IFO	Institute for Fermentation, Osaka, Japan
CBS	Centraalbureau voor Schimmel-cultures, Baarn, The Netherlands
ICPB	International Collection of Phytopathogenic Bacteria, University of Caliornia at Davis, California

BY STREPTOMYCES SPECIES

Northienamycin

A.J. Kempf and K.E. Wilson; U.S. Patent 4,247,640; January 27, 1981; assigned to Merck & Co., Inc. describe a fermentation process for preparing 6-hydroxymethyl-2-(2-aminoethylthio)-1-carbadethiapen-2-em-3-carboxylic acid which is useful as an antibiotic, and which has the formula:

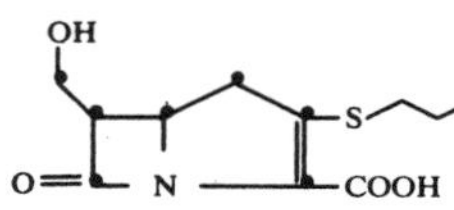

The antibiotic of the above formula will be recognized as northienamycin. Thienamycin is described in U.S. Patent 3,950,357, April 13, 1976 which fully describes the fermentation process.

The process comprises cultivating a thienamycin-producing strain of *Streptomyces cattleya* in an aqueous nutrient medium containing assimilable sources of carbohydrate, nitrogen and inorganic salts under submerged aerobic conditions and recovering the northienamycin so produced in substantially pure form. The organism cultivated is *Streptomyces cattleya* NRRL 8057.

8-Epi-Thienamycin Using *Streptomyces cattleya*

In the same fermentation broths that yield thienamycin and northienamycin (described in the previous patent), there is also found the substance (5R,6S,8S)-3-(2-aminoethylthio)-6-(1-hydroxyethyl)-7-oxo-1-azabicyclo[3.2.0] hept-2-ene-2-carboxylic acid which is useful as an antibiotic:

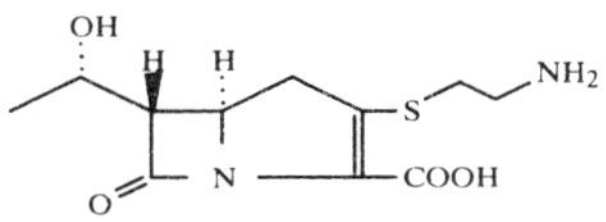

The antibiotic of the above formula will be recognized as 8-epi-thienamycin (also known as desacetyl 890 A_3).

The organism cultivated to produce these three thienamycins is a strain of *Streptomyces cattleya* NRRL 8057. The complete cultivation of the thienamycin-producing Streptomyces strain is detailed in three patents. The first, U.S. Patent 3,950,357, also details the means of recovering and separating thienamycin. The previous Patent 4,247,640, details the recovery and separation of northienamycin, and this patent contains the specifics for recovering and separating 8-epi-thienamycin.

Antibiotics A6888C and A6888X

S.M. Nash, K.F. Koch and M.M. Hoehn; U.S. Patent 4,252,898; February 24, 1981; assigned to Eli Lilly and Company have found that two new macrolide antibiotics plus cirramycin A_1 and B can be coproduced by a strain of *Streptomyces flocculus,* designated NRRL 11459. The new antibiotic substances are related structurally to cirramycin A_1 and cirramycin B and have the structural formulas shown below:

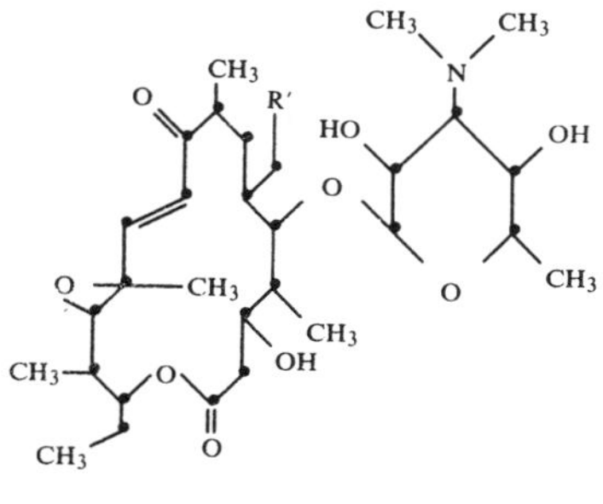

wherein (a) R' is –CHO and (b) R' is $-CH_2OH$, or a nontoxic, pharmaceutically acceptable acid addition salt thereof.

The new antibiotic compounds of the above formulas are arbitarily designated as Antibotic A6888C (R' is –CHO) and Antibiotic A6888X (R' is $-CH_2OH$). Antibiotics A6888C and A6888X exhibit antibacterial activity, particularly against Gram-positive bacteria, in standard test procedures.

This method of preparing the A6888 complex comprises cultivating *Streptomyces flocculus* NRRL 11459 in a culture medium containing assimilable sources of carbon, nitrogen, and inorganic salts under submerged aerobic conditions until a substantial amount of each compound is produced.

The term "complex" as used in the fermentation art refers to a mixture of the coproduced individual antibiotic factors. As will be recognized by those skilled in the art, the number and ratio of individual factors present in an antibiotic complex produced by a particular organism will vary depending upon the fermentation conditions used. "A6888 complex" means the complex comprising Antibiotic A6888C, Antibiotic A6888X, cirramycin A_1 and cirramycin B.

The organism used to produce the A6888 complex is a strain of *Streptomyces flocculus* isolated from a soil sample from Wyoming, U.S.A. The strain has been deposited with the permanent culture collection of the Northern Utilization Research and Development Division, U.S.D.A. NRRL 11459.

20-Dihydro-20-Deoxy-23-Demycinosyltylosin

Two other new macrolide antibiotics have been made by *R.H. Baltz, H.A. Kirst, G.M. Wild and E.T. Seno; U.S. Patent 4,304,856; December 8, 1981; assigned to Eli Lilly and Company.* These have the structures:

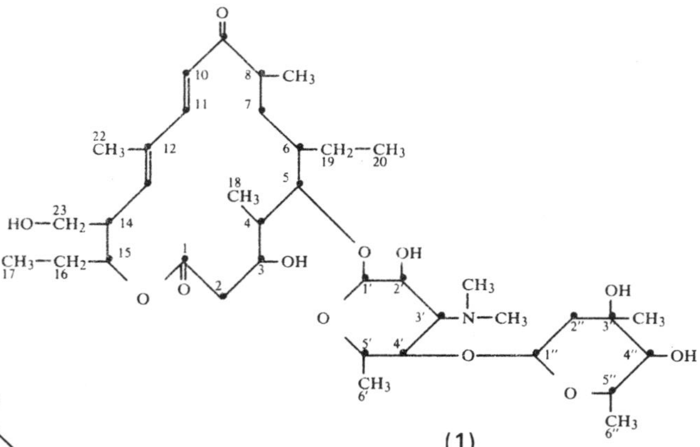

(1)

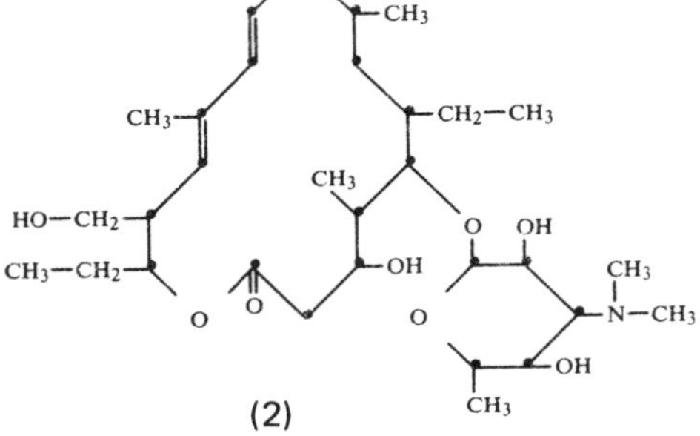

(2)

Formula 1 is 20-dihydro-20-deoxy-23-demycinosyltylosin, which will be called DH-DO-DMT for convenience herein, and Formula 2 is 20-dihydro-20-deoxy-5-O-mycaminosyltylonolide, which will be called DH-DO-OMT for convenience.

DH-DO-DMT and DH-DO-OMT inhibit the growth of organisms which are pathogenic to animals. More specifically, DH-DO-DMT and DH-DO-OMT are antibacterial agents which are active against Gram-positive microorganisms and Mycoplasma species.

DH-DO-DMT can be esterified on the 2', 4", 3", 23 and 3-hydroxyl groups and DH-DO-OMT can be esterified on the 2', 4', 23 and 3-hydroxyl groups to form useful acyl ester derivatives. Esterification of the 2'- and 4'-hydroxyl groups is easiest. Typical esters are those of a monocarboxylic acid or hemi-esters of a dicarboxylic acid having from 2 to 18 carbon atoms.

DH-DO-DMT and DH-DO-OMT are prepared by culturing a strain of *Streptomyces fradiae,* such as *S. fradiae* ATCC 31733, which produces these compounds under submerged aerobic conditions in a suitable culture medium until substantial antibiotic activity is produced. The culture medium used to grow *Streptomyces fradiae* ATCC 31733 can be any one of a number of media. For economy in production, optimal yield, and ease of product isolation, however, carbon sources in large-scale fermentation include carbohydrates such as dextrin, glucose, starch, and cornmeal, and oils such as soybean oil and the preferred nitrogen sources include cornmeal, soybean meal, fish meal, amino acids and the like. The conventional nutrient inorganic salts are incorporated in the culture media. For production of substantial quantities of DH-DO-DMT or DH-DO-OMT, submerged aerobic fermentation in tanks is preferred. Optimum antibiotic production appears to occur at temperatures of about 28°C.

Following its production under submerged aerobic fermentation conditions, DH-DO-DMT or DH-DO-OMT can be recovered from the fermentation medium by methods used in the art. Recovery of DH-DO-DMT or DH-DO-OMT is accomplished by an initial filtration of the fermentation broth. The filtered broth can then be further purified to give the desired antibiotic. A variety of techniques may be used in this purification. A preferred technique for purification of the filtered broth involves adjusting the broth to about pH 9; extracting the broth with a suitable solvent such as ethyl acetate, amyl acetate, or methyl isobutyl ketone; extracting the organic phase with an aqueous acidic solution; and precipitating the antibiotic by making the aqueous extract basic. Further purification involves the use of extraction, chromatographic and/or precipitation techniques.

The microorganism which produces DH-DO-DMT and DH-DO-OMT was obtained by chemical mutagenesis of a *Streptomyces fradiae* strain which produced tylosin. The microorganism produces only minimal amounts of tylosin, but produces DH-DO-DMT and DH-DO-OMT in almost equal amounts as major components.

The DH-DO-DMT- and DH-DO-OMT-producing microorganism is also classified as *Streptomyces fradiae.* A culture of the microorganism has been deposited in the American Type Culture Collection, under the accession number ATCC 31733.

Antibiotic Cephamycin C

Y. Kawamura, J. Shoji and K. Matsumoto; U.S. Patent 4,256,835; March 17, 1981; assigned to Shionogi and Company, Japan have developed a process for preparing the antibiotic cephamycin C characterized by cultivating a strain of *Streptomyces todorominensis* in a suitable medium and recovering cephamycin C from the culture broth.

Cephamycin C is one of the cephalosporins showing antimicrobial activity against Gram-positive and Gram-negative bacteria. More precisely, it is 7-(5-amino-5-carboxyvaleramido)-3-carbamoyloxymethyl-7-methoxy-3-cephem-4-carboxylic acid and strongly active against Gram-negative bacteria as well as β-lactamase-producing Proteus, with low toxicity. Cephamycin C is an excellent antibiotic and can be used as a human and veterinary drug.

Cephamycin C is prepared by the cultivation under controlled conditions of a hitherto undescribed strain of microorganism which was isolated from a soil sample and was tentatively called Streptomyces PA-30177. This strain has been designated *Streptomyces todorominensis* sp. nov. and the subculture has been deposited under the accession numbers FERM-P 4366 and ATCC 31489.

The process for producing the antibiotic cephamycin C comprises cultivating the *Streptomyces todorominensis,* ATCC 31489 in an aqueous nutrient medium at a temperature of about 20° to 40°C for about 20 to 100 hours under aerobic or submerged condition and isolating the accumulated antibiotic cephamycin C from the fermentation broth by filtering the broth and adsorbing the filtrate with a suitable adsorbent.

In work done by a different group of researchers, *T. Kamogashira, T. Nishida, M. Sugawara, T. Nihno and S. Takegata; U.S. Patent 4,332,891; June 1, 1982; assigned to Otsuka Pharmaceutical Co., Ltd., Japan,* another method of making cephamycin C is detailed, this one using another strain of Streptomyces. This strain, Streptomyces sp. OFR 1022, isolated from soil in Kenya, produces a great amount of cephamycin C in a culture medium and enables production of cephamycin C on a commercial scale. Not only does Streptomyces sp. OFR 1022 have excellent cephamycin C productivity in comparison with known cephamycin C producing strains, but the optimal cultural temperature used is high in comparison with those for the known strains. In accordance with this process, therefore, it is possible to produce cephamycin C without applying any cooling operation and by use of simplified apparatus at a low cost and in good yield.

Streptomyces sp. OFR 1022 has been deposited as FERM-P 4985 and ATCC 31666. An example of the cultivation of this Streptomyces to produce cephamycin C is as follows.

Example: Streptomyces sp. OFR 1022, which has been incubated on an oatmeal agar medium, was inoculated on a liquid culture medium containing 3% starch, 0.5% sucrose, 1% soybean flour and 0.3% dry yeast and having a pH of 7 and incubated with shaking at 37°C for 48 hours to give a seed culture solution.

In a 30 ℓ volume jar fermenter was placed 20 ℓ of culture medium containing 3% starch, 1% sucrose, 2% cottonseed flour, 1% dry yeast, 0.05% magnesium sulfate, 0.02% potassium dihydrogenphosphate, 0.05% disodium monohydrogenphosphate and 0.5% silicone (produced by Shinetsu Chemical Co., Ltd.) as a defoaming agent (after sterilization, pH 6.0), on which the above seed culture solution was then inoculated at a ratio of 1% and incubated with aeration at 37°C. The amount of the air aerated was 20 ℓ/min and the number of rotations of a propeller was 300 rpm. After cultivation for 90 hours, the amount of cephamycin C produced reached 2 mg/ml as determined by use of high-speed liquid chromatography.

Upon completion of the cultivation, the culture solution was subjected to centrifugal separation to remove the mycelia and the filtrate in the amount of 18 ℓ was adjusted to pH 7 to 8 and adsorbed on 3 ℓ of Diaion PA 406. Then it was eluted (with 0.5 M of sodium chloride aqueous solution) to obtain 2 ℓ of an antimicrobial active fraction. This fraction was concentrated under reduced pressure at about 30°C. The concentrated solution in the amount of 200 ml was passed through 400 ml of Silica Gel ODS (Waters Co.) and then again concentrated. The thus-concentrated solution was subjected to reverse phase chromatography with 0.01 M acetic acid on a 5.35 cm ϕ x 120 cm long column of Silica Gel ODS. Active fractions were collected and freeze-dried to obtain 18 g of white powdery cephamycin C.

Acanthomycin

The antibiotic acanthomycin is provided in a process by *A.D. Argoudelis, T.F. Brodasky and F. Reusser; U.S. Patent 4,259,450; March 31, 1981; assigned to The Upjohn Company.* It is obtained by culturing the microorganism *Streptomyces espinosus* ssp. *acanthus,* NRRL 11081, in an aqueous nutrient medium under aerobic conditions. Acanthomycin and its base addition salts have the property of adversely affecting the growth of Gram-positive bacteria, e.g., *Staphylococcus aureus* and *Streptococcus hemolyticus.* Accordingly, acanthomycin and its base addition salts can be used alone or in combination with other antibiotic agents to prevent the growth of or reduce the number of such bacteria in various environments.

The microorganism used for the production of acanthomycin is *Streptomyces espinosus* ssp. *acanthus,* NRRL 11081, which was studied and characterized by Alma Dietz of the Upjohn Research Laboratories. This antibiotic is produced when the organism, NRRL 11081, is cultivated in a conventional aqueous nutrient solution. It is recovered by a process comprising:

(a) filtering the acanthomycin-containing fermentation beer to obtain a clear beer;

(b) extracting the clear beer with a solvent for acanthomycin to obtain a crude preparation of acanthomycin;

(c) contacting the crude preparation of acanthomycin with ethyl acetate to give a precipitate containing acanthomycin;

(d) distributing the precipitate between water and 1-butanol to give a butanol phase containing acanthomycin;

(e) subjecting the butanol phase to chromatographic procedure on an anion exchanger to give fractions containing acanthomycin; and

(f) subjecting the fractions to chromatographic procedure on a nonionic macroporous copolymer of styrene crosslinked with divinylbenzene resin to give a purified preparation of acanthomycin.

The anion exchange resin may be Amberlite IRA-904 in the chloride form and the nonionic resin may be Amberlite XAD-7.

Acanthomycin is active against *S. aureus* and can be used to disinfect washed and stacked food utensils contaminated with these bacteria; it can also be used as a disinfectant on various dental and medical equipment contaminated with *S. aureus*. Further, acanthomycin and its salts can be used as a bacteriostatic rinse for laundered clothes, and for impregnating papers and fabrics; and, they are also useful for suppressing the growth of sensitive organisms in plate assays and other microbiological media. Since acanthomycin is active against *B. subtilis*, it can be used in petroleum product storage to control this microorganism which is a known slime and corrosion producer in petroleum products storage.

Antibiotics $890A_9$ and $890A_{10}$

The discovery of the remarkable antibiotic properties of penicillin stimulated great interest in this field which has resulted in the finding of many other valuable antibiotic substances. In general, the antibacterial activity of each of these antibiotics does not include certain clinically important pathogenic bacteria. For example, some are principally active against only Gram-positive types of bacteria. Acquired resistance over the course of widespread use of existing antibiotics in the treatment of bacterial infection has caused a serious resistance problem to arise.

Accordingly, the deficiencies of known antibiotics have stimulated further research to find other antibiotics which will be active against a wider range of pathogens as well as resistant strains of particular microorganisms.

P.J. Cassidy, S.B. Zimmerman, J.B. Tunac and S. Hernandez; U.S. Patent 4,264,735; April 28, 1981; assigned to Merck & Co., Inc. are concerned with an antibiotic substance, designated $890A_9$, which is highly effective in inhibiting the growth of various Gram-negative and Gram-positive microorganisms. Antibiotic $890A_9$ has a molecular structure as follows:

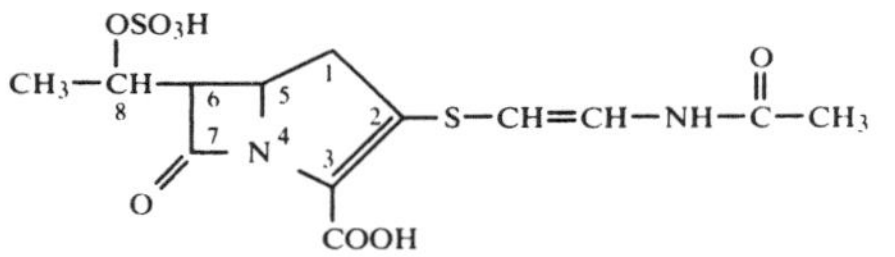

The new antibiotic substance is produced by growing under controlled conditions a new strain of *Streptomyces flavogriseus*, designated MA-4638 in the culture collection of Merck & Co., Inc. A culture has been placed on permanent deposit under NRRL 11,020 in the NRRL collection. *Streptomyces flavogriseus* MA-4638 produces antibiotic $890A_9$ which is isolated in substantially pure form from the fermentation broth.

$890A_9$ is produced during the aerobic fermentation, under controlled conditions, of suitable aqueous nutrient media inoculated with a strain of the organism, *Streptomyces flavogriseus*. Aqueous media, such as those employed for the production of other antibiotics, are suitable for producing $890A_9$. Such media contain sources of carbon, nitrogen and inorganic salts assimilable by the microorganism.

For optimum results it is preferable to conduct the fermentation at temperatures of from about 23° to 28°C. The initial pH of the nutrient media suitable for growing strains of the *Streptomyces flavogriseus* culture and producing antibiotic

$890A_9$ can vary from about 6.0 to 9.0. Although the antibiotic $890A_9$ can be produced by both surface and submerged cultures, it is preferred to carry out the fermentation in the submerged state.

This fermentation process, using *Streptomyces flavogriseus* MA-4638 as described above, also provides an antibiotic which has been named $890A_{10}$. It was developed by the same authors, in U.S. Patent 4,264,736, issued on the same date as the above and also assigned to Merck & Co., Inc. Antibiotic $890A_{10}$ has the molecular structure:

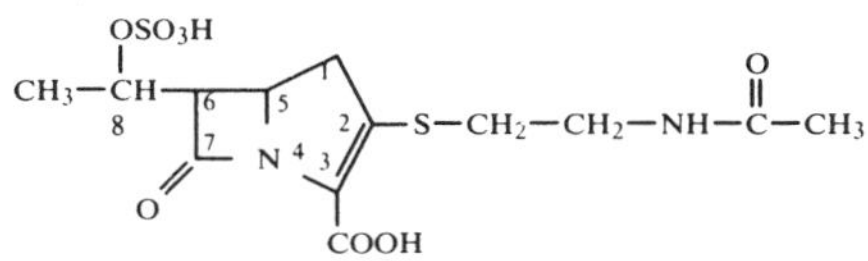

Lincomycin

The antibiotic lincomycin, formerly known as lincolnensin, can be produced by the microorganism *S. lincolnensis* var. *lincolnensis,* NRRL 2936, as described in U.S. Patent 3,086,912. The incubation temperature range used in that patent for the production of lincomycin is 18° to 40°C, and preferably 26° to 30°C. Also produced during the lincomycin fermentation is the compound known as lincomycin B. Though lincomycin and lincomycin B have activity against essentially the same spectrum of microorganisms, it is known that lincomycin B is significantly less active against the microorganisms than is lincomycin. Accordingly, lincomycin is the preferred antibiotic of the two.

In conducting the above fermentation, it is necessary to use a large amount of cooling water in most fermentation equipment to maintain the desired fermentation temperature. Further, the maintenance of a temperature within the range of 18° to 40°C, though essential for antibiotic production as disclosed above, is conducive to the development and proliferation of contaminating microorganisms in the fermentation vessel.

M.E. Bergy, J.H. Coats and V.S. Malik; U.S. Patent 4,271,266; June 2, 1981; assigned to The Upjohn Manufacturing Company describe the fermentation preparation of lincomycin by the new microorganism *Streptomyces vellosus* var. *vellosus,* NRRL 8037, at a temperature range of 18° to 45°C. It has been found, unexpectedly, that the titer of lincomycin produced at 45°C is comparable to that which is produced at 28°C.

A distinct advantage in using this microorganism to prepare lincomycin is the need for less fermentor cooling capacity. The need for less cooling capacity is especially significant in high temperature climates and in areas having limited water supplies since water is the generally used means for cooling and maintaining fermentation temperatures. A further distinct advantage in the process is that lincomycin is produced without the concomitant production of lincomycin B.

The actinomycete used according to this process for the production of lincomycin is *Streptomyces vellosus.* A subculture of this organism has been deposited at Northern Regional Research Laboratories, U.S.D.A. Its accession number in this repository is NRRL 8037. This microorganism was studied and characterized by Alma Dietz of the Upjohn Research Laboratories.

Lincomycin is produced by the described microorganism when the microorganism is grown in a conventional aqueous nutrient medium under submerged aerobic conditions. Production of lincomycin by this process can be effected at temperatures of from 18° to about 45°C. Ordinarily, optimum production of lincomycin is obtained in about two to ten days.

When growth is carried out in large vessels and tanks, it is preferable to use the vegetative form, rather than the spore form, of the microorganism for inoculation to avoid a pronounced lag in the production of lincomycin and the attendant inefficient utilization of the equipment. Accordingly, it is desirable to produce a vegetative inoculum in a nutrient broth culture by inoculating this broth culture with an aliquot from a soil or a slant culture. When a young, active vegetative inoculum has thus been secured, it is transferred aseptically to large vessels or tanks. The medium in which the vegetative inoculum is produced can be the same as, or different from, that utilized for the production of lincomycin, as long as it is such that a good growth of the microorganism is obtained.

In a preferred recovery process, lincomycin is recovered from its culture medium by separation of the mycelia and undissolved solids by conventional means, such as by filtration and centrifugation. Lincomycin is then recovered from the filtered or centrifuged broth by passing the broth over a resin which comprises a nonionic macroporous copolymer of styrene crosslinked with divinylbenzene. Exemplary of this type of resin is Amberlite XAD-2. Lincomycin is eluted from the resin with a solvent system consisting of methanol water (95:5 v/v). Bioactive eluate fractions are determined by a standard microbiological disc plate assay using the microorganism *Sarcina lutea.* Biologically active fractions are combined, concentrated to an aqueous solution which is then freeze dried.

The freeze-dried material is then triturated with methylene chloride. The methylene chloride extract is concentrated to dryness and the residue triturated with acetone. The filtrate is mixed with ether to give a precipitate which is separated. The remaining filtrate is mixed with methanolic hydrogen chloride (1 N) to precipitate colorless lincomycin hydrochloride. This precipitate is isolated by filtration and crystallization from water-acetone to give crystalline lincomycin hydrochloride.

Antibiotics WS-3442 A, B, C, D and E and Their Acyl Derivatives

The process of *H. Imanaka, J. Hosoda, K. Jomon, H. Sakai, I. Ueda and D. Morino; U.S. Patent 4,283,492; August 11, 1981; assigned to Fujisawa Pharmaceutical Co., Ltd., Japan* is one for the production of antibiotics WS-3442 A, B, C, D and E by the culture of a species of Streptomyces named *Streptomyces wadayamensis* in a nutrient medium, and to the acyl derivatives of these antibiotics and a method for their production. WS-3442 A, B, C, D and E are known compounds having chemical structures shown below:

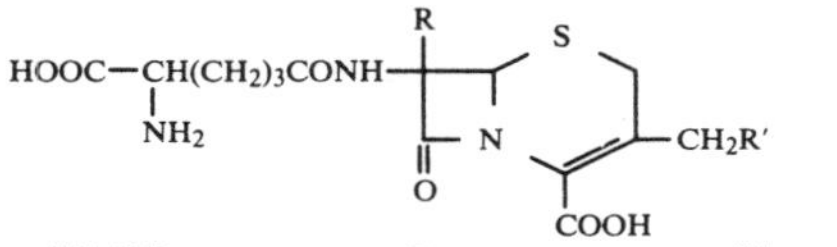

WS-3442	R	R'
A	H	H
B	H	$OCONH_2$
C	OCH_3	$OCONH_2$
D	OCH_3	H
E	H	OH

The microorganism useful for a method for production of antibiotics WS-3442 A, B, C, D and E is a new species of Streptomyces isolated from a soil sample collected in Wadayamacho, Japan and has the ATCC number 21948.

The antibiotics WS-3442 A, B, C, D and E are produced when a strain belonging to *Streptomyces wadayamensis* is grown in a nutrient medium containing the conventional assimilable sources of carbon and of nitrogen, and an inorganic salt under controlled submerged aerobic conditions. The nutrient medium may be any one of a number of media which can be utilized by a strain belonging to *Streptomyces wadayamensis.* These antibiotics WS-3442 A, B, C, D and E, produced by this method have activities against *Bacillus subtilis, Staphylococcus aureus, Escherichia coli, Proteus vulgaris,* etc.

Antibiotic X-14766A

C.-M. Liu and J. Westley; U.S. Patent 4,283,493; August 11, 1981; assigned to Hoffmann-La Roche Inc. describe a polyether ionophore antibiotic of the formula below and its pharmaceutically acceptable salts. (The shorthand expression Me and Et are utilized below to represent methyl and ethyl respectively.)

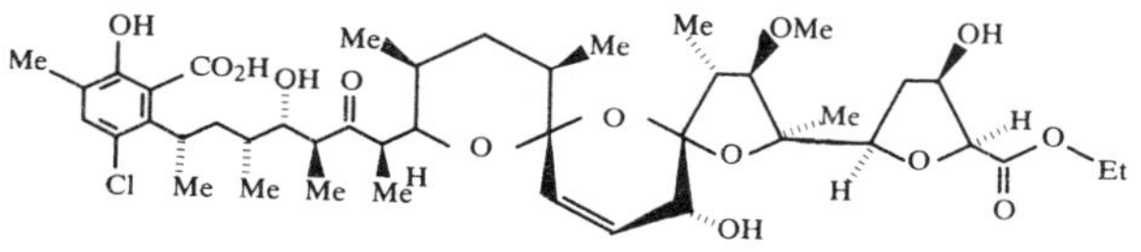

Antibiotic X-14766A is the designation given to a crystalline antibiotic produced by a Streptomyces organism isolated from a sample of soil collected at Playa Blanca, Mexico. Lyophilized tubes of the culture bearing the laboratory designation X-14766 were deposited with the U.S.D.A., Northern Regional Research Laboratories (NRRL). The culture was given the identification number NRRL 11335.

Antibiotic X-14766A is a polyether antibiotic and forms a variety of pharmaceutically acceptable salts. These salts are prepared from the free acid form of the antibiotic by methods well known for compounds of the polyether type in the art; for example, by washing the free acid in solution with a suitable base or salt. Examples of such pharmaceutically acceptable basic substances capable of forming salts for the purpose of this process include alkali metal bases, such as sodium hydroxide, potassium hydroxide, lithium hydroxide and the like; alkaline earth metal bases, such as calcium hydroxide, barium hydroxide and the like; and ammonium hydroxide. Alkali metal or alkaline earth metal salts suitable for forming pharmaceutically acceptable salts can include anions such as carbonates, bicarbonates and sulfates.

Examples of organic bases forming pharmaceutically acceptable salts with the polyether compounds are lower alkyl amines, primary, secondary and tertiary hydroxy-lower alkylamines such as ethylamine, isopropylamine, diethylamine, methyl-n-butylamine, ethanolamine and diethanolamine.

An amine especially preferred is N-methylglucamine. Salts of N-methylglucamine are of special value because of their water solubility which makes them amenable to parenteral use.

X-14766 is a variety of *S. malachitofuscus* which was named *S. malachitofuscus* ssp. *downeyi.* The Streptomyces X-14766A is cultivated in an aqueous carbohydrate solution containing a conventional nitrogenous nutrient under submerged aerobic conditions. The antibiotic is then isolated in a manner well known in the art.

Antibiotic X-14766A and its salts possess the property of adversely affecting the growth of certain Gram-positive bacteria. It is useful in wash solutions for sanitary purposes as in the washing of hands and the cleaning of equipment, floors or furnishings of contaminated rooms or laboratories. It is useful also for suppressing the growth of sensitive organisms in plate assays and other microbiological media. Antibiotic X-14766A has also exhibited antimalarial activity with an ED_{50} of 2.5 mg/kg against *Plasmodium bergei* in mice.

Antibiotic X-14766A also exhibited activity against swine dysentery and anticoccidial activity against *Eimeria tenella.* It was found to prevent and treat ketosis in ruminants and swine and it increases the feed utilization of ruminants.

Modification of Antibiotic A23187 Esters

B.J. Abbott and D.S. Fukuda; U.S. Patents 4,287,302; September 1, 1981 and 4,247,703; January 27, 1981; both assigned to Eli Lilly and Company describe derivatives of antibiotic A23187. This compound has the formula:

(1)

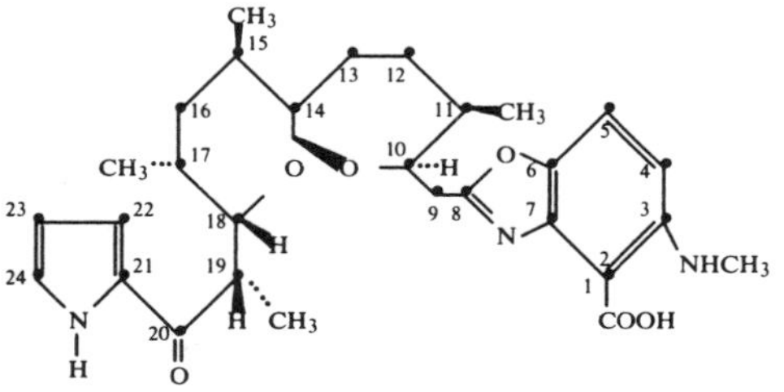

A23187 is one of the few naturally occurring compounds which have the capability of transporting divalent cations across biological membranes. [Refer to P. Reed et al, *J. Biol. Chem.,* 247, 6970 (1972).] It elicits a wide range of pharmacological responses, e.g., platelet aggregation [see P. Worner et al, *Thrombosis Res.,* 6, 295 (1975)], insulin release [see C. Wollheim et al, *J. Biol. Chem.,* 250, 1354 (1975)], histamine release [J. Foreman et al, *Nature,* 245, 249 (1973)], increased cardiac contractility [see D. Holland et al, *Proc. Soc. Exptl. Biol. Med.,* 148, 1141 (1975)], arrest of sperm motility [see D. Babcock et al, *J. Biol. Chem.,* 251, 3881 (1976)], and release of slow-reacting substances [M. Bach, *J. Immunol.,* 113, 2040 (1974)].

Antibiotic A23187 is prepared by culturing the microorganism *Streptomyces chartreusis,* NRRL 3882, as described by Gale et al, U.S. Patents 3,923,823 and 3,960,667.

The derivatives described by Abbott and Fukuda are produced by the biotransformation of an A23187 (lower) alkyl ester. The products formed by the hydrolysis of such derivatives are also of interest. The alkyl ester derivatives of A23187 have the following differences from Formula 1 above.

In Formula 2, the $-NHCH_3$ group in position 3 is replaced by an $-NH_2$ group; the carboxylic acid group of position 1 is replaced by a $-COOR$ group in which R is methyl, ethyl, n-propyl, isopropyl, n-butyl, isobutyl or tert-butyl.

In Formula 3, the same changes are made as in Formula 2, and an –OH group is added at position 16.

In Formula 4, the –OH group remains in position 16 and the ester group at position 1, but there is the original group of Formula 1 at position 3.

To these products are added the hydrolysis products of the compounds 2, 3 and 4, by which the esters again become free carboxylic acid compounds or sodium, potassium, or lithium salts thereof; or dimeric complexes thereof with a divalent cation selected from the group consisting of beryllium, magnesium, calcium, strontium, barium, manganese, cadmium, iron, zinc, lead, and mercury.

The monovalent cation salts of these compounds are prepared by conventional means such as by treating the acid with an appropriate base containing the desired monovalent cation. The complexes are prepared by conventional means such as by treating an alkaline metal salt of the compound in water with the desired divalent cation. The complexes may also be made by adding the divalent cation to a water solution of the compound at neutral pH. Suitable sources of the divalent and monavalent cations will be apparent to those skilled in the art.

The esters of Formulas 2, 3 or 4 are prepared by the biotransformation of a (lower) alkyl ester of antibiotic A23187. The preferred microorganism for the biotransformation is a strain of *Streptomyces chartreusis,* Calhoun and Johnson. This strain is different from the strain of *Streptomyces chartreusis* NRRL 3882 used for the preparation of antibiotic A23187. The strain employed for the biotransformation of antibiotic A23187 has been deposited with Northern Utilization Research and Development Division, U.S. Department of Agriculture. Its accession number is NRRL 11407. The strain was isolated from a soil sample collected in Venezuela by suspending portions of the sample in sterile deionized water and streaking the suspension on nutrient agar in Petri dishes. After incubation at 25° to 35°C, until growth was attained, colonies of the organism were transferred to agar slants with a sterile platinum loop. The agar slants were then incubated to provide a suitable inoculum for the culture of the organism used for the biotransformation.

The biotransformation of a A23187 lower alkyl ester to obtain a compound of Formulas 2, 3 or 4, is accomplished by cultivating *Streptomyces chartreusis* NRRL 11407 in a culture medium containing assimilable sources of carbon, nitrogen and inorganic salts under submerged aerobic conditions; adding the A23187 lower alkyl ester to the culture medium; and incubating the culture medium until substantial amounts of the desired transformation products are formed. Under these conditions, a mixture of the following lower alkyl esters is produced: 16-hydroxy A23187 (Formula 4); N-demethyl A23187 (Formula 2); and 16-hydroxy-N-demethyl A23187 (Formula 3). Starting from A23187 methyl ester the transformation products are isolated in a yield of from 4 to 6% (by weight) from substrate. The mixture of transformation products is recovered from the culture medium by methods conventional in the fermentation art, e.g., by solvent extraction. The transformation products are separated, isolated, and purified by chromatography.

It should be recognized that A23187 as the free carboxylic acid cannot be used as a substrate for cultivation with *Streptomyces chartreusis* NRRL 11407. An ester derivative of A23187 is essential for utilization by the organism. Thus, in order to obtain transformation products of A23187 having ionophoric properties the preparation must include formation of the alkyl ester of A23187, incubation of the ester with *Streptomyces chartreusis* NRRL 11407, and hydrolysis of the ester.

It should also be noted that a number of organisms other than *Streptomyces chartreusis* NRRL 11407 may be employed for the transformation of an A23187 ester. However, *Streptomyces chartreusis* NRRL 11407 is preferred because it produces the best yield of transformation products.

Antibiotic A-33853

A large variety of pathogenic microorganisms, such as bacteria and protozoa, are causative agents in producing diseased states in man and animals. Included in the list of causative agents are such organisms as *Staphylococcus aureus, Salmonella typhosa,* and *Pasteurella multocida,* the last being epidemiologically associated with pneumonia in sheep and cattle. *Mycoplasma gallisepticum* and other Mycoplasma cause respiratory problems in chickens and turkeys.

Although a number of antibiotics have been developed, some of which possess activity against one or more pathogenic organisms, there remains a need for more effective agents to combat the many diseases caused by these organisms.

M.M. Hoehn and K.H. Michel; U.S. Patent 4,293,649; October 6, 1981; assigned to Eli Lilly and Company describe an antibiotic substance A-33853 and its tetraacetyl derivative. This antibiotic is produced by culturing Streptomyces sp. NRRL 12068, or an A-33853-producing mutant or variant thereof, under submerged aerobic fermentation conditions.

Antibiotic A-33853 inhibits the growth of certain pathogenic microorganisms, as well as a number of viral organisms. The antibotic A-33853 and its tetraacetyl derivative both show in vitro activity against *Trichomonas vaginalis.* It has also shown antibacterial activity against penicillin-resistant Staphylococcus and Streptococcus strains. This antibiotic substance has the following formula:

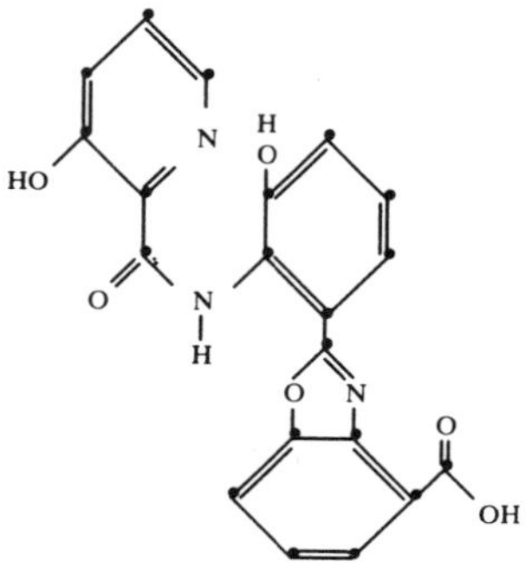

The compound represented by the above structural formula is named, according to *Chemical Abstracts* nomenclature rules, 2-[3-hydroxy-2-[[(3-hydroxy-2-pyridinyl)carbonyl] amino] phenyl] -4-benzoxazolecarboxylic acid. For convenience, it is designated herein as A-33853.

This A-33853 antibiotic is produced by culturing the previously undescribed microorganism Streptomyces sp. NRRL 12068, or an A-33853-producing mutant or variant thereof, in a culture medium containing assimilable sources of carbon, nitrogen, and inorganic salts, under submerged aerobic fermentation conditions until a substantial amount of antibiotic activity is produced. The fermentation mixture is filtered to remove the biomass, i.e., the mycelia, and the pH of the filtrate adjusted to a pH of about 6 to 7, preferably about pH 6.5. The antibiotic is then isolated from the filtered fermentation broth preferably by extraction with an organic solvent, such as chloroform, and concentration of the extract. Other solvents suitable for use in this extraction step include ethyl acetate, methyl acetate, n-butanol, and related water-immiscible polar solvents.

Antibiotic A-33853, on treatment with acetic anhydride in pyridine, forms a compound hereinafter referred to as the tetraacetyl derivative for convenience in terminology.

The Streptomyces sp. culture which is useful for the production of the A-33853 antibiotic, and which was initially isolated from a soil sample from Alaska, has been deposited at the Northern Regional Research Center, from which it is available under the number NRRL 12068.

Antibiotic 37,454 RP

J. Florent, J. Lunel and D. Mancy; U.S. Patent 4,293,650; October 6, 1981; assigned to Rhone-Poulenc Industries, France have provided a new antibiotic, designated 37,454 RP, which has particular interest for the anticoccidial activity displayed by itself and its salts with metals and nitrogen-containing bases. It has the formula:

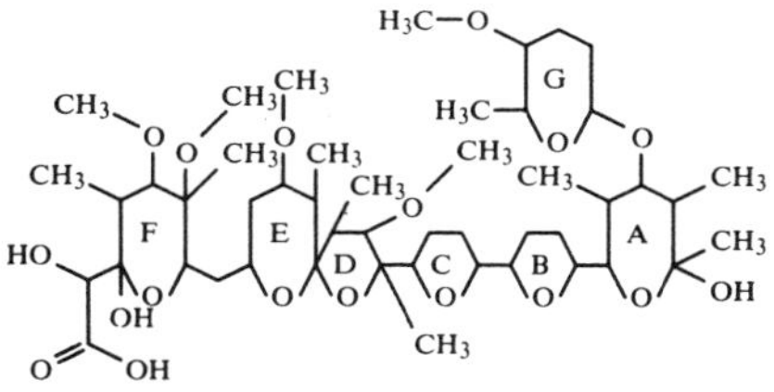

37,454 RP can be obtained by cultivating a new microorganism which belongs to the genus Streptomyces which is referred to by the name *Streptomyces gypseus* DS 27,461 (NRRL 11,168). This strain of Streptomyces was isolated from a sample of soil taken in Algeria, and the number DS 27,461 has been given to it. A sample of this organism was deposited with the Northern Regional Research Laboratory, U.S.D.A., as NRRL 11,168.

The culture is carried out under submerged aerobic conditions at a temperature between 23° and 33°C, commencing at a pH between 6.2 and 7.8 for 4 to 8 days. 37,454 RP is isolated from the culture medium by filtering the culture medium at an acid pH in the presence of a filtration aid, extracting 37,454 RP from the filter cake by means of a lower alcohol or a chlorinated solvent and isolating the 37,454

RP from the extract by crystallization after concentration under reduced pressure, if necessary adding a poor solvent or a nonsolvent for 37,454 RP and cooling the solution.

The anticoccidial activity of the antibiotic was determined in chicken infested in particular with *Eimeria tenella* or *Eimeria acervulina* and manifested itself when the sodium salt of the antibiotic was incorporated into the chicken feed at concentrations of between 0.005 and 0.04% by weight.

Monensin Urethane Derivatives

C.-M. Liu and J. Westley; U.S. Patents 4,263,427; April 21, 1981 and 4,294,925; October 13, 1981; both assigned to Hoffmann-La Roche Inc. describe the polyether antibiotics which are monensin urethane derivatives and their pharmaceutically acceptable salts. These compounds have the following formula:

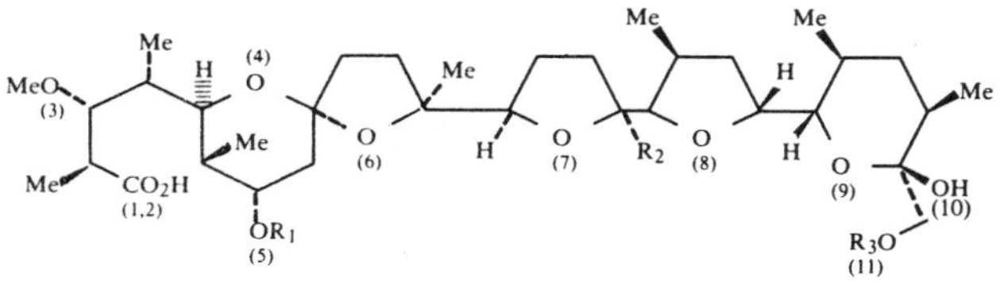

wherein R_1 is hydrogen, R_4 is alkyl, aryl, alkylaryl, arylalkyl, haloaryl, nitroaryl, haloarylalkyl, alkoxyaryl, aryloxyaryl, arylcycloalkyl, acylaryl and cycloalkyl; R_2 is methyl or ethyl and R_3 is $-CONHR_4$. Me stands for the methyl group.

These compounds and their salts exhibit activity as antibacterial agents, growth promotive agents in ruminants, coccidiostats, antihypertensives, antimalarial agents and as agents in the treatment of swine dysentery.

Certain of the monensin urethane derivatives, viz those of the above formula wherein R_4 is phenethyl and their pharmaceutically acceptable salts, are produced by Streptomyces organisms designated as Strains X-14667, X-14573 and X-14575. Streptomyces sp. X-14667 was isolated from a soil sample collected from Aesculapius temple, Epidaurus, Greece. Streptomyces sp. X-14573 was isolated from a soil sample collected at the University of Arizona, Tempe, Arizona. Streptomyces sp. X-14575 was isolated from a soil sample collected in a corn field in the Catskills, New York. Cultures of these three strains were deposited at the U.S.D.A. Northern Regional Research Laboratory and the American Type Culture Collection, and have the following accession numbers:

X-14667 = NRRL 11336 and ATCC 31551
X-14573 = NRRL 11337 and ATCC 31552
X-14575 = NRRL 11338 and ATCC 31553

An organic base especially preferred for making a pharmaceutically acceptable salt of these derivatives is N-methylglucamine. Salts of N-methylglucamine are of special value because of their water solubility which makes them amenable to parenteral use.

Streptomyces X-14667, X-14573 and X-14575 when grown under suitable conditions, produce monensin urethane derivatives. A fermentation broth containing Streptomyces X-14667, X-14573 or X-14575 is prepared by inoculating spores or mycelia of the organism producing the derivatives into an aqueous carbohydrate solution containing a nitrogenous nutrient under submerged aerobic conditions and thereafter isolating the end product from the solution.

The compounds of this process show in vitro activity against certain Gram-positive bacteria and therefore find utility as antibacterial agents useful in wash solutions for sanitary purposes as in the washing of hands and the cleaning of equipment, floors or furnishings of contaminated rooms or laboratories.

Further, it has been found that certain of these monensin urethane derivatives, viz those of the above formula wherein R_4 is aryl, haloaryl, nitroaryl, alkylaryl or cycloalkyl and their pharmaceutically acceptable salts, exhibit anticoccidial activity versus the organism *Eimeria tenella.* They are also useful for the purposes stated above.

Production of Rachelmycin

The known antibiotic, rachelmycin, has been found to be identical with antibiotic CC-1065 described in U.S. Patent 4,169,888. *D.E. Nettleton, Jr., J.A. Bush and W.T. Bradner; U.S. Patent 4,301,248; November 17, 1981; assigned to Bristol-Myers Company* have provided a new process for making this antibiotic. They cultivate certain rachelmycin-producing strains of Streptomyces in an aqueous nutrient medium under submerged aerobic conditions until a substantial amount of rachelmycin is produced and, optionally, isolate the rachelmycin from the culture medium.

These two rachelmycin-producing strains of Streptomyces sp. are ATCC 31128 and *Streptomyces anandii* ssp. *arraffinosus* designated ATCC 31431. One of the preferred rachelmycin-producing strains is a variant of a culture isolated from a Manlius, New York manure sample. The parent culture was designated strain C-329 and used to prepare a variant-designated C-329 variant 70 which gave high productivity of rachelmycin. Strain C-329 variant 70 has the accession number ATCC 31128.

The other preferred rachelmycin-producing strains, designated strain C-22,437, has been isolated from a Katpadi, Madras, India soil sample. It has the accession number ATCC 31431.

Rachelmycin inhibits the growth of various Gram-positive bacteria, e.g., *Staphylococcus aureus,* and can thus be used alone or in combination with other antibacterial agents to prevent the growth of, or reduce the number of, sensitive pathogenic bacteria. In addition to usefulness as an antibacterial agent for treatment of infectious diseases in animals (including man), the rachelmycin antibiotic may be employed in wash solutions for sanitation purposes, e.g., for washing hands and disinfecting various laboratory, dental and medical equipment, and as a bacteriostatic rinse for laundered clothes.

Rachelmycin has also been found to induce bacteriophage production in lysogenic strains of bacteria, thus indicating antitumor activity. Additionally, in tests against various transplantable rodent tumor systems, rachelmycin exhibited significant tumor inhibitory effects against P-388 leukemia, L-1210 leukemia, Sarcoma 180, Walker 256 carcinosarcoma, B16 melanoma and Lewis lung carcinoma.

Example 1: *Fermentation of rachelmycin with strain C-329 by shake-flask fermentation* — Streptomyces sp. strain C-329 variant 70 is grown in a test tube on a sterile agar slant medium consisting of 2.0 g glucose, 20.0 g oatmeal, 2.0 g soy peptone and 20.0 g agar, made up to 1.0 ℓ with distilled water. After 7 days of incubation at 27°C, growth from the surface of the slant is transferred to a 500 ml Erlenmeyer flask with 100 ml of sterile medium consisting of 50.0 g cornstarch, 10.0 g Pharmamedia (cottonseed meal; Traders Oil Mill Co.), 10.0 g Mellasoy (soy flour; Swift Chemical Co.) and 30 g $CaCO_3$ made up to 1.0 ℓ with distilled water. This vegetative culture is incubated at 27°C for 48 hours on a Gyrotory Tier shaker (Model G53, New Brunswick Scientific Co., Inc.) set at 210 rpm describing a circle with a 5.1 cm diameter.

Four milliliters of vegetative culture is transferred to a 500 ml Erlenmeyer flask containing 100 ml of sterile production medium consisting of 50 g glucose, 20.0 g Pharmamedia, 10.0 g Fermo 30 autolyzed yeast (Yeast Products Inc.) and 10.0 g $CaCO_3$, made up to 1 ℓ with distilled water. This production culture is incubated at 30°C on the previously described shaker for 8 days at which time rachelmycin is at a maximum level found both in mycelium and extracellular fluid.

Example 2: *Extraction of rachelmycin from strain C-329* — Whole broth from fermentation of strain C-329 variant 70 was stirred with an equal volume of methyl isobutyl ketone at broth pH (generally 8 to 8.5) for 1 hour. A large amount of diatomaceous earth filter aid was thoroughly stirred into the mixture, and the latter was then filtered on a filter aid mat with vacuum. The organic phase in the filtrate was separated, concentrated to a small volume under vacuum, and diluted with diethyl ether. The crude rachelmycin precipitate which formed was collected and air-dried.

Antibiotic A73A

R.S. Dewey, J.E. Flor, S.B. Zimmerman, P.J. Cassidy, S. Omura and R. Oiwa; U.S. Patent 4,304,859; December 8, 1981; assigned to Merck & Co., Inc. describe the production of an efrotomycin-like antibiotic, A73A. In its structure shown below, the pyran has the following configuration at its asymmetric centers: S at the hemiketal carbon 2, S at the hydroxyl carbon 4 and S at the pentadienyl side chain bearing carbon 6.

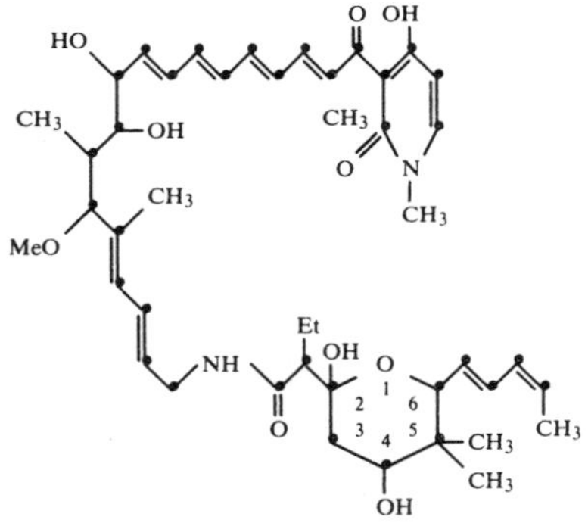

A73A is obtained by growing under controlled conditions the microorganism, *Streptomyces viridifaciens,* in a fermentation broth. By extensive taxonomic studies, *S. viridifaciens* was identified as an actinomycete and designated MA-4864 in Merck's culture collection. A culture has been deposited as ATCC 31495.

A preferred method for obtaining the antibiotic is by growing, under controlled conditions, the microorganism, *Streptomyces viridifaciens* MA-4864 in a medium containing suspended nutrient matter or in a clear medium substantially free of suspended nutrient matter and adsorbing the filtered broth on the appropriate resin and eluting. The eluate is concentrated. The concentrate is extracted by adjusting to acid pH and adding a water-immiscible organic solvent. The solvent layer is drawn off and evaporated in vacuo. The residue is redissolved in an appropriate polar organic solvent and added dropwise to nonpolar organic solvent such as hexane. The precipitate formed is chromatographed over silica gel to yield the antibiotic A73A.

In the process described above wherein extraction is carried out with water-immiscible polar organic solvents, representative examples of such solvents include alkyl esters of lower alkanoic acids such as methyl formate, ethyl formate, methyl acetate, ethyl acetate, n-butyl acetate, isobutyl acetate, ethyl propionate; a ketone such as cyclohexanone; or a halogenated lower hydrocarbon such as chloroform, methylene chloride, carbon tetrachloride, ethylene dichloride, 1-chloro-2,2-dimethylpropane, tetrachloroethylene, or bromoform.

A73A shows activity against Gram-positive bacteria, but only limited activity against Gram-negative bacteria. In addition to its use as an antibiotic, A73A is useful as a feed additive to permit the growth of animals such as chickens, sheep and cattle. The use of A73A shortens the time required for bringing animals up to marketable weight.

Antibiotic SM-173B

A. Fujiwara, M. Fujiwara, T. Hoshino, Y. Sekine and M. Tazoe; U.S. Patent 4,304,861; December 8, 1981; assigned to Hoffmann-La Roche Inc. have produced the antibiotic SM-173B which is antibiotically active against a large number of Gram-positive bacteria and against mycobacteria. It can accordingly be used as a disinfection agent and also as a therapeutic agent, to be administered orally or parenterally. The compound belongs to the anthracylinone series. It differs, however, in its physico-chemical properties from the hitherto known antibiotics of this series and accordingly represents a new substance to which can be ascribed the following formula:

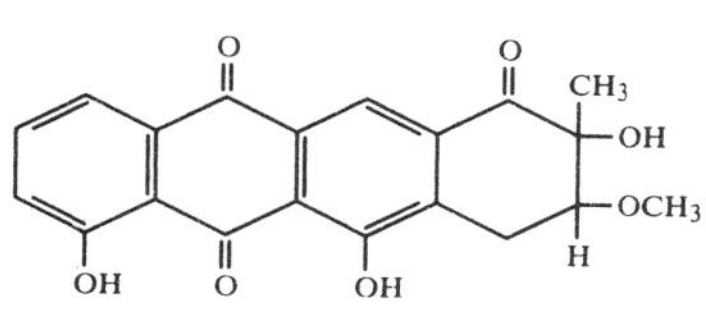

Antibiotic SM-173B is antibiotically active against a large number of Gram-positive bacteria and against mycobacteria. It can accordingly be used as a disinfection agent and also as a therapeutic agent.

According to the process, antibiotic SM-173B is manufactured by cultivating an antibiotic SM-173B-producing microorganism of the genus Streptomyces under aerobic conditions in an aqueous culture medium and isolating antibiotic SM-173B from the fermentation broth. Preferred strains are *Streptomyces chromofuscus* SM-173 as well as variants thereof. *Streptomyces chromofuscus* SM-173 was isolated from soil from Kumamoto-ken, Japan, and identified as a strain belonging to *Streptomyces chromofuscus.*

The strain denoted as *Streptomyces chromofuscus* SM-173 has been deposited with the Fermentation Research Institute, Japan, under the number FERM-P-3824. A subculture of this deposited strain has been deposited in the microorganism collection of the U.S. Department of Agriculture, Northern Utilization Research and Development Division, under NRRL 11092.

The cultivation can be carried out in a culture medium which contains customary nutrients usable by the microorganism being cultivated. It is carried out under aerobic conditions in an aqueous medium, preferably by submerged fermentation. The cultivation is suitably carried out at a temperature of 25° to 35°C, the optimal temperature being 28°C and is preferably carried out at a pH of 5 to 8. The cultivation time depends on the conditions under which the cultivation is carried out. In general, it is sufficient to carry out the cultivation for 20 to 100 hours.

The isolation of antibiotic SM-173B from the fermentation broth can be carried out according to methods known per se. For example, the mycelium can be separated from the fermentation broth by centrifugation or filtration and antibiotic SM-173B can be extracted from the filtrate with a water-immiscible organic solvent such as ethyl acetate, butyl acetate, etc.

Example: Spores of *Streptomyces chromofuscus* SM-173 (FERM 3824; NRRL 11092) were inoculated into six 500 ml flasks each containing 100 ml of nutrient medium [2% glucose, 1% Pharmamedia (cottonseed oil, Traders Oil Mill), 0.5% yeast powder (Ebios Pharmaceutical Co. Ltd., Japan), 1% wheat gluten] and cultivated at 28°C for 3 days while shaking. The culture broth was used to inoculate 30 ℓ of nutrient medium having the same composition in a 50 ℓ fermenter. The cultivation was carried out at 27°C while stirring (500 rpm) and aerating (30 ℓ/min). After 48 hours, the culture was centrifuged, there being obtained 22 ℓ of filtrate. The filtrate was extracted twice with the same volume of ethyl acetate. The extracts were dried over sodium sulfate and concentrated in vacuo.

There were thus obtained 12 ml of crude antibiotic SM-173B in the form of a yellow syrup. This syrup was added to a silica gel column (diameter 4 cm, height 60 cm, Kieselgel 60, Merck) and antibiotic SM-173B was eluted with chloroform. The fractions which contained only antibiotic SM-173B [determined with thin-layer chromatography plates, silica gel, benzene/acetone (4:3)] were combined and concentrated to 10 ml under reduced pressure. A small amount of n-hexane was added to the concentration which was then stored in the cold overnight. There was thus obtained antibiotic SM-173B in the form of orange-colored crystals.

On the other hand, the mycelium obtained and collected in the process described earlier was extracted with 15 ℓ of acetone. After separating the mycelium cake and the solvent, the extract was extracted twice with 3.5 ℓ of ethyl acetate each time and dried over sodium sulfate. After concentration under reduced pressure, there was obtained crude antibiotic SM-173B in the form of a syrup. This syrup was absorbed on silica gel (silicic acid, Mallinckrodt) in a glass column (diameter 4 cm, height 50 cm). Antibiotic SM-173B was eluted from the column with a mixture of benzene and acetone (95:5). The fractions which contained only antibiotic SM-173B (determined as described in the preceding paragraph) were pooled and evaporated to dryness in vacuo. The concentrate was dissolved in a small amount of hot benzene and left to stand overnight. There was thus obtained further antibiotic SM-173B in the form of orange-colored crystals.

Antibiotic U-60,394

L.A. Dolak and L.E. Johnson; U.S. Patent 4,306,021; December 15, 1981; assigned to The Upjohn Company have produced an antibiotic U-60,394, which has similarities to matchamycin, a known antibiotic [see A. Kimura and H. Nishimura, *J. Antibiotics* 33, 461 (1970)]. Matchamycin has the molecular formula $C_{20}H_{13}O_6N_3Cu$ and is produced by *S. amagasakensis*. It can be differentiated from U-60,394 by melting point, UV spectrum, IR spectrum and mass spectrum. *S. amagasakensis* grown in U-60,394 medium does not produce U-60,394, nor does *S. woolensis* when grown in the Japanese medium produce matchamycin.

Antibiotic U-60,394 is producible in a fermentation under controlled conditions using a biologically pure culture of the new microorganism *Streptomyces woolensis,* Dietz and Li sp. nov. NRRL 12113.

Antibiotic U-60,394 is strongly active against various Gram-positive bacteria. Further, the base addition salts of antibiotic U-60,394 are also active against these bacteria. Thus, antibiotic U-60,394 and its salts can be used to disinfect washed and stacked food utensils contaminated with *S. aureus.* They can also be used as disinfectants on various dental and medical equipment contaminated with *S. aureus.* Still further, antibiotic U-60,394 and its salts can be used as a bacteriostatic rinse for laundered clothes, and for impregnating papers and fabrics; and, they are also useful for suppressing the growth of sensitive organisms in plate assays and other microbiological media.

Antibiotic Narasin

Narasin is an antibiotic which is active against Gram-positive bacteria, anaerobic bacteria, and fungi, and is useful as an anticoccidial agent and as an agent for increasing feed utilization in ruminants.

Production of narasin by fermentation of *Streptomyces aureofaciens* NRRL 5758 or *Streptomyces aureofaciens* NRRL 8092, has been described by Berg et al, U.S. Patent 4,038,384.

R.E. Kastner and R.L. Hamill; U.S. Patent 4,309,504; January 5, 1982; assigned to Eli Lilly and Company have devised a process for the preparation of the antibiotic narasin which consists of cultivating a newly discovered strain designated herein as *Streptomyces lydicus* DeBoer et al, NRRL 12034, or a narasin-producing mutant or variant thereof, in a conventional culture medium containing assimilable sources of carbon, nitrogen, and inorganic salts under submerged aerobic fermentation conditions.

The microorganism used is a biologically pure culture derived from a soil sample collected near the Surinam River, Surinam, South America, and the culture was given the number A-39861.3 for identification purposes.

Culture A-39861.3 is classified as a strain of *Streptomyces lydicus* DeBoer et al, based upon a simultaneous culturing of *Streptomyces hygroscopicus, Streptomyces endus, Streptomyces platensis,* and *Streptomyces lydicus* using the methods and media recommended by Shirling and Gottlieb [*Intern. Bull. of Systematic Bacteriol.* 16, 313-340 (1966)], along with certain supplementary tests. From the characteristics obtained, *Streptomyces lydicus* was selected as being the most closely related species. The principal differences between culture A-39861.3 and the *S. lydicus* type culture are in the carbon utilization pattern, a higher level of NaCl tolerance by *S. lydicus,* and the production of a white aerial mycelium by culture A-39861.3 on several media.

In further work by *R.E. Kastner and R.L. Hamil; U.S. Patent 4,342,829; August 3, 1982; assigned to Eli Lilly and Company,* a different Streptomyces species is used to produce narasin. This strain is designated as *Streptomyces granuloruber* NRRL 12389. This microorganism is a biologically pure culture derived from a soil sample collected near the Surinam River, Surinam, South America. The culture was given the number A-39912 for identification purposes.

Culture A-39912 is classified as *Streptomyces granuloruber,* based upon a simultaneous culturing of *Streptomyces ruber; Streptomyces griseoruber;* and *Streptomyces griseoaurantiacus,* using the methods and media recommended by Shirling and Gottlieb.

A number of different media may be used to produce narasin with *Streptomyces granuloruber,* NRRL 12389. Conventional media are used, containing assimilable sources of carbon, nitrogen, and inorganic salts. The pH of the uninoculated fermentation medium varies with the medium used for production, but should fall in the range of from about pH 6.5 to 7.5.

This narasin-producing organism can be grown over a broad temperature range of from about 25° to 43°C. Optimum production of narasin with NRRL 12389 appears to occur at a temperature of about 30°C.

The narasin antibiotic can be recovered from the fermentation medium by methods known in the art and described by Berg et al in U.S. Patent 4,038,384. *Streptomyces granuloruber* NRRL 12389 produces substantial amounts of narasin factor A, as well as slight amounts of narasin factors B and D. The components may, as desired, be obtained as single antibiotics by further purification of the complex, for example by column chromatographic techniques as described in the Berg patent.

Antibiotic A-32256

The new antibiotic substance A-32256 has been developed by *K.H. Michel and M.M. Hoehn; U.S. Patent 4,316,959; February 23, 1982; assigned to Eli Lilly and Company.* It is produced by culturing Streptomyces sp. NRRL 12067, or an A-32256-producing mutant or variant thereof, in a culture medium containing assimilable sources of carbon, nitrogen, and inorganic salts, under submerged aerobic fermentation conditions.

A-32256 inhibits the growth of certain pathogenic microorganisms, in particular those within the genera Staphylococcus and Streptococcus, with genera being Gram-positive microorganisms. The A-32256 antibiotic also inhibits the growth of Gram-negative anaerobic organisms of the genus Bacteroides, has shown antitrichomonal activity in vitro, and inhibits the growth of Gram-positive organisms of the genus *Propionibacterium acnes.* Antibiotic A-32256 has shown utility in improving ruminant feed efficiency and as a mosquito larvicide.

The Streptomyces sp. culture which is useful for the production of the A-32256 antibiotic was isolated from a soil sample from Yellowstone National Park, and has been deposited and made a part of the stock culture collection of the Northern Regional Research Center, from which it is available to the public under the number NRRL 12067.

A number of different media may be used to produce antibiotic A-32256 with Streptomyces sp. NRRL 12067. For economy in production, optimal yield, and ease of product isolation however, certain culture media are preferred. Thus, for example, preferred carbon sources are glucose, mannitol, maltose, starch, and tapioca dextrin. The optimum level of carbon source is about 2 to 4%.

Suitable nitrogen sources include soybean meal, meat solubles, peanut meal and pork blood meal. A preferred nitrogen substrate is Bacto-peptone (Difco Laboratories) at a level of 0.7 to 1.0%.

Essential trace elements necessary for the growth and development of the organism may occur as impurities in other constituents of the media in amounts sufficient to meet the growth and biosynthetic requirements of the organism. However, it may be beneficial to incorporate in the culture media additional soluble nutrient inorganic salts capable of yielding sodium, potassium, magnesium, calcium, ammonium, chloride, carbonate, phosphate, sulfate, nitrate and like ions.

The Streptomyces sp. NRRL 12067 culture will produce antibiotic in media that do not contain molasses. However, molasses increases antibiotic yields. Although levels of up to about 4% or more of molasses can be used, an optimum level is about 2%. Molasses can be partially replaced by manganese and several other divalent cations.

Polypropylene glycol having a molecular weight of about 2,000 to 4,000, used at a level of from about 0.1 to 0.5%, preferably at about 0.25%, increases production of the antibiotic by as much as 100%.

Although small quantities of the A-32256 antibiotic may be obtained by shake-flask culture, submerged aerobic fermentation in tanks is preferred for producing substantial quantities of the A-32256 antibiotic. For tank fermentation, it is preferable to use a vegetative inoculum. The vegetative inoculum is prepared by inoculating a small volume of culture medium with the spore form, mycelial fragments, or a lyophilized pellet of the organism to obtain a fresh, actively growing culture of the organism. The vegetative inoculum is then transferred to a larger tank where, after a suitable incubation time, the A-32256 antibiotic is produced in optimal yield.

The A-32256-producing organism can be grown over a broad temperature range of from about 25° to 37°C. Optimum production of A-32256 antibiotic appears to occur at a temperature of about 25°C.

Antibiotics KA-7038 and Their Acid Addition Salts

T. Deushi, A. Iwasaki, K. Kamiya, T. Mizoguchi, M. Nakayama, H. Itoh and T. Mori; U.S. Patent 4,329,426; May 11, 1982; assigned to Kowa Company, Ltd., Japan have succeeded in isolating an antibiotic-producing strain belonging to the genus Streptomyces from the soil at Sannan-cho, Hikami-gun, Hyogo Prefecture, Japan. From its morphological, cultural and physiological characteristics, the strain was assumed to be a new species belonging to the genus Streptomyces, and termed Streptomyces sp. KC-7038. This strain KC-7038 was deposited as FERM No. 4388 and ATCC No. 31530.

It was ascertained that the antibiotics produced by the strain KC-7038 had not previously been described. They have antibacterial action against Gram-positive bacteria and Gram-negative bacteria. This substance was termed substance KA-7038. Further investigations led to the discovery that the substance KA-7038 can be further separated into seven antibiotics, KA-7038 I through VII, and that they can be readily converted to their acid addition salts by treatment with acids. The structures of two of the antibiotics are shown below, as examples.

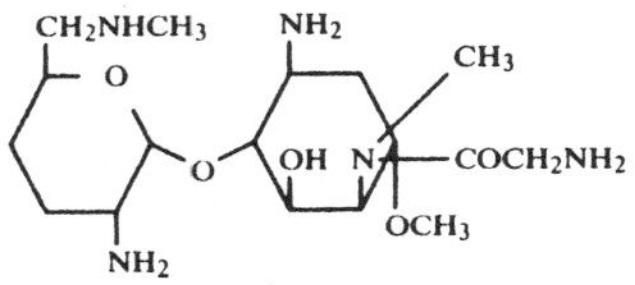

KA-7038 I

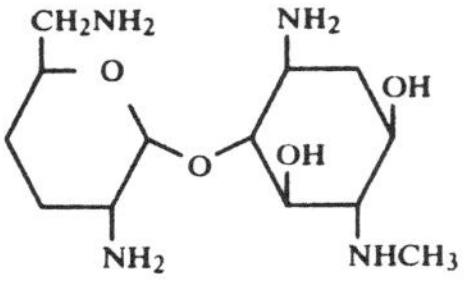

KA-7038 II

Suitable culture media for use in fermenting the KA-7038-producing strain of the genus Streptomyces comprise the conventional carbon and nitrogen sources and as optional ingredients, inorganic salts (minerals), very small amounts of heavy metals, etc.

Since the substance KA-7038 is a water-soluble basic substance soluble in water but difficultly soluble in common organic solvents, it can be separated from the culture broth by utilizing the procedures which are customarily used in isolating and purifying water-soluble basic antibiotics. For example, there can be used an adsorption-desorption method using an ion exchange resin, active carbon, etc.; column chromatographic method using cellulose, silica gel, alumina, etc.; and a method for extracting with butanol, amyl alcohol, etc., using a higher fatty acid as an adjuvant.

Examples of the weak acidic cation exchange resin used to recover the substance KA-7038 are Amberlite IRC-50, IRC-84 and CG-50 (Rohm & Haas Co.); and Diaion WK-10 and WK-20 (Mitsubishi Chemical Co., Ltd.).

The substance KA-7038 that can be isolated by the methods described above can be separated into KA-7038 I through VII by dissolving it in water, charging it on a column of an adsorbent such as a weak acidic ion exchange resin of the type described above or a weak acidic ion exchanger such as CM-Sephadex or CM-cellulose to cause the substance to be adsorbed on the adsorbent, and then eluting it with an alkaline aqueous solution such as dilute ammonium hydroxide, or an aqueous solution of ammonium carbonate or ammonium formate by a gradient method or a stepwise method. According to this separating procedure, substance KA-7038 IV, substance KA-7038 VII, substance KA-7038 I, substance KA-7038 II, substance KA-7038 VI, substance KA-7038 III and substance KA-7038 V as free bases are separated successively.

The resulting substances KA-7038 I through VII thus separated can be obtained in powder form by concentrating the eluate and lyophilizing the condensate. These substances KA-7038 obtained as free bases can be converted to their acid addition salts by treatment with pharmaceutically acceptable inorganic or organic acids, such as sulfuric acid, hydrochloric acid, phosphoric acid, carbonic acid, acetic acid, maleic acid, citric acid, etc.

Antibiotic compositions provided by these preparations are useful for both humans and animals such as poultry, domestic animals and cultivated fish and have a broad antibacterial spectrum.

Antibiotic Tunicamycin

Tunicamycin is a well-known antibiotic which was first described by Takasuki et al in *Journal of Antibiotics* of April 1971, pages 215-223. Although initially reported as a single compound, it was discovered subsequently that tunicamycin actually is a complex in which the common structural units are uracil, N-acetyl glucosamine, a unique 11-carbon sugar and a fatty acid. The structure of tunicamycin and its four major components is shown below.

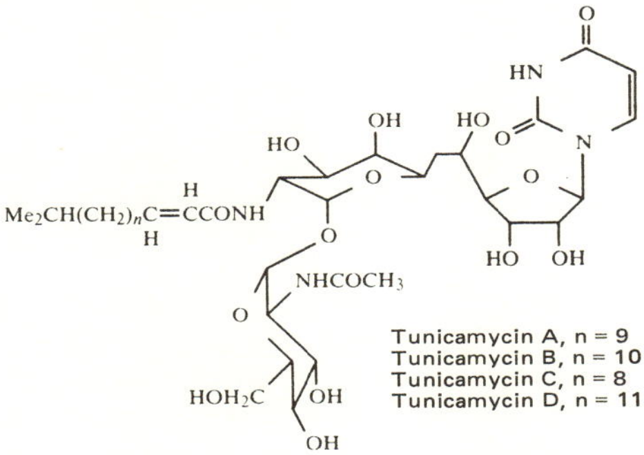

Tunicamycin A, n = 9
Tunicamycin B, n = 10
Tunicamycin C, n = 8
Tunicamycin D, n = 11

Tunicamycin is an antibiotic active against Gram-positive bacteria, yeasts, fungi and plant and animal viruses. Further, since tunicamycin acts by inhibiting the formation of lipid-linked intermediates in the biosynthesis of complex carbohydrates, it has become a valuable probe for molecular biologists.

E. Tejera, S.A. Currie, J.E. Flor and R.L. Monaghan; U.S. Patent 4,330,624; May 18, 1982; assigned to Merck & Co., Inc. have found that the antibiotic, tunicamycin, may be prepared by cultivating a heretofore unknown strain of *Streptomyces griseus.* The antibiotic is produced by growing this microorganism under controlled conditions in a fermentation medium followed by isolation of the antibiotic. The specific *Streptomyces griseus* actinomycete strain used for the preparation of tunicamycin has been designated MA-4729 in the Merck culture collection and ATCC 31591 in the American Type Culture Collection.

Tunicamycin is produced according to this process during the aerobic fermentation of suitable aqueous nutrient media under controlled conditions via the inoculation with the organism *Streptomyces griseus* MA 4729 ATCC 31591. Such aqueous media contain sources of carbon, nitrogen and inorganic salts assimilable by the microorganism. The choice of media is not critical and the fermentation may be carried out in media containing suspended nutrient matter or predominantly clear media wherein the media is substantially free of suspended nutrient matter.

For optimum results it is preferable to conduct the fermentation at temperatures of 27° to 28°C. The pH of the nutrient media for growing the *Streptomyces griseus* MA 4729 ATCC 31591 culture and producing the antibiotic tunicamycin should be in the range of from about 6.0 to 8.0.

A small-scale fermentation of the antibiotic is conveniently carried out by inoculating a suitable nutrient medium with the antibiotic-producing culture and, after transfer to a production medium, permitting the fermentation to proceed at a constant temperature of about 28°C on a shaker for several days. At the end of the incubation period, the antibiotic activity is isolated from the fermentation broth by fixation through diatomaceous earth and extraction with 60% aqueous acetone. The pooled extracts are evaporated in vacuo to remove the acetone. The pH of the concentrate is adjusted to 9 with 1 N aqueous sodium hydroxide and passed over a column of XAD-2 (Amberlite, Rohm & Haas Co.) at a broth-to-resin ratio of 10:1 (v/v).

The antibiotic activity is eluted with 85% aqueous acetone in one column volume. The eluate is evaporated to remove the acetone and the pH of the concentrate is adjusted to 3 with 1 N hydrochloric acid. The antibiotic activity is extracted into n-butanol twice and the combined extracts are evaporated to dryness. The residue is dissolved in methanol and chromatographed on a LH-20 (Pharmacia Fine Chemicals) column in methanol. Active fractions are determined by silica gel thin layer chromatography developed in a 4:1:1 n-butanol-acetic acid-water system.

The resulting chromatogram is visualized under UV (254 μm) light and is found to correlate with the bioautograph against *Saccharomyces cerevisiae.* Fractions thus selected are pooled and rechromatographed on a LH-20 column in n-butanol saturated with water. The eluate is evaporated and lyophilized to obtain purified tunicamycin complex.

In another preparation of tunicamycin by different researchers, *R.L. Hamill, M.M. Hoehn and L.D. Boeck; U.S. Patent 4,336,333; June 22, 1982; assigned to Eli Lilly and Company* have designated four tunicamycin factors, from A through D, each with subgroups 1 and 2. A_1 has an isofatty acid, the chain length of n in the formula (shown in the previous patent) being 8; A_2 has a normal fatty acid, n being 8; B_1 has an isofatty acid; B_2 a normal one, both with n = 9, etc.

The process described in this patent relates to methods of preparing the tunicamycin complex, and the tunicamycin isomers noted above. In particular, it constitutes a method of preparing tunicamycin which comprises cultivating *Streptomyces chartreusis* Calhoun and Johnson, 1956, NRRL 12338 in a culture medium containing assimilable sources of carbon, nitrogen, and inorganic salts under submerged aerobic conditions until a substantial amount of tunicamycin is produced, separating the desired factor from other coproduced factors, and recovering the desired factor. It is understood that each of the individual factors is isolated in a form substantially free of other coproduced tunicamycin factors.

The new microorganism useful for producing the antibiotic tunicamycin was isolated from a soil sample collected in the Netherlands Antilles.

After completion of the fermentation using *Streptomyces chartreusis* Calhoun and Johnson, 1956, NRRL 12338, tunicamycin is present both in the mycelia and in the broth. Most of the tunicamycin is present, however, in the mycelia. The mycelia are collected by conventional means, such as by filtration (using a filter aid, if desired) or by centrifugation. Tunicamycin can be recovered from the mycelial filter cake and from the filtered broth by methods well known in fermentation technology, such as solvent extraction, precipitation, and chromatography.

Reverse-phase high pressure liquid chromatography using silica gel/C_{18} adsorbent is a preferred method for the final purification of tunicamycin. In this method, tunicamycin (as obtained, for example, after chromatography using silica gel or Diaion HP-20, available from Mitsubishi Chemical Industries), dissolved in water or methanol-water (2:1), is placed on a column equilibrated with methanol-water (2:1) at a pressure of 60 to 85 psi. The column is then eluted at the above pressure with methanol-water (2:1) to remove further impurities and with methanol-water (4:1) to remove tunicamycin. The fractions, as they are eluted, are monitored with an ultraviolet spectrometer at 254 nm.

When it is desired to obtain the individual factors, tunicamycin can be separated by chromatographic techniques, such as reverse-phase high performance liquid chromatography.

Tunicamycin complex and the individual tunicamycin factors inhibit the growth of certain pathogenic organisms, particularly Gram-positive bacteria, fungi, and yeasts, as demonstrated in vitro by standard paper-disc or agar dilution inhibition tests.

De(Mycinosyloxy)Tylosin

The process developed by *R.H. Baltz, G.M. Wild and E.T. Seno; U.S. Patent 4,334,019; June 8, 1982; assigned to Eli Lilly and Company* relates to a new macrolide antibiotic, 23-de(mycinosyloxy)tylosin, hereafter called DMOT and to its 20-dihydro derivative. DMOT has the following structure:

(1)

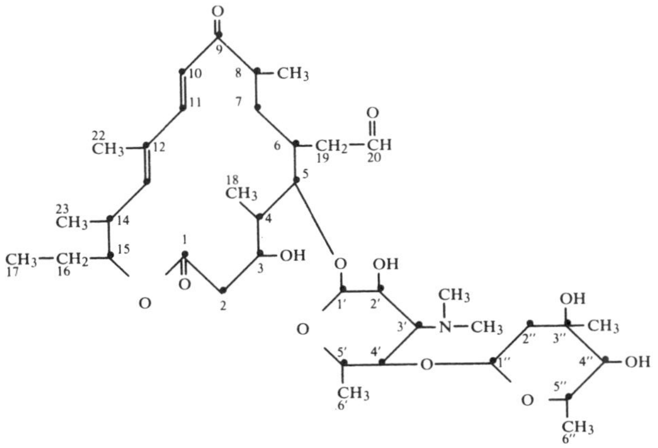

The dihydro derivative of DMOT has the same structure as that of formula 1 except that the –CHO group in position 20 is a $-CH_2OH$ group.

DMOT and dihydro-DMOT inhibit the growth of organisms which are pathogenic to animals. More specifically, they are antibacterial agents which are especially active against Gram-positive microorganisms and Mycoplasma species.

The hydroxyl groups of DMOT and dihydro-DMOT can be esterified on the 2',4",3" and 3-hydroxyl groups to form useful acyl ester derivatives. In addition, dihydro-DMOT can be esterified on the 20-hydroxyl group. Esterification of the 2'-hydroxyl group is most facile. Typical esters are those of a monocarboxylic acid or hemi-esters of a dicarboxylic acid having from 2 to 18 carbon atoms.

DMOT, dihydro-DMOT and their acyl ester derivatives are basic compounds which, when treated with acids, are converted to acid addition salts, which are also useful, when pharmaceutically acceptable.

DMOT and dihydro-DMOT are prepared by culturing a strain of *Streptomyces fradiae* which produces these compounds under submerged aerobic conditions in a suitable culture medium until substantial antibiotic activity is produced. As will be appreciated by those skilled in the art, DMOT is produced first in the fermentation process. Dihydro-DMOT is produced when the fermentation is carried out for a longer time, thus permitting the DMOT present to be reduced enzymatically.

The culture medium used to grow *Streptomyces fradiae* NRRL 11271 can be any one of a number of media. For economy in production, optimal yield, and ease of product isolation, however, preferred carbon sources in large-scale fermentation include carbohydrates such as dextrin, glucose, starch and cornmeal and oils such as soybean oil. Preferred nitrogen sources include cornmeal, soybean meal, fish meal, amino acids and the like.

The new microorganism used in this process was obtained by chemical mutagenesis of a *Streptomyces fradiae* strain which produced tylosin. The microorganism obtained by mutagenesis produces only minimal amounts of tylosin, but produces DMOT as a major component.

For characterization purposes, the new organism was compared with *Streptomyces fradiae* strain M48-E 2724.1, a tylosin-producing strain derived from *S. fradiae* NRRL 2702 described by Hamill et al in U.S. Patent 3,178,341.

This organism which produces DMOT and dihydro-DMOT, NRRL 11271, is classified as a new strain of *Streptomyces fradiae.* It has been deposited and made part of the stock culture collection of the Northern Regional Research Center from which it is available to the public under the accession number NRRL 11271.

Antibiotic Nanaomycin E

Nanaomycin A is a known compound having antibiotic activity against Gram-positive bacteria, Trichophyton and Mycoplasma and may be used, for example, as antibacterial and therapeutic agents for humans and animals. Conventionally, nanaomycin A is produced by culturing a microorganism belonging to the genus Streptomyces and capable of producing nanaomycin A in a medium under aerobic conditions to accumulate nanaomycin A in the fermented liquor and nanaomycin A is recovered therefrom. A preferred strain for this purpose is Streptomyces *rosa variant notoensis* (FERM P-2209; ATCC 31135).

S. Omura, H. Tanaka, I. Takahashi, S. Ishii, K. Mineura, K. Shirahata and M. Kasai; U.S. Patents 4,353,986; October 12, 1982 and 4,296,040; October 20, 1981; both assigned to Kyowa Hakko Kogyo KK, and Kitasato Kenkyusho, Japan have found that a large amount of another substance having interesting properties is accumulated in this fermentation. This substance, designated as nanaomycin E, also possesses antibiotic activity. Moreover, nanaomycin E may also serve as a starting material for the preparation of nanaomycin A, general formula (1) below. This new antibiotic, nanaomycin E is represented by the formula (2):

(1)

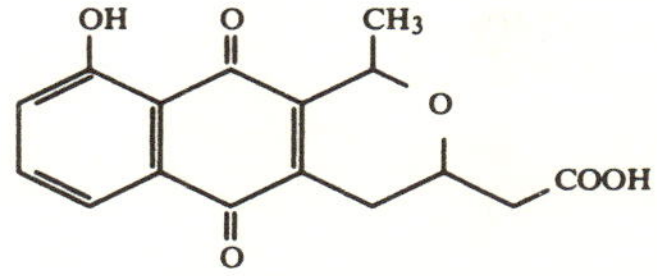

(2)

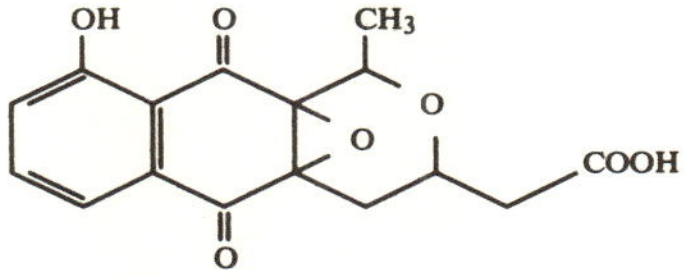

Nanaomycins B, C and D, structurally related to nanaomycins A and E, also possess antibiotic activity. When the fermentation of *S. rosa* ATCC 31135 referred to above is performed under conventional conditions, nanaomycin E is the major product by a considerable margin.

The recovery and isolation of nanaomycin E may be effected as follows. The fermented liquor is extracted with a suitable solvent such as ethyl acetate under acidic conditions. The extract is concentrated to dryness under reduced pressure to give crude powders containing nanaomycin E in association with other nanaomycin-type compounds. The crude powders are subjected to silica gel column chromatography using a solvent system of benzene/ethyl acetate. The eluted fractions contain various nanaomycin-type compounds, and those containing nanaomycin E are collected and combined. The combined fractions containing nanaomycin E are concentrated under reduced pressure to obtain crude powders of nanaomycin E which are then subjected to further silica gel column chromatography carried out in a similar manner to the first column chromatography and so on, resulting in yellow-white powders of nanaomycin E. These powders are dissolved in a suitable solvent (e.g., n-hexane/methylene chloride) and concentrated to dryness under reduced pressure to obtain orange pillar crystals of nanaomycin E.

BY NOCARDIA SPECIES

Antibiotics C-14919 E-1 and E-2

E. Higashide, M. Asai and T. Hasegawa; U.S. Patent 4,245,047; January 13, 1981; assigned to Takeda Chemical Industries, Ltd., Japan have prepared two antibiotics, C-14919 E-1 and C-14919 E-2 by the cultivation of certain new microorganisms belonging to the genus Nocardia, in conventional nutrient systems. The microorganisms C-14919 were deposited as FERM 3991 and ATCC 31280. The aforementioned microorganism is actinomycete organism discovered in the course of screening for antibiotic producers and is thought to belong to Group III of the genus Nocardia.

The culture medium employed in the process may be a liquid or a solid medium, if it contains nutrients which the strain may utilize, although a liquid medium is preferred for high-production runs. The medium may contain carbon and nitrogen sources which Strain No. C-14919 may assimilate and digest, inorganic matter, trace nutrients, etc.

The cultivation may be conducted by any of the stationary, shake, aerobic submerged and other cultural methods. For high-production runs, submerged aerobic culture is preferred. While the conditions of culture, of course, depend on the condition and composition of the medium, the strain used, cultural method and other factors, it is normally preferred to carry out incubation at 20° to 32°C with an initial pH near neutral. Particularly desirable is a temperature from 23° to 26°C in an intermediate stage of cultivation, with an initial pH of 6.0 to 7.0. While the incubation time also is variable, the time required normally ranges from about 72 to 192 hours.

Because C-14919 E-1 and E-2 which are produced in the fermentation broth are lipophyl and neutral substances, they can be conveniently recovered by separation and purification procedures which are normally employed for the harvest of such microbial metabolites. C-14919 E-1 and E-2 display activity against Gram-positive bacteria, fungi and yeasts. Therefore, the antibiotics are of use as germicides or disinfectants against pathogenic bacteria or fungi of the same species as the assay or organisms.

The therapeutic activity (intraperitoneally administered for consecutive days) of C-14919 E-1 and E-2 against P388 leukemia in mice (1 x 10^6 cells/animals, mouse, intraperitoneally transplanted) was investigated. The antitumor activity in terms of the extension of life spans was 144% at the dose level of 5 mg/kg/day.

Thus, since antibiotics C-14919 E-1 and E-2 have a life-extending effect upon tumor-bearing mammals (e.g., mouse), they are expected to be useful as antitumor agents.

Rifamycin Compounds P, Q, R and U

It has previously been reported that during fermentation in normal growth media, *S. mediterranei* synthesizes a family of antibiotics collectively referred to as the rifamycin complex (P. Sensi, et al, *Antibiotics Annual,* 1959-1960, page 262). Subsequent work revealed that the addition of sodium diethyl barbiturate to the culture medium resulted essentially in the formation of a single fermentation product, rifamycin B (P. Margalith and H. Pagani, *Applied Microbiology,* 9, 325, 1961). According to a further discovery, a mutant strain of *S. mediterranei* was found which elaborates essentially rifamycin B only, irrespective of the presence or absence of sodium diethyl barbiturate in the fermentation medium, so that an improved process for making rifamycin B results. The rifamycin B producing strain was identified as *S. mediterranei* ATCC 21789 (see U.S. Patent 3,871,965).

Work by *R.J. White, G. Lancini and P. Antonini; U.S. Patent 4,263,404; April 21, 1981; assigned to Gruppo Lepetit SpA, Italy* has found that mutant strains of *S. mediterranei* ATCC 13685 produced new rifamycin compounds designated rifamycin P, Q, R and U. These strains are obtained by treating ATCC 13685 with common chemical mutagenic agents such as nitrous acid or nitroso-guanidine derivatives or with physical mutagenic agents such as x-rays and UV radiation. The new mutant strains rather than being assigned to the genus *S. mediterranei* are assigned to the genus Nocardia in accordance with the proposals by J.E. Thiemann et al, *Arch. Mikrobiol.,* 67, 147-155 (1969). Samples of these microorganisms are identified respectively as *Nocardia mediterranea* ATCC 31064, ATCC 31065 and ATCC 31066.

The procedure for fermentation is the normal one, the microorganism being cultivated in an aqueous nutrient medium under stirred and aerated submerged conditions, preferably at room temperature.

The crude extract of rifamycins can be purified chromatographically on a column of silica gel. Prior to chromatography it is convenient to dissolve the crude extract in a phosphate buffer pH 7.0 to 8.0 and to treat with a mild oxidizing agent. The buffer solution is then extracted with a water-immiscible solvent, this organic extract containing a new rifamycin designated rifamycin R. The extracted buffer solution is acidified to pH 2 to 4 and then again extracted with a water-immiscible solvent. This organic extract contains three new compounds of the rifamycin family denominated rifamycin P, Q and U.

The further purification of rifamycin R is effected by chromatography on a column of suitable adsorbing material such as silica gel and eluting with an appropriate mixture of organic solvents.

The second organic extract containing rifamycins P, Q and U is purified in a similar way to that described for rifamycin R.

Derivatives of Rifamycin S

Antibiotically active compounds which are derivatives of rifamycin S have been prepared by *T. Schupp, P. Traxler and J. Nüesch; U.S. Patent 4,298,692; November 3, 1981; assigned to Ciba-Geigy Corporation.* These have the basic formula shown below:

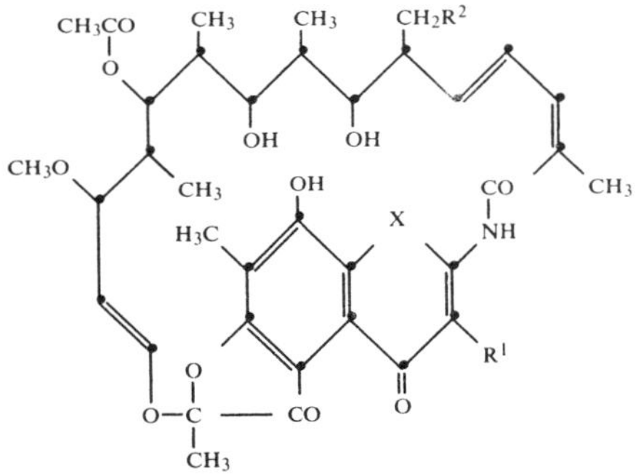

In 3-hydroxyrifamycin S, X = >C=O, R^1 = OH and R^2 = H. In 3,31-dihydroxyrifamycin S, X = >C=O, R^1 and R^2 = OH. In 1-desoxy-1-oxarifamycin S, X = —O—, R^1 and R^2 = H.

These rifamycin analogues exhibit a very good antibiotic activity similar to that of rifamycin S, especially an antimicrobial activity against Gram-positive and Gram-negative bacteria. They prove to be superior to rifamycin S in particular in their antibiotic activity against Gram-negative bacteria, such as *Escherichia coli* and Pseudomonas sp.

The compounds are obtained by isolation from fermentation material, in which they are formed, individually or together in any combination, by cultivating of certain strains of *Nocardia mediterranei.*

The isolation is carried out physico-chemically by means of known methods of separation, especially by centrifugation, filtration, solvent extraction precipitation, crystallization and chromatography, especially adsorption chromatography and partition chromatography. In a typical isolation process, the fermentation material (culture broth) is freed of mycelium by filtration, optionally using filter auxiliaries, such as diatomaceous earth, and the culture filtrate is subjected to extraction with an organic solvent which is water-miscible only to a limited extent, such as ethyl acetate. Before extraction, the culture filtrate is preferably adjusted to a pH of approximately 2 to 3. Volatile fractions, especially solvents, are removed from the organic solution by evaporation and the residue remaining (crude extract) is subjected to further processing. In order to isolate the compounds and to remove accompanying substances and also in order to separate the individual compounds from one another, the crude extract is further purified by chromatography, e.g., by column chromatography.

The rifamycin derivatives are produced by cultivating, under aerobic conditions in an aqueous nutrient medium containing a carbon source, a nitrogen source and mineral salts, a strain of *Nocardia mediterranei,* which is derived from *S. mediterranei* ATCC 13 685 as the parent strain, characterized in that a recombinant strain *Nocardia mediterranei* which is resistant to streptomycin and autotrophic with respect to amino acids is cultivated.

An advantageous recombinant strain is obtained, e.g., by crossing of and selective gene exchange between two *Nocardia mediterranei* mutants A and B which are derived from *S. mediterranei* ATCC 13 685 as the parent strain, [compare T. Schupp et al, *J. of Bacteriology* 121, 128-136 (1975)]. The mutant strains A and B are in their turn obtained in a known manner by the mutagenic action of UV light or N-methyl-N'-nitro-N-nitrosoguanidine on the mycelium suspension of the parent strain *S. mediterranei* ATCC 13 685 and are isolated by selection in accordance with their specific properties.

The mutant strain A produces chiefly rifamycin B, and is in this respect similar to the parent strain, but compared therewith has, on the one hand, a resistance towards streptomycin that is 50 times as great, but, on the other hand, is auxotrophic with respect to cysteine, lysine and leucine, that is, needs these three amino acids to grow. The mutant strain B is distinguished especially because it is sensitive to streptomycin, and particularly because it does not produce rifamycin B, but only an intermediate of rifamycin synthesis, rifamycin W [compare R.J. White et al, *Proc. Nat. Acad. Sci.,* 71, 3260-3269 (1974)]. The recombination is carried out in a known manner by cultivating both mutant strains together, and isolating the recombinant strain in accordance with its specific properties by selection in accordance with known methods.

Antibiotic C-15003

E. Higashide, M. Asai and S. Tanida; U.S. Patent 4,320,200; March 16, 1982; assigned to Takeda Chemical Industries, Ltd., Japan have developed the antibiotic C-15003. This compound has the general formula below:

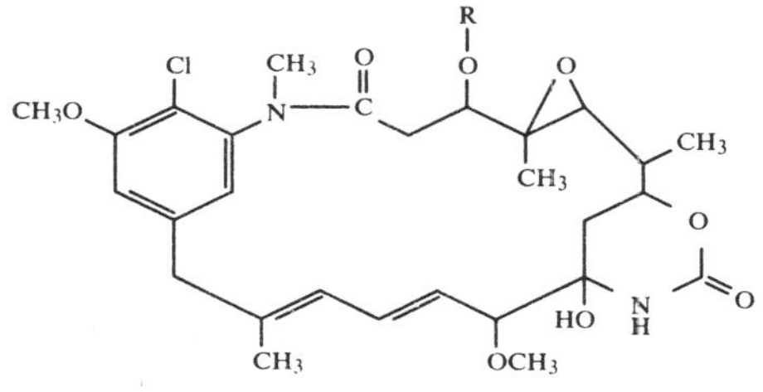

In the above formula R represents $-CO-CH(CH_3)_2$, $-CO-CH_2-CH_2-CH_3$ or $-CO-CH_2-CH(CH_3)_2$.

The production of this antibiotic has been accomplished by the cultivation of a C-15003-producing strain of the genus Nocardia, having an ATCC number 31281 and an FERM number of C-15003. The culture of the Nocardia strain is performed under the usual submerged aerobic conditions in an appropriate nutrient solution at an initial pH of 6.5 to 7.5 and temperature of 23° to 30°C.

Antibiotic C-15003 has strong inhibitory activity aganst fungi and protozoa and, therefore, is of value as an antifungal or antiprotozoan agent. Furthermore, because antibiotic C-15003 displays a life span-extending action upon tumor-bearing mammalian animals (e.g., mice), it is also expected that the compound will be of use as an antitumor drug.

Antibiotic C-15003, as an antifungal and antiprotozoan agent, can be used with advantage for an assessment of the bacterial ecology in the soil, active sludge, animal body fluid or the like. Thus, when valuable bacteria are to be isolated from soil samples or when the actions of bacteria are to be evaluated independently of those of fungi and protozoa in connection with the operation and analysis of an active sludge system used in the treatment of wastewater, the antibiotic may be utilized to obtain a selective growth of the bacterial flora without permitting growth of the concomitant fungi and protozoa in the specimen. In a typical instance, the sample is added to a liquid or solid medium and 0.1 ml of a 10 to 100 μg/ml solution of the antibiotic in 1% methanol-water is added per ml of the medium, which is then incubated.

The antibiotic C-15003 can also be used as an antimicrobial agent for the treatment of certain plant diseases. In the typical application, antibiotic C-15003 is used in the form of a 1% methanolic aqueous solution containing 0.5 to 5 μg/ml of the antibiotic for the control of reddish brown sheath rot, blast, Helminthosporium leaf spot and the blight of rice plants.

Antibiotic EM 4940

According to *R.B. Sykes, J.S. Wells and W.-C. Liu; U.S. Patent 4,339,535; July 13, 1982; assigned to E.R. Squibb & Sons, Inc.* cultivation of a strain of the microorganism Nocardia sp. 11,340 which has been deposited as ATCC 31531, yields an antibiotic substance EM 4940. The chemical structure of EM 4940 shown below has been elucidated through the use of x-ray crystallography and is as follows, that is, 3,4-dihydro-4-hydroxy-5-(3-hydroxy-2-pyridinyl)-4-methyl-2H-pyrrole-2-carboxyamide. EM 4940 is a mixture of two antibiotics, a 2,4-trans isomer and a 2,4-cis isomer, which have been designated EM 4940A and EM 4940B, respectively.

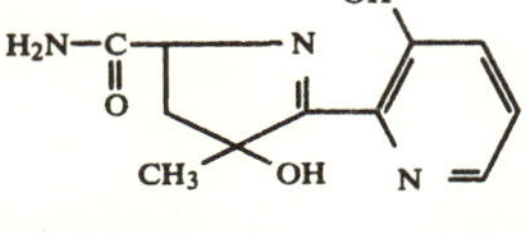

The antibiotic EM 4940 can be produced by cultivating Nocardia sp. SC 11,340 ATCC 31531 at or about 28°C under submerged aerobic conditions on an aqueous nutrient medium containing an assimilable carbohydrate and nitrogen source. The fermentation is carried out until substantial antibiotic activity is imparted to the medium, usually about 120 to 144 hours, preferably about 144 hours. EM 4940 can be separated from the fermentation medium and purified using art-recognized techniques.

The antibiotics have activity against a range of Gram-positive and Gram-negative bacteria, yeasts, fungi, Acholeplasma, and the protozoan *Trichomonas vaginalis.* The antibiotics can, therefore, be used as (1) an environmental disinfectant, (e.g., as a spray or dust in a conventional carrier) and (2) an agent to combat infections due to the above-enumerated bacteria, yeasts, fungi, Acholeplasma and *Trichomonas vaginalis* in various mammalian species (e.g., topically, orally or parenterally using pharmaceutically acceptable carriers, excipients, etc.).

Antibiotic C-15003 P-3

Antibiotic C-15003 P-3 (hereinafter P-3) is a compound obtained by cultivating a microorganism of the genus Nocardia. The compound and its production are described in U.S. Patent 4,162,940 in detail. The known process produces several components of antibiotic C-15003 simultaneously. For separating each component from the cultured broth, very complicated processes are required, and the process is not an advantageous one in view of the yield of the desired compound.

With the purpose of overcoming this drawback, *K. Hatano, M. Nakamichi and S. Akiyama; U.S. Patent 4,356,265; October 26, 1982; assigned to Takeda Chemical Industries, Ltd., Japan* devised a process for recovering P-3 specifically and have found that the addition of isobutyl alcohol and/or isobutyl aldehyde into the culture medium in which Nocardia is cultivated accomplishes the purpose.

Antibiotic C-15003 refers generically to the four compounds having the following general formula as a group, or a mixture of two or more of the compounds.

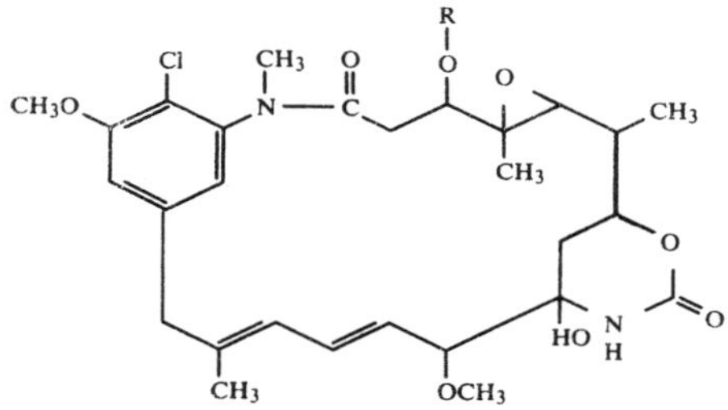

In the formula when R is $-COCH_2CH_3$, the compound is referred to as P-2; when R is $-COCH(CH_3)_2$, the compound is P-3; and when R is $-COCH_2CH(CH_3)_2$, the compound is P-4.

The microorganism used in the process is a strain of Nocardia, strain number C-15003, and has the following accession numbers: FERM-P 3992, IFO 13726 and ATCC 31281.

The isobutyl alcohol, its fatty acid ester (e.g., formate, acetate, propionate, butyrate or isobutyrate) and/or isobutyraldehyde are added in amounts not less than 0.005 v/v % based on the volume of culture medium, preferably from 0.01 to 0.5 v/v %. The compound may be added before or during the production of the antibiotic.

The medium used for the cultivation is the conventional one, an aerobic submerged cultivation being preferred, with an initial pH of 6.5 to 7.5 and a temperature of 20° to 30°C. Recovery and purification are accomplished in the usual way.

Antibiotic P-3 shows antiprotozoan, antifungal and antitumor activity.

BY ACTINOPLANES SPECIES

Gardimycin Antibiotics

F. Parenti, C. Coronelli, G. Tamoni and G. Lancini; U.S. Patent 4,276,382; June 30, 1981; assigned to Gruppo Lepetit SpA, Italy have produced and isolated a new family of antibiotic substances obtained by fermentation of strains belonging to the genus Actinoplanes. These substances will hereafter be referred to as metabolite A, metabolite B and metabolite C. Metabolite B is also named gardimycin.

The strains were isolated from soil samples collected at Temossi, Italy and Garbadi Bridge, India. Collection numbers are A/6353 for the strain isolated from the Italian soil sample and A/10889 for the strain isolated from the Indian soil sample; both have been deposited and made part of the stock culture collection of ATCC where they were assigned the numbers 31048 and 31049 respectively.

In the preparation of the antibiotic substances the selected organism is cultivated under aerobic conditions in a conventional aqueous nutrient medium.

Metabolite A, metabolite B (gardimycin) as well as their salts with alkali and alkali earth metals and metabolite C show good antibacterial in vitro and in vivo activities. More particularly, gardimycin exhibits an outstanding in vitro antimicrobial action especially against Gram-positive bacteria at concentration levels between 1 and 50 μg/ml.

Another favorable characteristic of the antibiotic substance gardimycin is that it is active against clinically isolated Streptococcus strains. Furthermore, gardimycin displays also a very interesting in vivo activity against experimental infections in mice caused by pathogenic bacteria of the genus Diplococcus and Streptococcus. These favorable antimicrobial properties are coupled with a very low toxicity. The Actinoplanes strains useful for cultivation of these antibiotics have been given the names *Actinoplanes liguriae* ATCC 31048 and *Actinoplanes garbadinensis* ATCC 31049.

Antibiotic A/16686

B. Cavalleri, H. Pagani and G. Volpe; U.S. Patents 4,303,646; December 1, 1981 and 4,328,316; May 4, 1982; both assigned to Gruppo Lepetit SpA, Italy have developed a new antibiotic, arbitrarily designated A/16686, and a process for its cultivation from a hitherto undescribed strain which has been characterized taxonomically as a new strain of the Actinoplanes genus.

Antibiotic A/16686 is a glycopeptide antibiotic with a basic character which is capable of forming acid addition salts. Therefore the physiologically acceptable acid addition salts of antibiotic A/16686 are part of this process.

Antibiotic A/16686 inhibits in vitro the growth of certain pathogenic bacteria, especially those which are Gram-positive. Moreover parenteral administration of antibiotic A/16686 gives a high degree of protection against experimental infections in mice.

As stated above, antibiotic A/16686 is produced by culturing a new strain of the Actinoplanes genus. A culture of this strain, which was isolated from a soil sample collected at Vaghalbod (India), has been deposited as ATCC 33076.

For producing antibiotic A/16686 the strain Actinoplanes sp. ATCC 33076 is cultivated under aerobic conditions in the usual type of aqueous nutrient medium containing an assimilable source of carbon, an assimilable source of nitrogen and inorganic salts.

The A/16686-producing strain can be grown at temperatures between about 20° and 37°C and preferably at temperatures of about 28° to 30°C.

During the fermentation, antibiotic production can be followed by testing samples of the broth or of extracts of the mycelial solids for antibiotic activity. Organisms known to be sensitive to antibiotic A/16686 are useful for this purpose. One especially useful assay organism is *Sarcina lutea.* The bioassay is conveniently performed by the agar diffusion method on agar plates. Maximum production of antibiotic activity generally occurs between about the third and fifth day. The antibiotic produced during fermentation of the strain Actinoplanes sp. ATCC 33076 is found both in the broth and in the mycelial mass. Recovery of antibiotic A/16686 may therefore be carried out by separated extraction of broth and mycelium.

Extraction of the mycelial mass is best accomplished with methanol, but other lower alcohols and chloroform are also suitable. Antibiotic A/16686 is recovered as a raw product from the extracting solvent by routine procedure. Analogously, the broth is extracted also, preferably with n-butanol, and a further amount of raw antibiotic A/16686 is obtained by precipitation from this solution. Purification of the raw antibiotic A/16686 is then achieved by treating the product recovered from the extracting solvents with a mixture chloroform:ethanol:water 4:7:2, separating the oily product which forms and pouring it in water. This treatment causes solidification of the product which is recovered by filtration, and further purified by silica gel column chromatography eluting with a mixture of acetonitrile and 0.01 N HCl 1:1. Antibiotic A/16686, which according to this procedure is recovered in the form of the hydrochloride, is then desalted by chromatography on a crosslinked dextran gel column.

Antibiotic A/16686 has been found to possess a high order of activity in vitro and in vivo (in mice), being particularly effective against Gram-positive microorganisms. It has also been found to possess a relatively low level of toxicity when used on test animals.

Mycoplanecin Antibiotic

M. Arai, A. Torikata, R. Enokita, T. Haneishi and M. Nakajima; U.S. Patent 4,332,902; June 1, 1982; assigned to Sankyo Company Limited, Japan describe a new antibiotic substance, named mycoplanecin; a new microorganism, Actinoplanes nov. sp. strain number 41042 capable of producing mycoplanecin, and a process for producing mycoplanecin by cultivating the mycoplanecin-producing microorganism of the genus Actinoplanes.

The antibiotic has the following physical and chemical properties: (1) It is a white powder, with a MP of 161° to 167°C. (2) Its elemental analysis is: C, 61.78%; H, 8.48%; and N, 11.75%. (3) It is soluble in methanol, ethanol, ethyl acetate, acetone and chloroform; sparingly soluble in benzene; and insoluble in water. (4) It has strong antibacterial activity against various bacteria of the genus Mycobacterium. Additional properties are given in the patent, including figures showing the substance's UV and infrared spectrums.

The microorganism used in the process was isolated from a sample of soil collected in Sumoto City, Japan, and are identified by the numbers FERM 4504 and NRRL 11462.

The cultivation of the Mycoplanecin-producing microorganism can be performed under the conditions conventionally employed for the cultivation of Actinoplanes species. Shaken culture or submerged culture with aeration and agitation in a liquid medium are preferred.

Mycoplanecin is present in both the liquid portion and the mycelial portion of the culture broth produced by the cultivation. In order to recover the antibiotic from the culture broth on completion of the cultivation, the mycelium and other solids are first removed from the liquid phase by filtration, for example, using diatomaceous earth as a filter aid, or by centrifugation. The Mycoplanecin which is present in the mycelial portion on the filtrate or the supernatant can then be isolated and purified by conventional techniques suited to its physico-chemical properties.

BY MICROMONOSPORA SPECIES

Gentamycin

A.P. Da Luz; U.S. Patent 4,275,155; June 23, 1981 has devised a method of manufacturing aminoglycoside antibiotics, particularly gentamycin, by fermentation using a new variety of microorganisms of the genus Micromonospora; more particularly, this process deals with methods of producing gentamycins with a new species *Micromonospora scalabitana.* Cultures of the living microorganism were deposited as ATCC 31435.

In the course of screening soil samples of Portugal, a beautiful red microorganism was isolated from a soil sample from the plains of Santarem, a city that in Roman times was called Scalabis. It was clear from the first tests that this microorganism was an antibiotic producer and that the metabolite was active against Gram-positive and Gram-negative microorganisms. *Micromonospora scalabitana* is closely related to *Micromonospora narashinoensis.*

Micromonospora scalabitana grows very well in many culture media. Such media must contain certain assimilable substances as sources of carbon and nitrogen and there must be an adequate supply of oxygen. The sources of carbon and nitrogen may be selected from the group consisting of dextrose, sucrose, lactose, dextrin, starch and starch-containing materials; and soybean meal, bean meal, pea meal, corn steep liquor and ammonium sulfate.

There may also be included in the nutrient solution 0.05% calcium carbonate and a synergistic mixture of soluble salts of trace elements consisting of 0.0001%

nickel acetate, 0.0001% cobalt chloride, 0.0005% zinc sulfate, 0.0005% manganese sulfate, 0.0025% magnesium sulfate and 0.0025% ferrous sulfate.

Under the term gentamycin all possible variations of gentamycin are possible, such as gentamycin A, B, the whole group of gentamycins C, antibiotics JI-20A and JI-20B, sisomycin and all other so far described gentamycin-related substances. Since there are great morphological and physiological variations among variants and mutants of *M. scalabitana* there are also great variations in the number, quality and potency of the substances produced. This can be easily demonstrated by paper chromatography or thin layer chromatography.

Production of Aminoglycoside Antibiotics Related to Kanamycin

Y. Oka, H. Ishida, M. Morioka, Y. Numusaki, T. Yamafuji, T. Osono and H. Umezawa; U.S. Patent 4,279,997; July 21, 1981; assigned to Yamanouchi Pharmaceutical Co., Ltd., Japan have found that cultivation of gentamycin-producing strains of Micromonospora species or their mutants in a medium containing kanamycin A or B (known aminoglycoside antibiotics) provides compounds with the following structure.

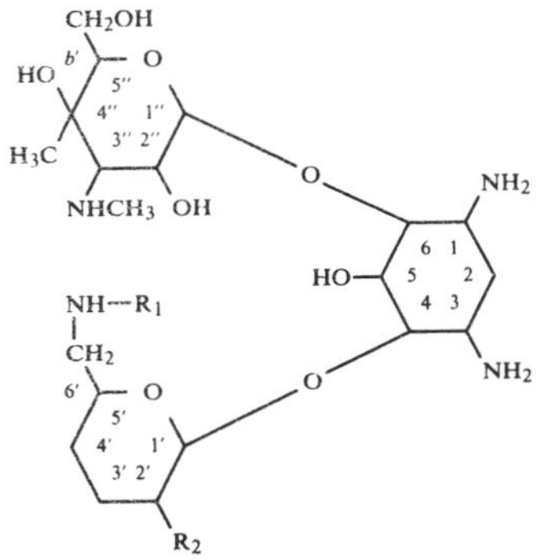

In the formula R_1 represents a hydrogen atom or a methyl group and R_2 represents an amino group or a hydroxyl group, and a process for production thereof.

The compounds have chemical structures in which kanamycins A and B have been deoxylated both at 3'- and 4'-positions, C-methylated at the 4"-position, and their amino group at the 3"-position has been methylated and their amino group at the 6'-position has optionally been methylated. These compounds possess superior antibiotic activities to those of kanamycins A and B.

The compounds of the general formula above can further be converted into acid addition salts with pharmacologically acceptable nontoxic inorganic or organic acids, such as hydrochloric acid, sulfuric acid, phosphoric acid, acetic acid, propionic acid, stearic acid, tartaric acid, maleic acid, fumaric acid, aspartic acid, glutamic acid, etc. These salts can be obtained by lyophilization of their aqueous solution or crystallization from a water-miscible solvent. When they are aseptically prepared, they can be used for injection.

The compound of the general formula above can be used for various purposes both in vivo and in vitro. For in vitro application, they can be used for sterilizing and disinfecting laboratory equipment and medical instruments, and for in vivo application, they are effective for treatment and prevention of various bacterial infections in animals and humans. They possess broad and strong antibacterial activities against various Gram-positive and Gram-negative bacteria, and show much stronger activity than those of kanamycins A and B, that is, the starting materials. They also show potent antibacterial activities against various kanamycin A- and B-resistant *Escherichia coli.*

Any gentamycin-producing strains of Micromonospora species or their mutants can be used for cultivation if they can deoxylate the 3'- and 4'-positions of KM-A or KM-B, methylate the amino grop at the 3"-position, C-methylate the 4"-position, and if desired, methylate the amino group at the 6'-position. Useful strains of known gentamycin-producing organisms are *Micromonospora echinospora* NRRL 2985(IFO-13149) and *Micromonospora purpurea* NRRL 2953(IFO-13150). Micromonospora sp. K-6993, belonging to a new Micromonospora species has been isolated from a soil sample collected at Ishigaki island, Okinawa prefecture, Japan, and confirmed to produce gentamycin. These strains have been deposited as ATCC 31348, ATCC 31349 and ATCC 31350, respectively.

In order to convert kanamycin A or kanamycin B into the compounds of the general formula above, the abovementioned strains are cultivated in the presence of kanamycin A or kanamycin B by conventional cultivation methods for producing antibiotics.

Antibiotics XK-62-3 and XK-62-4

In accordance with the process developed by *T. Iida, K. Shirahata, S. Ishii, R. Okachi and T. Nara; U.S. Patent 4,298,690; November 3, 1981; assigned to Kyowa Hakko Kogyo Co., Ltd., Japan* the antibiotics XK-62-3 and XK-62-4 are produced by fermentation of a microorganism belonging to the genus Micromonospora, which is capable of producing one or both of these antibiotics in a nutrient medium until substantial antibacterial activity is detected in the culture liquor. At the completion of culturing, the active fractions containing XK-62-3 or XK-62-4 are isolated from the culture liquor by means such as ion exchange resin treatment. The compositions are represented by the following general formula:

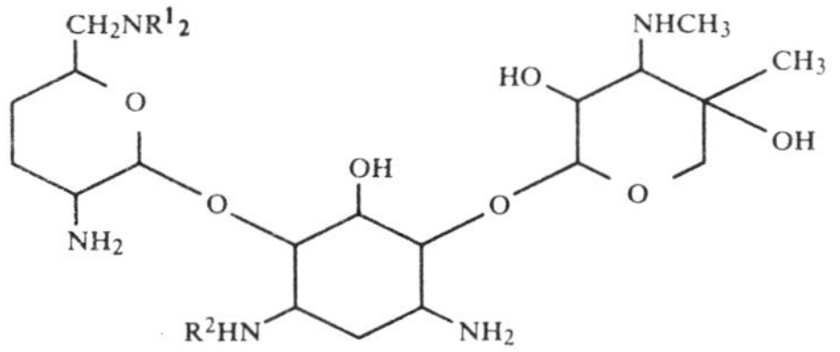

In the formula R^1 is H when R^2 is CH_3 and R^1 is CH_3 when R^2 is H.

XK-62-3 and XK-62-4 exhibit broad antibacterial activity, and are, therefore, useful to clean and sterilize laboratory glassware and surgical instruments, and may also be used in combination with soaps, detergents and wash solutions for sanitation purposes. Further, they are expected to be useful as therapeutic compounds in connection with infections induced by various bacteria.

The antibiotics XK-62-3 and XK-62-4 have a very strong antibacterial activity against a broad range of Gram-positive and Gram-negative bacteria. The compounds are particularly effective upon microorganisms of the genera Proteus and Pseudomonas upon which only a few antibiotics have been known to be effective. Furthermore, XK-62-3 has a strong antibacterial activity against certain strains of *Escherichia coli* and *Pseudomonas aeruginosa* which are resistant to various known antibiotics.

XK-62-3 and XK-62-4 are produced by fermentation of a microorganism belonging to the genus Micromonospora, namely *Micromonospora sagamiensis* var nonreducans KY 11504 (NRRL-11101) (FERM-P 3962). The strain has been deposited as *Micromonospora sagamiensis* KY 11504 with the U.S. Deaprtment of Agriculture and with the Fermentation Research Institute Agency and has been accorded the accession numbers noted above.

The microorganism was obtained by double mutation of *Micromonospora sagamiensis* var nonreducans MK-62, ATCC 21803, a known XK-62-2 producing organism. The microbiological properties of the parent strain are set forth in detail in U.S. Patent 4,045,298. Generally, conventional methods for culturing Actinomycetes may be employed in this process.

Aminocyclitol Antibiotics

In separate work, *H. Yamamoto; U.S. Patent 4,288,547; September 8, 1981* has produced aminocyclitol antibiotics of the gentamycin class, useful as antibacterials. The compounds have one of the below structures:

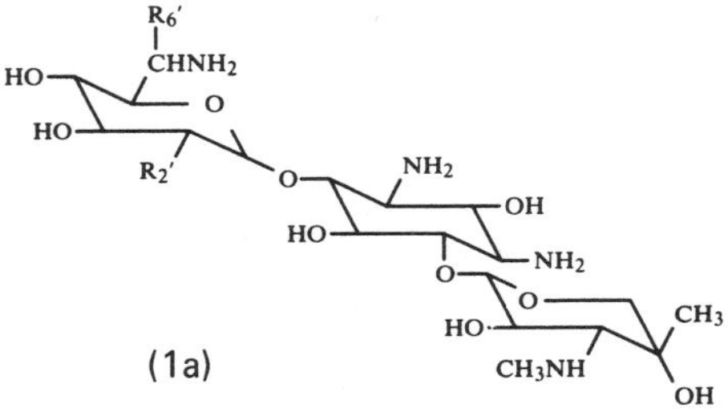

(1a)

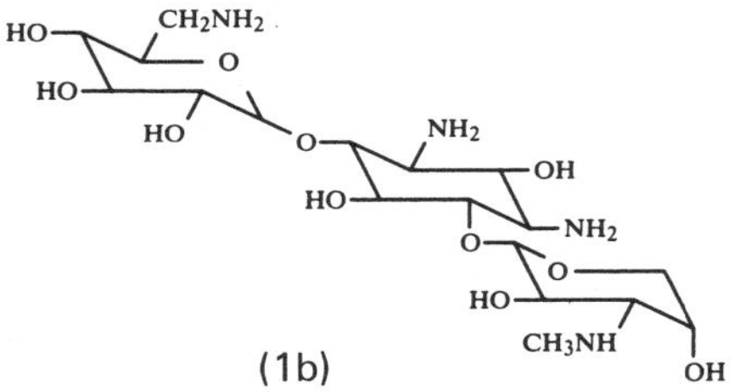

(1b)

In formula (1a), $R_{2'}$ is hydroxy or amino and $R_{6'}$ is hydrogen or methyl. The method for producing these compounds comprises culturing a nutrient medium containing carbohydrates, a source of assimilable nitrogen, essential salts and D-streptamine with *Micromonospora purpurea* ATCC 31,536.

The fermentation products are isolated from the nutrient medium and acylated with an ester of an ω-(N-benzyloxycarbonyl)amino-α-hydroxy-lower-alkanoic acid, after first blocking the 6'- and/or 2'-amine group with an amine-protecting group, followed by catalytic hydrogenolysis of the benzyloxycarbonyl group and removal of the amine-protecting groups to prepare the 1-N-(ω-amino-α-hydroxy-lower-alkanoyl) derivatives.

7-Deazaadenosine and 7-Deazainosine

Due to the usefulness of the known antibiotics, 7-deazaadenosine and 7-deazainosine, new processes for their production are in demand. *T. Nara, R. Okachi, I. Kawamoto, T. Sato and T. Oka; U.S. Patent 4,316,957; February 23, 1982; assigned to Abbott Laboratories* have found a way to accomplish this.

In accordance with this process, at least one of the compounds 7-deazaadenosine and 7-deazainosine represented by the general formula

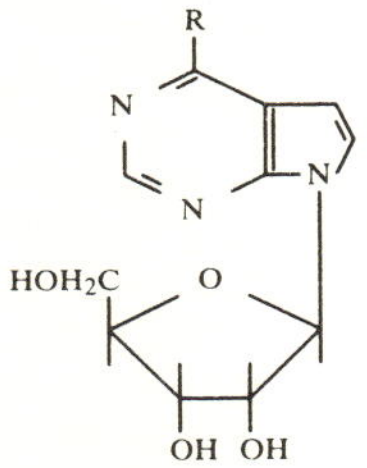

in which R is NH_2 or OH, are produced by culturing a microorganism belonging to the genus Micromonospora and having the identifying characteristics of *Micromonospora chalcea* ssp. *tubercidica* NRRL 11,107 in a nutrient medium until substantial antibacterial activity is detected in the culture liquor; and thereafter isolating at least one of the compounds from the culture liquor. *Micromonospora chalcea* ssp. *tubercidica* is a strain isolated from a soil sample from Machida City, Tokyo, Japan. Cultures of this Micromonospora have been deposited also with the Fermentation Research Institute of Japan as accession number FERM-P-3963.

Generally, conventional methods for culturing microorganisms of the Actinomycetes may be employed in this process. Various nutrient sources may be employed for the culture medium although natural sources are preferred.

A liquid culturing method, and especially a submerged stirring culturing method, is most suitable for this process. It is desirable to carry out the culturing step at a temperature of 25° to 40°C and at approximately neutral pH. Under such conditions substantial antibacterial activity is detected in the culture liquor usually after 2 to 7 days. When the antibacterial activity in the culture liquor reaches a maximum, culturing is discontinued and the desired active compounds, namely 7-deazaadenosine and/or 7-deazainosine are isolated and purified from the culture liquor after the microbial cells have been removed, such as by filtration.

Since the desired antibiotics are basic and readily soluble in water, but poorly soluble in the ordinary organic solvents, the antibiotics can be purified by a combination of adsorption and desorption from cation exchange resins; cellulose column chromatography; adsorption and desorption using a column of Sephadex LH-20; silica gel column chromatography; adsorption and desorption from active carbon and the like methods.

BY MICROORGANISMS OF OTHER GENERA

Antibiotic Bu-2349 by Use of a Bacillus Species

Various aminoglycoside antibiotics such as kanamycin, gentamycin, streptomycin, neomycin, tobramycin, amikacin and paromomycin are known in the art. There exists a need, however, for additional new broad-spectrum antibiotics, particularly those having activity against aminoglycoside-resistant organisms.

H. Kawaguchi, M. Konishi, T. Tsuno and T. Miyaki; U.S. Patent 4,260,683; April 7, 1981; assigned to Bristol-Myers Company have provided a new aminoglycoside antibiotic complex designated herein as Bu-2349. The complex is produced by fermentation of a microorganism belonging to the genus Bacillus. Any strain belonging to the genus Bacillus and capable of forming Bu-2349 in culture medium may be used. The preferred producing organisms are designated Bacillus sp. strain F173-B61 and Bacillus sp. strain F-b 262-B54 in the Bristol-Banyu culture collection. The above strains were isolated from soil samples collected in West Germany and India, respectively, and have been deposited with the American Type Culture Collection as ATCC 31429 and ATCC 31430, respectively.

This aminoglycoside complex comprises two bioactive aminoglycoside components which have been arbitrarily designated as Bu-2349A and B. The complex and each of its components exhibit a broad spectrum of antibacterial activity and are thus valuable as antibacterial agents, as nutritional supplements in animal feeds and as therapeutic agents for animals. In particular, they are useful in the treatment of infectious diseases in mammals (including man) caused by Gram-positive and Gram-negative bacteria including such diseases attributed to aminoglycoside-resistant bacteria. Additionally, the antibiotics are useful in cleaning and sterilizing laboratory glassware and surgical instruments and may be used in combination with soaps, detergents and wash solutions for sanitation purposes.

The Bu-2349 antibiotics are produced by conventional fermentation methods by cultivating a Bu-2349-producing strain of the genus Bacillus, most preferably Bacillus sp. ATCC 31429 or ATCC 31430 or a mutant thereof, under submerged aerobic conditions in an aqueous nutrient medium and recovering the Bu-2349 from the medium in substantially pure form. The structure of the Bu-2349 complex is as follows:

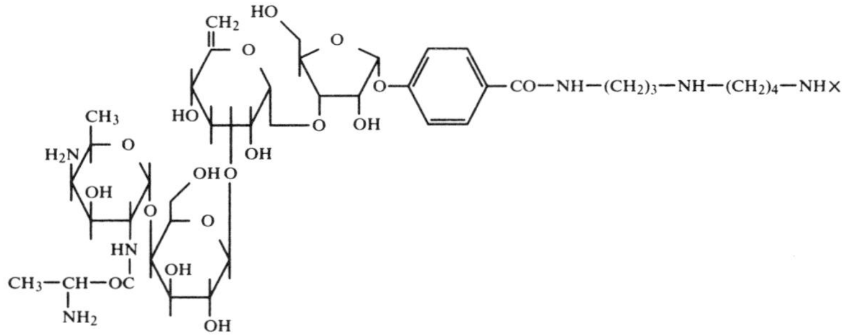

where X = $-(CH_2)_4-NH$ for compound Bu-2349A and X = $-H$ for compound Bu-2349B.

Prodigiosin by *Serratia marcescens* R-2

Prodigiosin is an antibiotic having antimicrobial activity against Gram-positive bacteria, fungi, protozoa and the like. It is also known that prodigiosin has antitumor activity.

Although prodigiosin has been produced by microbiological methods it has been difficult to get high yields. Problems have also been found in establishing a suitable medium for the cultures.

K. Nakamura and K. Kitamura; U.S. Patent 4,266,028; May 5, 1981; assigned to Kirin Beer KK, Japan have solved these problems in the microbiological production of prodigiosin by using a special strain of *Serratia marcescens* and a special medium for the fermentation. Their process consists of the use of a bacterial strain of *Serratia marcescens* R-2 obtained from soil at Takasakishi, Japan and having the ATCC accession number 31453. The synthetic culture medium contains a higher fatty acid having 12 to 18 carbon atoms, a salt thereof and/or an ester thereof with an alcohol as the sole or main assimilable source of carbon.

The esters of the higher fatty acid include such as lower alkyl esters, preferably C_{1-4} alkyl esters, e.g., methyl esters, ethyl esters, n-butyl esters, polyoxyethylenesorbitan esters (Tween) and the like. These esters are generally represented by the C_{1-4} alkyl esters of oleic acid.

These higher fatty acid compounds can be used in a reasonable concentration provided that the compounds of a given concentration do not inhibit the growth of the bacteria. A suitable concentration is generally in the range of 0.5 to 10% and preferably in the range of 1 to 3.5% by weight.

Ammonium sulfate is most preferred as the source of nitrogen. The effect of ammonium sulfate is considered to be partly due to the presence of sulfate ions. Thus, production of prodigiosin can be enhanced by the addition of sulfate ions to the culture medium when a source of nitrogen other than ammonium sulfate is used. The source of nitrogen is used in a concentration normally employed in conventional media and preferably in the range of 0.1 to 1%.

Other components which have been employed in conventional synthetic media such as phosphates and magnesium salts are essential or advantageous for use in this process. These salts can be used in concentrations normally employed in conventional synthetic media.

The starting pH of the culture medium is preferably in the range of 7.0 to 9.0. The medium is adjusted to a pH within these ranges with a suitable acid or alkali such as hydrochloric acid or sodium hydroxide.

Cultivation of the strains can be carried out in solid and liquid media. An aerobic cultivation in liquid media is most preferred to produce and obtain prodigiosin in large quantities.

The temperature of cultivation is generally in the range of about 16° to 34°C and most preferably in the vicinity of about 28°C. The cultivation is conducted with aeration and stirring and is generally terminated in 2 or 3 days.

U-53,946 by *Paecilomyces abruptus*

In the process of *L.J. Hanka and P.F. Wiley; U.S. Patent 4,282,327; August 4, 1981; assigned to The Upjohn Company,* a new antibiotic, U-53,946, is obtained by culturing *Paecilomyces abruptus* sp. nov. NRRL 11,110, in an aqueous nutrient medium under aerobic conditions. Antibiotic U-53,946 has the property of adversely affecting the growth of such Gram-positive bacteria as *Staphylococcus aureus, Sarcina lutea, Bacillus subtilis, Mycobacterium avium,* and *Streptococcus pyogenes.* It can be used, for instance, to disinfect washed and stacked food utensils contaminated with *S. aureus;* it can also be used as a disinfectant on various dental and medical equipment contaminated with *S. aureus.* Further, U-53,946 can be used for treating breeding places of silkworms, to prevent or minimize infections which are well known to be caused by *Bacillus subtilis.*

Still further, U-53,946 can be used to control *Mycobacterium avium* which is a known producer of generalized tuberculosis in birds and rabbits.

Antibiotic U-53,946, in its essentially pure form: (a) has the molecular formula $C_{61}H_{107}N_{11}O_{14}$; (b) has the following elemental analysis: C, 60.13%, 59.73%; H, 9.21%, 8.92%; N, 12.16%, 11.97%; O, 18.44%; (c) has a specific rotation of $[\alpha]_D^{25}$ = -26° (c, 0.9175 EtOH); (d) is soluble in lower alcohols, e.g., methanol, ethanol and butanol; ketones, e.g., methyl ethyl ketone, halogenated solvents, ethyl acetate, or other ester-type solvents; and is relatively insoluble in aliphatic hydrocarbons; (e) has a characteristic infrared absorption spectrum; and (f) a characteristic UV spectrum. (Both of these spectrums are shown in the patent.)

A-30912 Antibiotics Using ***Aspergillus nidulans***

L.D. Boeck and R.E. Kastner; U.S. Patent 4,288,549; September 8, 1981; assigned to Eli Lilly and Company have developed a new method for producing the A-30912 antibiotics. This method comprises culturing *Aspergillus nidulans* var. *roseus,* NRRL 11440 under submerged aerobic fermentation conditions until a substantial level of antibiotic activity is produced. The A-30912 antibiotics are separated initially as a complex (the A-42355 antibiotic complex). A-30912 factors A, B, D and H are isolated from the A-42355 complex by chromatography. The A-30912 antibiotics have antifungal activity.

The authors further describe the biologically pure culture of the microorganism *Aspergillus nidulans* var. *roseus* which is useful for the production of the A-30912 antibiotics.

The antibiotic complex produced is arbitrarily designated herein as the A-42355 antibiotic complex. To avoid confusion, however, the A-30912 designations will be used for those individual factors of the A-42355 complex which are identical to A-30912 factors.

As is the case with the A-30912 antibiotic complex, A-30912 factor A is the major factor of the A-42355 antibiotic complex; A-30912 factors B, D, and H continue to be minor factors in the A-42355 complex. The compound A-30912 factor A and its production by *Aspergillus nidulans* var. *echinulatus* and *Aspergillus rugulosus* has previously been described (U.S. Patents 4,024,245 and 4,024,246).

A-30912 factor H (A-30912H), the newly discovered component of the A-30912 antibiotic complex, is quite similar to A-30912 factor A. Under conditions known thus far, A-30912H is a minor factor in the A-30912 complex, being present in amounts in the range of from about 0.01 to 1.0% of the total complex. Another minor factor of the A-30912 complex has been recognized, but has not been isolated in an amount sufficient for characterization. A-30912 factor H is best separated from this factor by silica gel TLC using an ethyl acetate:methanol (3:2) or an acetonitrile:water (95:5) solvent system. In either system, the uncharacterized minor factor is more polar than the other A-30912 factors. A-30912 factor H is a white amorphous solid.

Amino acid anlysis of A-30912 factor H indicated the presence, after hydrolysis, of threonine and four other as-yet-unidentified amino acids. A-30912H has the structure shown on the following page, where R represents the linoleoyl radical.

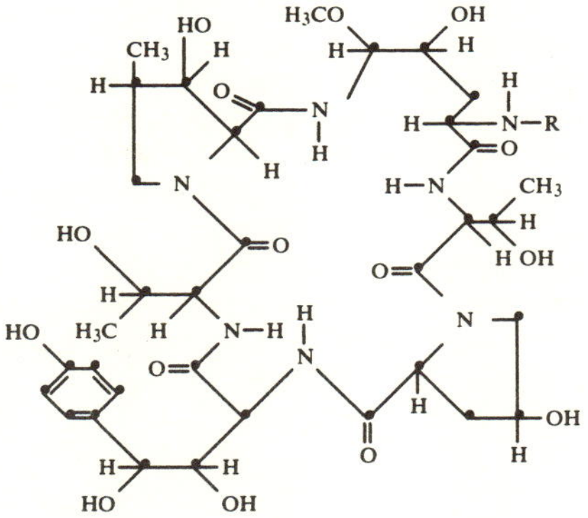

The method for producing the described A-30912 antibiotics comprises cultivating a new culture which is a variety of *Aspergillus nidulans.* The new culture has been named *Aspergillus nidulans* var. *roseus.* This microorganism is a biologically pure culture which was isolated from a soil sample from Greenfield, Indiana. One of its strain characteristics is production of the A-30912 antibiotics. A subculture of this microorganism has the number NRRL 11440.

The culture medium used to grow *Aspergillus nidulans* var. *roseus* can be any one of a number of media. For economy in production, optimal yield and ease of product isolation, however, certain culture media are preferred. Thus, for example, a preferred carbon source in large-scale fermentation is cottonseed oil or glucose. Preferred nitrogen sources are enzyme-hydrolyzed casein, soybean meal and soluble meat peptone. Nutrient inorganic salts can be incorporated in the culture media, including the customary soluble salts capable of yielding sodium, magnesium, zinc, iron, calcium, ammonium, chloride, carbonate, sulfate, nitrate, phosphate, and the like ions and essential trace elements necessary for the growth and development of the organism.

For production of a substantial quantity of the A-42355 antibiotic complex, submerged aerobic fermentation in tanks is preferred.

A. nidulans var. *roseus* NRRL 11440 can be grown at temperature between about 20° and 43°C. Optimum production of the A-30912 antibiotic complex occurs at temperatures below 30°C.

The A-30912 antibiotics can be recovered from the fermentation medium by methods known in the fermentation art and isolated from the separated A-42355 complex by chromatography using various adsorbents.

Reversed-phase high-performance, low-pressure liquid chromatography (HPLPLC) using silica gel/C_{18} adsorbent is a preferred method for final purification of the A-30912 antibiotics. Silica gel/C_{18} is available from E. Merck, Darmstadt, Germany.

Antibiotic U-59,760 by *Saccharopolyspora hirsuta*

H.A. Whaley and J.H. Coats; U.S. Patent 4,293,651; October 6, 1981; assigned to The Upjohn Company have utilized a culture of the new microbe *Saccharopoly-*

spora hirsuta strain 367, NRRL 12045, to produce the useful antibiotic U-59,760. The physical and chemical characteristics of this antibiotic cannot be distinguished from the antibiotic known as compound 47444 which is described in U.S. Patent 4,148,883. The production of the antibiotic in U.S. Patent 4,148,883 is accomplished by an entirely different microbe which is classified as a Nocardia. The microorganism used for this production of antibiotic U-59,760 is a biologically pure culture of *Saccharopolyspora hirsuta* strain 367, NRRL 12045.

This microorganism was studied and characterized by Alma Dietz and Grace P. Li of The Upjohn Company Research Laboratories and is an Actinomycete isolated in the Upjohn soils screening laboratory which was found to have the macroscopic, microscopic and whole cell hydrolysate properties of the genus Saccharopolyspora.

Unique properties of the genus Saccharopolyspora are its butyrous or gelatinous-type vegetative growth, sparse aerial growth which is best studied after 21 days incubation, yellow to orange to orange-tan vegetative growth and pigment production, spore chains developing in a sheath with unique hair tufts with truncated bases and smooth areas between the tufts, the presence of meso-diaminopimelic acid, arabinose and galactose in whole cell hydrolysates and in cell wall preparations. L-diaminopimelic acid was detected in addition to meso-DAP in cell wall preparations.

The antibiotic U-59,760 is produced when the elaborating organism is grown in an aqueous nutrient medium under submerged aerobic conditions at temperatures preferably between about 20° and 28°C. Ordinarily, optimum production of the compound is obtained in about 3 to 15 days.

The antibiotic is active against various microorganisms such as *Staphylococcus aureus, Streptococcus pyogenes,* and *Klebsiella pneumoniae.*

Since antibiotic U-59,760 is active against *S. aureus,* it can be used to disinfect washed and stacked food utensils contaminated with this bacteria. Also, it can be used as a disinfectant on various dental and medical equipment contaminated with *S. aureus.* Further, antibiotic U-59,760 can be used as a bacteriostatic rinse for laundered clothes, and for impregnating papers and fabrics; and, it is also useful for suppressing the growth of sensitive organisms in plate assays, and other microbiological media.

Nocardicin by Cultivation of *Microtetraspora caesia*

K. Tomita, H. Tsukiura and H. Kawaguchi; U.S. Patent 4,320,199; March 16, 1982; assigned to Bristol-Myers Company describe a process for the preparation of the known antibiotic nocardicin by cultivating a strain of *Microtetraspora caesia* sp. ATCC numbers 31724 or 31725 under submerged aerobic conditions in an aqueous medium until a substantial amount of nocardicin is produced in the culture medium, and, optionally, recovering nocardicin from the culture medium. As with prior art procedures, this process provides a mixture of nocardicin A (the major component) and nocardicin B (the minor component) which, if desired, may be readily separated by known procedures. Nocardicin has the structure shown on the following page and is produced by fermentation of certain new species of the genus Microtetraspora under submerged aerobic conditions in an aqueous culture medium containing assimilable sources of carbon and nitrogen.

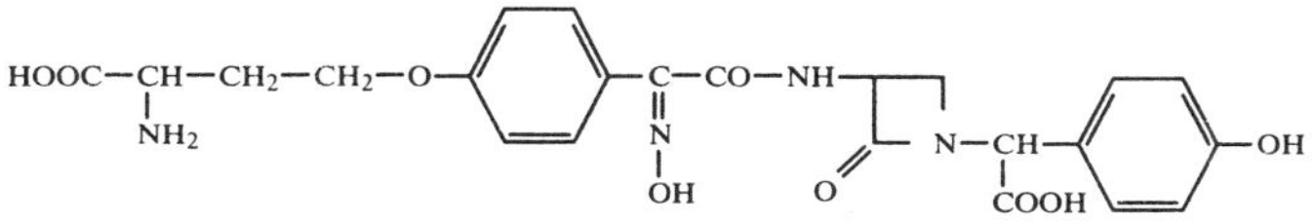

In the course of screening for β-lactam antibiotics, two unusual Actinomycetes strains, numbers G432-4 and G434-6, which produced nocardicin were isolated from soil samples collected in India. Both strains grow well on natural media or chemically defined organic media at temperatures between 20° and 50°C.

As the results of comparative studies with several genera of the Actinomycetales, strains numbers G-432-4 and G434-6 were determined to be a new species of genus Microtetraspora. The new species name *Microtetraspora caesia* is proposed for strains G432-4 and G434-6 in view of the bluish-gray colored aerial mycelium. Cultures of strains G432-4 and G434-6 have been deposited as ATCC 31724 and ATCC 31725, respectively.

Antibiotic EM-5400 Using *Agrobacterium radiobacter*

R.B. Sykes and W.L. Parker; U.S. Patent 4,321,326; March 23, 1982; assigned to E.R. Squibb & Sons, Inc. have found that cultivation of a strain of the microorganism *Agrobacterium radiobacter* ATCC 31700 in a conventional culture medium produces the antibiotic substance EM5400. This substance comprises a mixture of certain derivatives of azetidinesulfonic acid.

The components of EM5400 have been isolated; they are salts of: 3-[[2-(acetylamino)-3-(4-hydroxyphenyl)-1-oxopropyl]amino]-2-methoxy-2-oxo-1-azetidinesulfonic acid; 3-[[2-(acetylamino)-1-oxo-3-phenylpropyl]-amino]-3-methoxy-2-oxo-1-azetidinesulfonic acid; 3-[[2-(acetylamino)-1-oxo-3-(sulfoxy)-3-[4-(sulfoxy)-phenyl]propyl]amino]-3-methoxy-2-oxo-1-azetidinesulfonic acid; [3S(R*)]-3-[[2-(acetylamino)-3-(4-hydroxyphenyl)-1-oxopropyl]amino]-2-oxo-1-azetidinesulfonic acid; and 3-[[2-(acetylamino)-3-hydroxy-1-oxo-3-[4-(sulfoxy)-phenyl]propyl]-amino]-3-methoxy-2-oxo-1-azetidinesulfonic acid.

To form the antibiotic mixture EM5400 according to the preferred method, *Agrobacterium radiobacter* ATCC 31700 is grown at, or near, room temperature (25°C) under submerged aerobic conditions in an aqueous nutrient medium containing an assimilable carbohydrate and nitrogen source. The fermentation is carried out for at least about 18 hours.

After the fermentation is completed, the beer is acidified, preferably to about pH 4, the cells removed by either centrifugation or filtration. The supernate is extracted with 0.05 M cetyldimethylbenzylammonium chloride in methylene dichloride and the resulting extract, concentrated to a small volume, is then extracted with a 1.08 M aqueous solution of sodium thiocyanate, at pH 4.35. The aqueous layer, reduced to a small volume, is diluted with 4 volumes of methanol to remove methanol-insoluble impurities. The methanol-soluble material is chromatographed on Sephadex G-10 with aqueous methanol (2:1, v/v). Progress of the fractionation in this and subsequent steps is followed by high voltage electrophoresis on pH 7 buffered paper. When applicable, thin layer chromatography on silica gel, with methylene dichloride:methanol (4:1, v/v) as developing solvent, is also done.

The compounds which make up the antibiotic mixture EM5400 can be used as agents to combat bacterial infections (including urinary tract infections and respiratory infections) in mammalian species, such as domesticated animals (e.g., dogs, cats, cows, horses, and the like) and humans.

Preparation of 3-Trehalosamine with a Nocardiopsis

L.A. Dolak, A.L. Laborde and T.M. Castle; U.S. Patents 4,276,412; June 30, 1981; 4,306,028, December 15, 1981; and 4,310,627; January 12, 1982; all assigned to The Upjohn Company have developed a new aminoglycoside antibiotic, considered to be 3-trehalosamine. The microorganism used for the production of 3-trehalosamine is *Nocardiopsis trehalosei* sp. nov.

A subculture of this microorganism can be obtained from the permanent collection of the Northern Regional Research Laboratory under accession number NRRL 12026. The microorganism was studied and characterized by Alma Dietz and Grace P. Li of The Upjohn Research Laboratories.

An Actinomycete isolated in the Upjohn soils screening laboratory has been characterized and found to have the macroscopic, microscopic, and whole cell hydrolysate properties of the genus Nocardiopsis [Meyer, *Int. J. Syst. Bacteriol.* 26 (1976):487-493]. The new soil isolate was compared with this strain. The new soil isolate differs from *N. dassonvillei* in its color properties, its failure to grow on sucrose and D-fructose in the synthetic medium of Shirling and Gottlieb and in its cultural characteristics. It is further distinguished by the production of 3-trehalosamine.

This culture is considered a member of the distinctive Actinomycete genus Nocardiopsis based on its nocardioform substract mycelium, aerial mycelium with distinctive spore chain development; at first zigzag or twisted ribbon-like, subsequently constricting to form spores of irregular size (mostly elongated) with a smooth surface. Mature spore chains are not readily detected before 21 days. meso-Diaminopimelic acid is present in whole-cell hydrolysates and in cell-wall preparations. No diagnostic carbohydrates are present.

On the basis of the distinctions noted, the new culture is considered to be a new species of the genus Nocardiopsis. It is proposed that the culture be designated *Nocardiopsis trehalosei* Dietz and Li sp. nov.

The antibiotic 3-trehalosamine is produced when the elaborating organism is grown in a conventional aqueous nutrient medium under submerged aerobic conditions.

Production of the compound can be effected at any temperature conducive to satisfactory growth of the microorganism, preferably between about 20° and 28°C. Ordinarily, optimum production of the compound is obtained in about 3 to 15 days. The medium normally remains alkaline during the fermentation. The final pH is dependent, in part, on the buffers present, if any, and in part on the initial pH of the culture medium.

When growth is carried out in large vessels and tanks, it is preferable to use the vegetative form, rather than the spore form, of the microorganism for inoculation to avoid a pronounced lag in the production of the compound and the attendant inefficient utilization of the equipment. Accordingly, it is desirable to produce a vegetative inoculum in a nutrient broth culture by inoculating this broth culture with an aliquot from a soil, liquid N_2 agar plug, or a slant culture. When a young, active vegetative inoculum has thus been secured, it is transferred aseptically to large vessels or tanks. The medium in which the vegetative inoculum is produced can be the same as, or different from, that utilized for the production of the compound, so long as a good growth of the microorganism is obtained.

A variety of procedures can be employed in the isolation and purification of the compound produced by the process from fermentation beers.

Isolation can be accomplished by resin adsorption over a cationic exchange resin. The resin can be eluted with a suitable salt or buffer such as ammonium sulfate in water. Purification of the antibiotic can be accomplished by chromatography of the isolated material over a suitable chromatographic column, for example, IR-45 (OH–). Antibiotic 3-trehalosamine can be shown by the following structural formula:

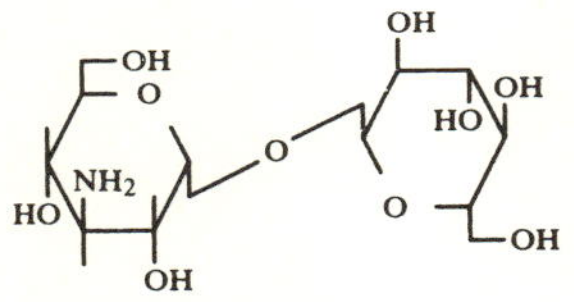

3-Trehalosamine is active against various Gram-positive and Gram-negative bacteria. Since it is active against *Staphylococcus aureus,* it can be used to disinfect washed and stacked food utensils contaminated with this bacterium; it can also be used as a disinfectant on various dental and medical equipment contaminated with *Staphylococcus aureus.* Further, 3-trehalosamine and its salts can be used as a bacteriostatic rinse for laundered clothes, and for impregnating papers and fabrics; and, they are also useful for suppressing the growth of sensitive organisms in plate assays and other microbiological media. Since 3-trehalosamine is active against *B. subtilis,* it can be used in petroleum product storage to control this microorganism which is a known slime and corrosion producer in petroleum products storage.

ANTIBIOTIC BIOCONVERSION

10-Dihydrosteffimycin

The process for preparing the antibiotic steffimycin, and the description of its various biological properties, are disclosed in U.S. Patent 3,309,273. The antibiotic at that time was known as steffisburgensimycin. The process for preparing steffimycin B and its characterization are disclosed in U.S. Patent 3,794,721.

V.P. Marshall, D.W. Elrod and P.F. Wiley; U.S. Patent 4,264,726; April 28, 1981; assigned to The Upjohn Company have found that antibiotic 10-dihydrosteffimycin can be prepared in a fermentation process using the known microorganism *Actinoplanes utahensis,* NRRL 5614. This microorganism is described in U.S. Patent 3,824,305. 10-Dihydrosteffimycin is prepared in the fermentation by the addition of steffimycin to the fermentation.

The antibiotic 10-dihydrosteffimycin B also can be prepared in a fermentation using *Actinoplanes utahensis,* NRRL 5614, or the microbe Chaetomium sp. (BB 427) NRRL 11442. This antibiotic is produced in the fermentations by the addition of steffimycin B to the fermentation.

The antibiotics of this process are recovered from the fermentation beers using a series of filtration, extraction, and thin layer chromatography (TLC) procedures. 10-Dihydrosteffimycin and 10-dihydrosteffimycin B have the following structures:

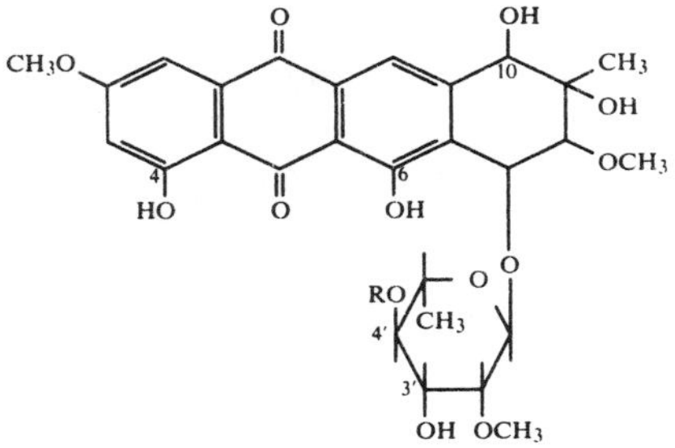

R = H (10-Dihydrosteffimycin)
R = CH_3 (10-Dihydrosteffimycin B)

Actinoplanes utahensis, NRRL 5614, is a known microorganism deposited at the Northern Regional Research Laboratory, U.S. Department of Agriculture.

Chaetomium sp. (BB 427) is a biologically pure culture of a new microorganism deposited in the permanent collection at the Northern Regional Research Laboratory as accession number NRRL 11442.

Steffimycin and steffimycin B can be added to their respective fermentations advantageously as dimethylformamide solutions (25 mg/ℓ), or as a milled aqueous suspension, to a final medium concentration of about 1 to 100 mg/ℓ, advantageously, to about 25 mg/ℓ. The addition of these compounds to the fermentations can be done at any time after suitable growth of the microbe is evidenced and preferably between about 36 and about 48 hours of fermentation time.

A variety of procedures can be employed in the isolation and purification of the compounds produced by this process, for example, solvent extraction, partition chromatography, silica gel chromatography, liquid-liquid distribution in a Craig apparatus, adsorption on resins, and crystallization from solvents.

These compounds are active against various Gram-positive bacteria. For example, they are active against *Bacillus subtilis, Bacillus cereus, Staphylococcus aureus, Sarcina lutea,* and *Mycobacterium avium.* Accordingly, the compounds can be used as a disinfectant on washed and stacked food utensils contaminated with *Staphylococcus aureus.* They can also be used in birds and rabbits to control the organism *Mycobacterium avium,* which is a known producer of generalized tuberculosis in these animals. They can also be used in papermill operations to control the contamination of wool by the organism *Bacillus cereus.* They can also be used in petroleum products storage to control the microorganism *Bacillus subtilis* which is a known slime and corrosion producer in petroleum products storage.

Allylic Methyl-Hydroxylated Novobiocins

Novobiocin is an antibiotic useful in the treatment of staphylococcal infections and in urinary tract infections caused by certain strains of Proteus. It shows no cross resistance with penicillin and is active against penicillin-resistant strains of *Staphylococcus aureus.* Novobiocin is produced through fermentation by streptomycetes. The methods for production, recovery and purification of novobiocin are described in U.S. Patent 3,049,534. A number of modifications of novobiocin have been patented; however, none of them relates to modification of the isopentyl side chain on the benzamide ring.

O.K. Sebek and L.A. Dolak; U.S. Patent 4,304,855; December 8, 1981; assigned to The Upjohn Company have developed a procedure for producing a hydroxynovobiocin-type compound (one in which the terminal methyl group in the formula below is replaced by a $-CH_2OH$ group) or a pharmaceutically acceptable salt of this compound.

This process comprises: (1) cultivating *Sebekia benihana* having the identifying characteristics of NRRL 11,111 and novobiocinhydroxylating mutants thereof in an aqueous nutrient medium under aerobic conditions; (2) contacting a novobiocin-type compound of the formula below with the *Sebekia benihana* culture; and (3) recovering the hydroxynovobiocin-type compound.

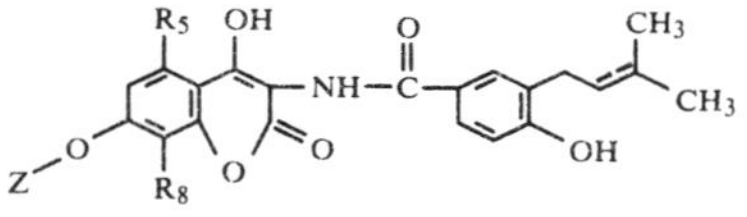

R_5 and R_8 may be the same or different and are hydrogen, alkyl of from 1 through 5 carbon atoms, alkenyl of from 1 through 5 carbon atoms, halogen, nitro, cyano, carboxyl, or $-NR_1R_2$, where R_1 and R_2 may be the same or different and are hydrogen or alkyl of 1 through 5 carbon atoms, ---- is a single or double bond, and Z is hydrogen or

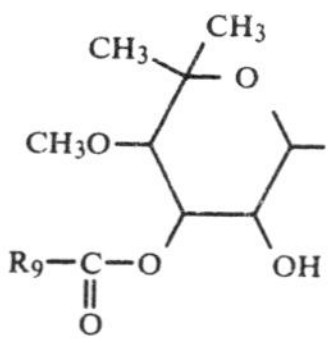

where R_9 is amino, 2-pyrryl, 2-(5-methyl)-pyrryl, 2-furyl, and 2-(5-methyl)-furyl.

The microorganism used in this process was studied and characterized by Alma Dietz and Grace P. Li of the Upjohn Research Laboratories, and is an unusual Actinomycete, isolated from a soil sample.

One of the characteristics of *Sebekia benihana* is the hydroxylation of the trans methyl group of the 3-methyl-2-butenyl side chain of ring A of novobiocin.

For the purpose of this process, the microorganism is grown in or on a sterile medium favorable to its development. Sources of nitrogen and carbon are present in the culture medium, the pH is properly adjusted and an adequate sterile air supply is maintained.

The preferred medium for this process is TYG medium. It is utilized for the growth of the microorganism prior to addition of the substrate and during the bioconversion process. It is adjusted to pH 7.2 in deionized water. Its composition is tryptone, 0.5%; yeast extract, 0.3%; and glucose, 2.0%.

It is preferred that the substitutes which undergo bioconversion by the described process are novobiocin compounds of the above formula or their salts where R_5 is hydrogen and R_8 is methyl or chlorine and R_9 is amino.

Following completion of the bioconversion, as measured by thin layer chromatography, the products are recovered and purified by methods well known to those skilled in the art. The fermentation beer is adjusted to pH 2 to 5 with an acid such as hydrochloric, sulfuric, phosphoric, etc. The solids are separated by centrifugation or by mixing the fermentation beer with approximately one-tenth volume of a filter aid such as Dicalite 4,200, or any other diatomaceous earth product.

The hydroxynovobiocins are useful in the same manner and in the same way as the corresponding novobiocin-type parent compounds, except that about ten times higher concentration than novobiocin should be used. The hydroxynovobiocin-type compounds are useful to sterilize glassware and utensils in the concentration range of 0.01 to 10.0%. Walls, bench tops and floors may be cleaned of susceptible organisms using the same concentration range. In addition, the hydroxy novobiocin-type compounds can be used to selectively destroy susceptible organisms in soil samples prior to screening for antibiotics. Further these hydroxy compounds may be used to destroy susceptible organisms in the bowels of animals for studies of digestion and excretion.

MISCELLANEOUS PREPARATIONS

Use of Phenoxyalkanes as Precursors for Penicillin V Production

Penicillin V, that is, 3,3-dimethyl-7-oxo-6-[(phenoxyacetyl)amino]-4-thia-1-azabicyclo[3.2.0]-heptane-2-carboxylic acid, also known as phenoxymethyl penicillin, is a widely prescribed antibacterial agent. Penicillin V is available in the free acid or various salt forms such as the potassium salt and is formulated primarily for oral administration in tablets or solutions.

Penicillin V is conventionally prepared by fermentation. A penicillin-producing microorganism is cultured in a medium containing a source of nitrogen, a source of carbon and energy, various inorganic salts, and a sidechain precursor. Phenoxyacetic acid and its salts such as potassium phenoxyacetic acid are conventionally employed as Penicillin V sidechain precursors.

L.J. Szarka and R.W. Eltz; U.S. Patent 4,250,258; February 10, 1981; assigned to E.R. Squibb & Sons, Inc. produce Penicillin V by a conventional fermentative biosynthetic process except that one or more phenoxyalkanes of the formula

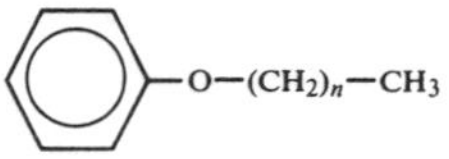

where n is an integer from 6 to 13 is included within the fermentation medium as the sidechain precursor.

By the term "conventional fermentation process" it is meant that the process conditions, the penicillin-producing microorganism, the nitrogen source, the carbon and energy source, and the salts present within the culture medium are those commonly employed in the production of Penicillin V. This process can employ various penicillin-producing cultures. In general those of the Penicillium type are preferred, with *Penicillium chrysogenum* being most preferred.

The phenoxyalkane sidechain precursors can be employed in this process in a similar manner to the phenoxyacetic acid salts. Thus, the phenoxyalkane or mixture of phenoxyalkanes can be added periodically in small amounts to the culture medium during the course of the fermentation process.

Preferably, the phenoxyalkanes employed in this process are one or more wherein n is an integer from 7 to 11. Most preferred are those wherein n is 8 or 9. These most preferred phenoxyalkanes are less toxic to the penicillin-producing microorganisms than the commercially employed potassium phenoxyacetic acid. Consequently, these most preferred phenoxyalkanes can be added in toto to the initial culture medium, thus eliminating the need for careful monitoring of the medium and the periodic introduction of additional sidechain precursor.

Zinc Complexes of Mononitrogen-Containing Divalent Pyrrole Ether Antibiotics

Previously known mononitrogen-containing divalent pyrrole ether antibiotics, such as that named X-14547, have generally been recovered and used as alkali metal or alkaline earth metal salts. The recovery process has been very complicated, requires the use of quite large quantities of various organic solvents (some of which are expensive), and has led to some antibiotic and solvent loss.

J.L. Martin; U.S. Patent 4,356,264; October 26, 1982; assigned to International Minerals & Chemical Corporation has developed a process for producing zinc complexes of this type of antibiotic, which complexes can be used for poultry to enhance growth and to combat coccidial infection and also for improving cardiovascular function in animals, particularly mammals, including man.

In accordance with this process, zinc complexes of pyrrole ether antibiotics can be advantageously formed by adding water-soluble zinc salts to the fermentation broth in which such antibiotics have been produced. When formed in a fermentation beer, the formation of these complexes facilitates the recovery of the pyrrole ether antibiotics from the fermentation beer in which the antibiotics can be produced by, among other things, avoiding the necessity of using recovery methods which involve extractions with organic solvents followed by their subsequent purification and reuse. The resulting broth-insoluble zinc complexes of the antibiotics can then be recovered from the broth and employed, for instance, as coccidiostatic, feeding efficiency improving and growth-promoting agents for poultry. Upon further purification of the recovered zinc complexes by suitable methods, they may be used to stimulate cardiovascular function in animals.

An antibiotic-containing fermentation broth can be prepared in conventional manner by fermenting a nutrient-containing liquid fermentation medium inoculated with a Streptomyces microorganism capable of producing the desired antibiotic such as Streptomyces sp. X-14547 (NRRL 8167). Characteristics, production and recovery of X-14547 are described in U.S. Patent 4,100,171 to Westley et al, U.S. Patent 4,161,520 to Osborne et al, and in articles by Liu et al, *J. Antibiotics* 32, 95-99 (1979) and Westley, *J. Antibiotics* 32, 100-107 (1979).

OTHER DRUGS PRODUCED BY MICROORGANISMS

ANTIBIOTICS HAVING ANTITUMOR OR ANTILEUKEMIC ACTIVITY

Anthracycline Glycosides

Anthracycline glycosides designated MA 144-G1, -G2, -L, -S1, -N1, -U1 and -Y which inhibit the growth of Gram-positive bacteria and mammalian tumors are described by *H. Umezawa, T. Takeuchi, T. Oki and T. Inui; U.S. Patent 4,245,045; January 13, 1981; assigned to Zaidan Hojin Biseibutsu Kagaku Kenkyu Kai, Japan.* These antibiotics may be produced either by fermentation of certain species of Streptomyces or by the chemical or enzymatic conversion of aclacinomycin A, cinerubin A, rhodirubin A, MA 144-M1, MA 144-M2, MA 144-G1, MA 144-G2, MA 144-U1, MA 144-U2, MA 144-Y or MA 144-N1. The so-produced antibiotics may be recovered, separated and purified by conventional methods used to isolate and purify water-insoluble antibiotics, the methods including at least one process selected from the group consisting of solvent extraction, solvent precipitation, concentration, gel filtration, countercurrent distribution, chelation with metal ions and adsorption followed by elution from an ion exchange resin, adsorbent siliceous earth material or synthetic adsorbent.

The MA 144 antibiotics may be prepared by fermentation of MA 144-producing strains of Streptomyces, such as *S. galilaeus* MA 144-M1 (FERM P-2455, ATCC 31133), *S. galilaeus* (ATCC 14969), *S. cinereoruber* (ATCC 19740), *S. niveoruber* (ATCC 14971), *S. antibioticus* (ATCC 8663), *S. purpurascens* (ATCC 25489), *S. sp.* ME 505-HE1 (ATCC 31273) and mutants thereof.

Production of the MA 144 compounds is carried out by cultivating the appropriate strain of Streptomyces as indicated above in a conventional aqueous nutrient medium containing known nutritional sources for actinomycetes, i.e., assimilable sources of carbon, nitrogen and inorganic salts. Submerged aerobic culture is preferably employed for the production of substantial amounts of the MA 144 components, much as for other fermentation antibiotics. The general procedures used for the cultivation of other actinomycetes are applicable.

As mentioned above, the components MA 144-G1, -G2, -L, -N1, -S1, -U1 and -Y are new antibiotics, useful in both human and veterinary medicine. They also possess marked inhibitory action against solid and ascitic-type mammalian malignant tumors. The antibiotics can also be used in the form of their nontoxic acid salts.

Figaroic Acid Complex

A new anthracycline antibiotic complex is reported by *W.T. Bradner, J.A. Bush, and D.E. Nettleton, Jr.; U.S. Patent 4,248,970; February 3, 1981; assigned to Bristol-Myers Company.* This antibiotic is designated a figaroic acid complex and it is prepared by fermentation of a new strain of *Streptosporangium* designated *S. sp.* strain C-31,751. The above organism was obtained from a soil sample taken from Seelyville, Indiana. A culture of the organism has been deposited without restrictions in the American Type Culture Collection as ATCC 31129.

Figaroic acid complex inhibits growth of various Gram-positive bacteria, for example, *Staphylococcus aureus* and *Mycobacterium tuberculosis*, and various protozoa and yeasts, for example, *Candida albicans, Histoplasma capsulatum, Trichomonas vaginalis* and Trichomonas faetus. The substance exhibits phage inducing properties and inhibits growth of various lymphatic and solid tumor systems in rodents including Sarcoma 180, L-1210 lymphatic leukemia, Walker 256 carcinosarcoma, P-388 lymphatic leukemia and B-16 melanoma. The figaroic acid complex may be used alone or in combination with other antibacterial agents to prevent the growth of, or reduce the number of, the sensitive Gram-positive bacteria, yeasts and protozoa mentioned above. It is useful to wash solutions for sanitation purposes, e.g., for washing hands and disinfecting various laboratory, dental and medical equipment or other contaminated materials and as a bacteriostatic rinse for laundered clothes. It is also useful in treating the abovementioned tumor systems in mice and rats.

This antibiotic is prepared by cultivation of the microorganism named above or a mutant of the Streptosporangium under submerged aerobic conditions at temperatures up to 43°C, conveniently around 27°C, for 170 to 210 hours in a conventional aqueous nutrient medium at a slightly alkaline pH. The antibiotic complex is extracted and isolated in the normal manner.

Cultivation of the Hemolytic ***Streptococcus pyogenes***

S. Inoue, M. Sotomura, H. Tanaka, S. Iwamoto, N. Takamatsu, A. Suzuki and I. Utsumi; U.S. Patent 4,306,024; December 15, 1981; assigned to Kanebo Ltd., Japan describe a process for cultivation of hemolytic *Streptococcus pyogenes* (hereafter *St. pyogenes*) and more particularly, to a process for cultivation of *St. pyogenes* by which to obtain bacterial cells having a high Streptolysin-S (hereafter SLS) producing ability and antitumor activity in high yield.

St. pyogenes are pathogenic bacteria of erysipelas, septicemia, puerperal fever and other various diseases, but some parts of *St. pyogenes* have long been known to have the antitumor activity, and in recent years they have been used clinically as anticancer agents. The antitumor activity of *St. pyogenes* has a close relation with its SLS-producing ability, and even among *St. pyogenes* present in nature it is only those strains having SLS-producing ability that have antitumor activity. It is also known that even with *St. pyogenes* having SLS-producing ability the antitumor activity will be lost if it is cultivated under such conditions as to lose the SLS-producing ability. For example, when cultivating *St. pyogenes* it is known that if glucose is added even in a small amount to the culture medium, the SLS-producing ability is rapidly lowered, and if the amount of glucose added reaches 0.3% or more the SLS-producing ability is nearly completely lost, entailing the simultaneous loss of its antitumor activity. This is true not only of glucose but also of lactose, fructose, mannose or glyceraldehyde.

In this process, high yields of SLS-producing *St. pyogenes* were produced, even when fermentable carbon sources such as glucose were used in the culture medium, by maintaining the pH of the culture medium at 5.6 or higher, preferably 6 to 7.

The culture medium, therefore, consists of a carbon source, preferably a saccha-ride such as glucose or sucrose in amounts of 0.3 to 2%, in soy peptone broth.

Class of Anthracycline Antibiotics

G. Cassinelli, A. Grein, S. Merli and G. Rivola; U.S. Patent 4,309,503; January 5, 1982; assigned to Farmitalia Carlo Erba SpA, Italy describe a new class of anthracycline antibiotics which are deoxy analogs of daunomycin and adriamycin (also known respectively as daunorubicin and doxorubicin) and their aglycones, and methods for preparing these compounds.

This class of anthracycline antibiotics, hereafter designated as glycosides A, B, C and D, are 11-deoxy derivatives of daunomycin and adriamycin, as well as their respective aglycones. These 11-deoxy derivatives have the following formula:

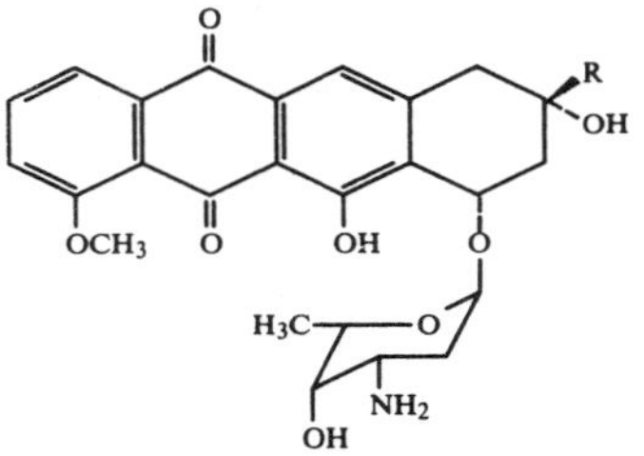

In glycoside A, R = $COCH_2OH$; in glycoside B, R = $CHOHCH_3$; in glycoside C, R = $COCH_3$; and in glycoside D, R = CH_2CH_3.

The aglycones are obtained from the glycosides by aqueous acid hydrolysis to remove the sugar moiety.

The method of preparing the glycosides A, B, C and D comprises culturing a new mutant strain of the microorganism *Streptomyces peucetius* var. *caesius* which produces the glycosides A, B, C and D, and recovering same from the culture medium.

In yet another aspect, there is provided a method for producing the mutant strain of this microorganism which comprises subjecting it to a mutagenic treatment with N-methyl-N'-nitro-N-nitrosoguanidine.

The new mutant strain of the microorganism has been designated B211 F.I. of the Farmitalia Collection of Microorganisms. Samples of the microorganism have also been deposited with the American Type Culture Collection as ATCC 31366 and with the Japanese Fermentation Research Institute as FRI 4363.

These anthracycline glycosides display antitumor and antibacterial activity. More particularly, glycosides A and C are useful as antitumor agents in experimental animals.

The production of these glycosides is carried out by conventional, well-known methods, and comprises culturing the microorganism in a previously sterilized liquid culture medium under aerobic conditions at a temperature of from 25° to 37°C (preferably at 28°C) for from 5 to 30 days (preferably 15 days) and at a pH which initially is from 6.5 to 7.0 and which, at the end of the fermentation period is from 6.5 to 8.0.

After the fermentation, the active compounds are contained in the mycelia and in the fermentation liquor. The anthracycline antibiotic complex can be extracted at pH 8.5 to 9.0 in the form of the free bases from the culture broth "in toto" with a water-immiscible organic solvent such as butanol, methyl isobutyl ketone, chloroform, methylene dichloride or ethyl acetate. Preferably, the mycelia and the fermentation liquor are separated by filtration at pH 4 with the aid of diatomaceous earth, and then extracted separately.

The filtration cake is extracted with a mixture of a water-soluble solvent, such as acetone, methanol or other lower alcohol, and an 0.1 N aqueous solution of an inorganic or organic acid, such as hydrochloric acid, sulfuric acid or acetic acid. Generally, a mixture of acetone: 0.1 N hydrochloric acid in a ratio 4:1 by volume is employed. The mycelia extracts are collected, adjusted to pH 4, and then concentrated under reduced pressure. The aqueous concentrate is combined with the filtered broth, adjusted to pH 8.5 to 9.0, then extracted with a water-immiscible organic solvent, preferably chloroform or n-butanol. The extracts are concentrated under reduced pressure and the anthracycline complex is precipitated by addition of five volumes of n-hexane. The constituents of the crude complex are then fractionated and purified by column chromatography.

Further purification of the antibiotic complex and its separation into its four components may be effected by silica gel column chromatography. The crude orange brown powder is dissolved in chloroform and the solution mixed with an equivalent of methanolic hydrogen chloride is chromatographed on silica gel with chloroform:methanol:water mixtures. The components D and C are eluted first, with a 94.8:5.0:0.2 mixture. The glycosides B and A follow with an 89.5:10.0:0.5 mixture. The components are usually separated as shown by paper and thin layer chromatography and the four components are obtained as their hydrochlorides in crystalline form.

Tallysomycin Derivatives

Although a number of glycopeptide antibiotics have been discovered, some of which are also effective in inhibiting the growth of tumors in mammals, there remains a need for additional antimicrobial and antitumor agents. Especially needed are antitumor agents which exhibit increased activity and/or broader spectrum relative to presently available agents or which show fewer undesirable side effects.

In an effort to find such agents, *T. Miyaki, O. Tenmyo, M. Konishi and H. Kawaguchi; U.S. Patent 4,314,028; February 2, 1982; assigned to Bristol-Myers Company* have developed a series of semibiosynthetic tallysomycin derivatives which have advantageous antimicrobial and antitumor properties.

An especially preferred compound is the tallysomycin B derivative having the formula:

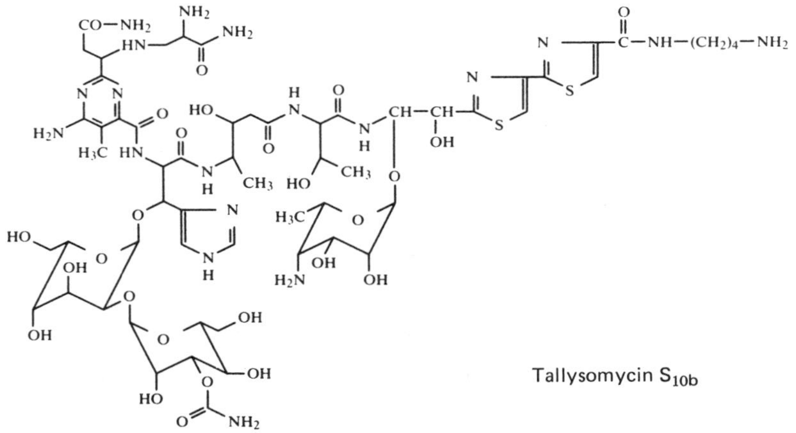

Tallysomycin S_{10b}

Initial studies using experimental animals indicate that tallysomycin S_{10b} (which may exist in both copper-free and copper-chelated forms and pharmaceutically acceptable salts thereof) has excellent antitumor activity against a broad spectrum of malignant tumors and at the same time shows less undesirable side effects (e.g., nephrotoxicity) than the naturally-occurring tallysomycin A and B.

In the preparation of these antibiotics, a tallysomycin-producing strain of *Streptoalloteichus hindustanus* (as disclosed in U.S. Patent 4,051,237), most preferably the strain *St. hindustanus* E 465-94, ATCC 31158 or a mutant thereof, is cultivated in a conventional aqueous nutrient medium.

In addition to the conventional nutrient constituents, there is added to the medium a precursor amine compound in the form of the free base or as an acid addition salt thereof. Generally, the precursor-amine compound is used in the form of a neutralized aqueous solution. The precursor-amine compound used to make tallysomycin S_{10b}, for example, is 1,4-diamino butane.

The amines, utilized as free base compounds or as acid addition salts with mineral acids such as HBr or HCl, are preferably added to the medium at a concentration of about 0.05 to 0.4% (w/v), more preferably about 0.1 to 0.2% (w/v), at the beginning of fermentation, although good results can also be achieved by portionwise addition of amine solution during the early stages of fermentation.

Fermentation is carried out following the procedure of U.S. Patent 4,051,237. The incubation temperature may be any temperature, preferably conducted at 27° to 32°C. A neutral or near neutral initial pH, e.g., pH ~6 to 7, is preferably employed in the medium, and production of antibiotic is generally carried out for a period of about 2 to 10 days. Ordinarily, optimum production is obtained in 3 to 7 days.

For preparation of relatively small amounts, shake flasks and surface culture can be employed, but for large scale production submerged aerobic culture in sterile tanks is preferred. When tank fermentation is to be carried out, it is desirable to produce a vegetative inoculum in a nutrient broth by inoculating the broth culture with a spore from the producing organism. When a young active vegetative inoculum has been obtained, the inoculum is transferred aseptically to the fermentation tank medium. Aeration in tanks and bottles may be provided by forcing air through or onto the surface of the fermenting medium. Agitation in tanks may be provided by a mechanical impeller, and an antifoaming agent such as lard oil may be added as needed.

Production of the desired tallysomycin derivative in the culture medium can be followed during the course of the fermentation by the paper disc-agar diffusion method using *Mycobacterium smegmatis* strain M6-3 as the test organism.

After optimum broth potency has been obtained (as determined, for example, by the assay procedure mentioned above), the mycelium and undissolved residues are separated from the fermentation broth by conventional means such as filtration or centrifugation. The antibiotic activity is in the filtrate and can be recovered therefrom by employing conventional adsorption techniques; see, for example, *J. Antibiotics* 30(10): 779-788 (1977).

Daunomycin and Baumycins

A number of anthracycline glycosides have been described in prior literature. Among them, daunomycin and adriamycin are particularly being watched with keen interest by those in the field of cancer chemotherapy and have already been applied clinically for human cancers.

In the continuation of studies on biosynthesis of anthracycline glycosides, particularly daunomycin and adriamycin, *T. Oki, A. Yoshimoto, K. Kouno, T. Inui, T. Takeuchi and H. Umezawa; U.S. Patent 4,337,312; June 29, 1982; assigned to Sanraku-Ocean Co., Ltd., Japan* have developed a process for producing daunomycin, baumycins and their related anthracycline glycosides with high yield from biologically inactive anthracyclinones by microbial glycosidation. They found that daunomycin and related-anthracycline-producing microorganisms, for example, *Streptomyces coeruleorubidus* ME130-A4 (FERM-P 3540, ATCC 13740), *Streptomyces peucetius* ssp *carneus* ATCC 21354, *St. peucetius* NRRL B-3826 (FERM-P 3989) and mutants therefrom, produce daunomycin and baumycins not from daunomycinone, but from aklavinone and ε-rhodomycinone by microbial glycosidation.

The starting materials of the process are biologically inactive anthracyclinones such as aklavinone and ε-rhodomycinone.

The anthracyclinones can be isolated directly from their culture medium or obtained by acid hydrolysis of the corresponding anthracycline glycosides, for example, from aclacinomycins A and B (U.S. Patent 3,988,315), MA 144 G1, G2, L, N1, S1, S2, U1, U2 (Japan Pat. Kokai No. SHO 53-44555), aclacinomycin Y (Japan Pat. Kokai No. SHO 54-63067), and rhodmycins produced by *Actinomyces roseoviolaceus, St. purpurascens, St. coeruleorubidus* (ATCC 13740) and *St. peucetius* (NRRL B-3826).

Fermentative production of daunomycin and baumycins is carried out as follows. The streptomyces culture, grown on YS agar slant (0.3% yeast extract, 1.0% soluble starch, 1.5% agar, pH 7.0) and stored at 6° to 7°C, was inoculated in a liquid medium consisting of starch, glucose, organic nitrogen sources, and inorganic nitrogen sources, as an example, and shake-cultured for 1 to 2 days at 25° to 32°C to prepare the seed culture. Then, the above seed culture was inoculated with 1 to 3% by volume to an aqueous medium, for example, consisting of sucrose, glucose, soybean meal and inorganic salts, and aerobically cultivated at 25° to 32°C for 36 to 100 hours. During cultivation, aklavinone and/or ε-rhodomycinone at the concentration of 10 to 200 μg/ml is added as substrate to the cultured medium in the logarithmic phase of the microbial growth, and the cultivation is further continued for 18 to 72 hours to complete the microbial conversion.

Optimal conditions for microbial conversion using *St. coeruleorubidus* ME 130-A4, parent strain, and ME130-A4 IU-222, anthracycline-pigment nonproducing mutant, are that the cultivation was carried out for 72 hours at 28°C in the production medium, and then 50 μg/ml of aklavinone and/or ε-rhodomycinone solution (2 mg/ml in methanol) were added to the cultured medium. After 48 hours cultivation, 43 μg/ml of the conversion products, 5 μg/ml of daunomycin, 15 μg/ml of baumycin A1 and 23 μg/ml of baumycin A2, were accumulated; recovery yield was over 80% on the weight basis of substrate. The parent strain can accumulate only 20 μg/ml as total products by direct fermentation procedure.

Antileukemics Tripdiolide, Triptolide and Celastrol

W.T. Chalmers, J.P. Kutney, P.J. Salisbury, K.L. Stuart, P.M. Townsley and B.R. Worth; U.S. Patent 4,328,309; May 4, 1982; assigned to the U.S. Secretary of the Department of Health & Human Services have developed a process for producing and recovering tripdiolide, triptolide and celastrol using cell cultures of the plant *Tripterygium wilfordii.*

In Formula (1) below, when R is OH, the compound is tripdiolide; when R is H, the compound is triptolide. Formula (2) is a quinone-methide compound known as celastrol. All three of these compounds show antitumor activity.

(1)

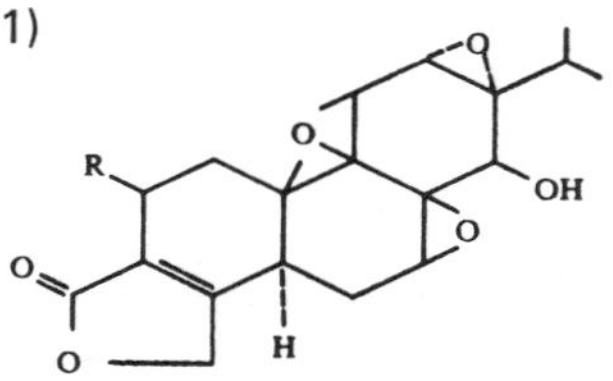

(2)

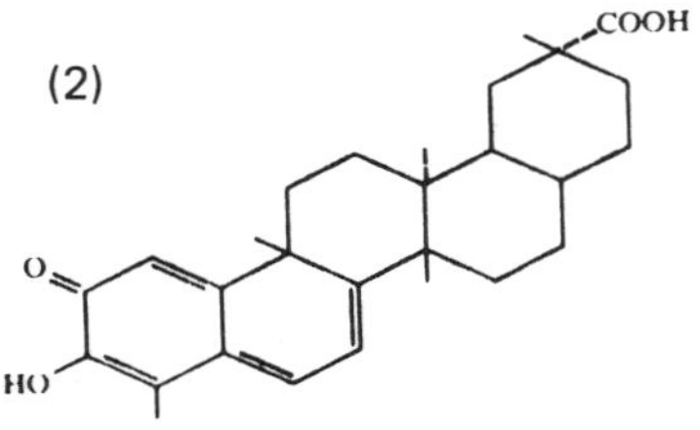

The process for obtaining tripdiolide, triptolide and celastrol comprises: (a) preparing a cellular inoculum from *Tripterygium wilfordii* Hook F; (b) inoculating a nutrient growth medium with the cellular inoculum and incubating the inoculated growth medium at 20° to 30°C for up to 8 weeks to produce a cellular product; (c) harvesting the cellular product from the inoculated growth medium; and (d) isolating tripdiolide, triptolide and celastrol from the cellular product and supernatant inoculated growth medium.

Tripterygium wilfordii Hook F is a plant of the Celastraceae family, which grows in Taiwan. A domestic source of plant material is available at the Plant Introduction Garden of the Bureau of Plant Industry, at Glenn Dale, Md.

Any part of the *Tripterygium wilfordii* Hook F plant can be used to provide the explant or initial cells from which an inoculum can be prepared. However, cells from the leaves are preferred. The cellular explant is maintained on suitable plant cell culture nutrient media, the product thereof serving as inoculum for the production of tripdiolide, triptolide and celastrol. In the initial phase of inoculum production, the cells are allowed to multiply in the presence of light.

The nutrient media employed for preparing the cellular inoculum and in which multiplication of the thus-produced cells to produce a cellular product is one which will contain sources of carbon, calcium, nitrogen, magnesium, phosphorus, sulfur and potassium. Preferably, the medium will also contain vitamins.

In a most preferred example, the process will be carried out as above, wherein the cellular inoculum is prepared from the leaves of *Tripterygium wilfordii* Hook F. Preparation of the cellular inoculum and incubation of the cellular inoculum to produce cellular product is done in media containing coconut milk, vitamins and sources of carbon, nitrogen, magnesium, calcium, phosphorus, sulfur and potassium. Preparation of the cellular inoculum is done in the presence of light in a medium containing indole-3-acetic acid to produce a callus and material from the callus is further incubated in the dark in a medium containing casein hydrolysate and 2,4-dichlorophenoxyacetic acid and incubation of the cellular inoculum in the nutrient growth medium to produce cellular product is done under subdued light. The pH of the nutrient growth medium is maintained below 7.

At the end of the incubation period, cellular product is removed from the supernatant nutrient growth medium, conveniently by filtration. The cellular product is crushed and extracted with ethyl acetate. The supernatant medium is also extracted with ethyl acetate and the combined ethyl acetate extracts are evaporated to dryness to produce a crude mixture of tripdiolide, triptolide and celastrol.

A preferred procedure for isolating the individual components of the crude product is by column chromatography on silica, using a gradient going from benzene to ethyl acetate.

Advantages of this process include not only higher yields of triptolide and related materials than obtainable from plant extracts but also that the proportions of triptolide, etc. are different than obtainable from the plant extracts.

HYPOCHOLESTEREMICS

MSD803 and Its Salts and Esters

Because of the possible connection between high blood cholesterol and atherosclerosis, many efforts have been made to find ways and substances which would reduce the cholesterol in the mammalian body. One of these ways is to inhibit in mammals the body's ability to synthesize cholesterol.

R.L. Monaghan, A.W. Alberts, C.H. Hoffman and G. Albers-Schonberg; U.S. Patent 4,294,926; October 13, 1981; assigned to Merck & Co., Inc. have developed a group of compounds which have excellent properties of inhibiting cholesterol biosynthesis and are useful as medicaments for hypercholesteremia and hyperlipemia. These compounds are MSD803 (Formula 1 below), its free hydroxyacid (Formula 2 below) and pharmaceutically acceptable salts of the acid and its lower alkyl and substituted alkyl esters.

(1)

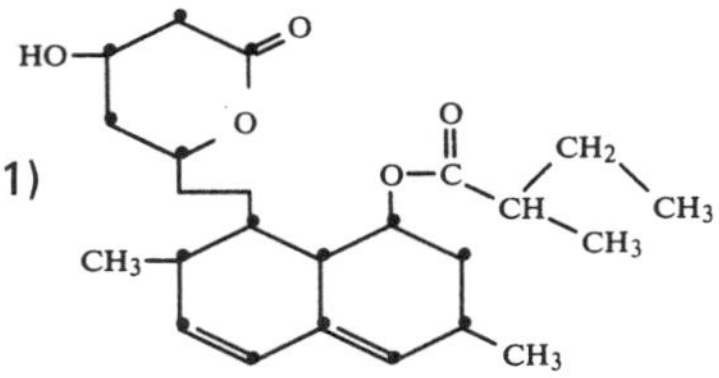

(2)

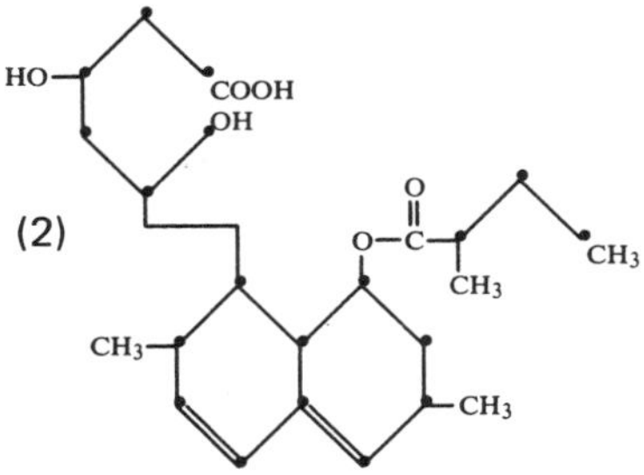

The microorganisms used to synthesize these compounds are members of the Aspergillus genus and are designated MF-4833 and MF-4845 in the Merck collection and ATCC 20541 and ATCC 20542, respectively, in the American Type Culture Collection. The latter gives the better yields. Both strains are *Aspergillus terreus.*

The culture of these organisms to produce MSD803 is carried out in aqueous media such as those employed for the production of other fermentation products. Such media contain sources of carbon, nitrogen and inorganic salts assimilable by the microorganisms. For optimum results it is preferable to conduct the fermentation at temperatures of from about 22° to 30°C. The pH of the nutrient media suitable for growing the Aspergillus culture and producing MSD803 can vary from about 6.0 to 8.0 and it is preferred to carry out the fermentation in the submerged state.

Monacolin K

High blood cholesterol levels are recognized as one of the main causes of heart disease, such as cardiac infarction or arteriosclerosis. As a result, considerable research has been undertaken with a view to discovering physiologically acceptable substances which are capable of inhibiting cholesterol biosynthesis and thus reducing blood cholesterol levels. One such compound is Monacolin K, which can be produced by cultivating microorganisms of the genus Monascus, especially *Monascus ruber* strain 1005 (FERM 4822).

The objective of *K. Tanzawa, S. Iwado, Y. Tsujita, M. Kuroda and K. Furuya; U.S. Patent 4,323,648; April 6, 1982; assigned to Sankyo Company Limited, Japan* is to provide a process for preparing Monacolin K using alternative new strains of the genus Monascus.

Thus, their process for preparing Monacolin K comprises cultivating one or more of the microorganisms *Monascus anka* SANK 10171 (IFO 6540), *Monascus purpureus* SANK 10271 (IFO 4513), *Monascus ruber* SANK 10671 (FERM 4958), *Monascus vitreus* SANK 10960 (FERM 4960), *Monascus paxli* SANK 11172 (IFO 8201), *Monascus ruber* SANK 13778 (FERM 4959), *Monascus ruber* SANK 15177 (FERM 4956) or *Monascus ruber* SANK 18174 (FERM 4957) in a culture medium therefor and extracting the Monacolin K from the resulting culture medium.

Monacolin K may be produced by cultivating the chosen microorganism in a culture broth under aerobic conditions using well-known techniques.

Any culture medium well-known in the art for the cultivation of fungi may be employed, provided that it contains, as is well-known, the necessary nutrient materials, especially an assimilable carbon source and an assimilable nitrogen source. Glucose, glycerin and starch are particularly preferred carbon sources for the production of Monacolin K. Examples of suitable sources of assimilable nitrogen are peptone, corn steep liquor, rice bran and inorganic nitrogen sources. Peptone is particularly preferred. When producing Monacolin K, an inorganic salt and/or a metal salt may, if necessary, be added to the culture medium. Furthermore, if necessary, a minor amount of a heavy metal may also be added.

The microorganism is preferably cultivated under aerobic conditions using cultivation methods well-known in the art, for example solid culture, shaken culture or culture under aeration and agitation. The microorganism will grow over a wide temperature rnage, especially for the production of Monacolin K, the more preferred cultivation temperature is within the range from 20° to 30°C.

During in vivo tests on rabbits and rats, Monacolin K was found to reduce blood cholesterol levels, and its low toxicity would seem to make its potential for treatment of human hypercholesteremia very attractive. The acute oral toxicity of Monacolin K in the mouse is 1 g/kg of body weight.

ANTIFUNGAL ANTIBIOTICS

Antibiotic A43F

The objective of *J.C. Onishi, G.L. Rowin and J.E. Miller, Jr.; U.S. Patent 4,254,224; March 3, 1981; assigned to Merck & Co., Inc.* is to provide an antifungal antibiotic which is highly effective in inhibiting the growth of a wide variety of plant pathogens. Another objective is to provide a process for preparing the antibiotic substance by the fermentation of a nutrient media with an as yet unclassified species of fungus initially isolated from a fluff sample taken from a chicken incubator.

The antibiotic is characterized by having the following structural formula:

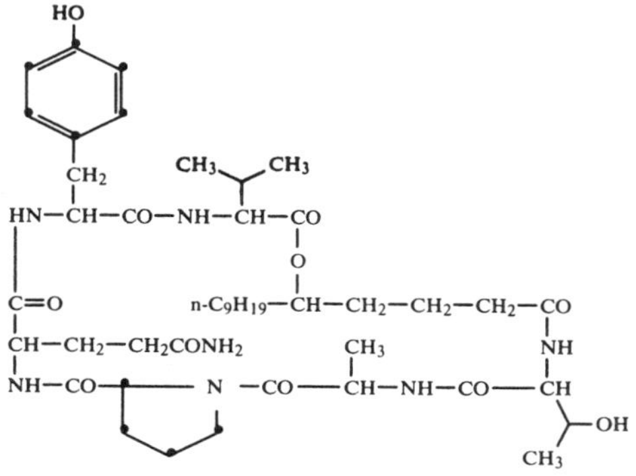

Antibiotic A43F inhibits the growth of a variety of fungi which are pathogenic to plant life including *Puccinia recondita* f. sp. *tritici, Piricularia oryzae, Phytophthora infestans, Alternaria solani* and *Erysiphe polygoni*. It is contemplated, therefore, that antifungally effective amounts of antibiotic A43F could be applied as an agricultural fungicide in the treatment and control of pathogenic fungus infestations of plants. The antifungal activity of antibiotic A43F has been confirmed in vivo employing standard greenhouse bioassay techniques.

The antibiotic A43F is prepared by growing under controlled conditions an as yet unclassified species of fungus designated MF 4683 in the culture collection of Merck & Co., Inc. A viable culture thereof has been placed on permanent deposit with the culture collection of the American Type Culture Collection as ATCC 20529.

The method of producing antibiotic A43F comprises cultivating fungus species ATCC 20529 in a culture medium containing assimilable sources of carbohydrate, nitrogen and inorganic salts until a substantial amount of antibiotic activity is produced by the fungus.

The separation of antibiotic A43F from the culture medium is achieved by: (a) adding an equal volume of methanol to the whole broth and filtering the resulting mixture; (b) concentrating the filtrate from (a) to one-tenth the volume of the whole broth and extracting the filtrate with an equal volume of ethyl acetate; (c) concentrating the ethyl acetate extract from (b); chromatographing over silica gel, eluting with methanol in ethyl acetate and separating the active fraction; (d) chromatographing the active fraction from (c) over Sephadex LH20 in a mixture of methylene chloride, hexane and methanol (10:10:1); and (e) evaporating the active fraction to dryness.

Mildiomycin

Mildiomycin is an antibiotic substance, Antibiotic B-98891, which can be used (1) to treat and prevent mildew in a wide variety of plants and (2) as a miticide.

As a result of studies on methods for producing mildiomycin in commercial quantities and at low cost, *T. Suzuki, H. Sawada and T. Asai; U.S. Patent 4,334,022; June 8, 1982; assigned to Takeda Chemical Industries, Ltd., Japan* have found a method for producing the compound by cultivation of a microorganism of Actinomycetes in a culture medium containing an N-methyl compound in an amount of at least 3 mM.

The microorganisms of Actinomycetes having an ability to produce mildiomycin used in this method include, for example, Actinomycetes belonging to the genus Streptoverticillium, in particular, *Streptoverticillium rimofaciens* FERM-P 2549 (IFO 13592, ATCC 31120). This organism was originally identified as *Streptomyces rimofaciens*, but, in view of subsequent modification of criteria for classification, is now identified as belonging to the genus Streptoverticillium.

The N-methyl compounds which can be used in the culture medium are those having at least one nitrogen atom substituted with 1 to 4 methyl groups in the molecule thereof. Of these compounds, quaternary ammonium salts having trimethylammonio group in which the nitrogen atom is substituted with 3 methyl groups, i.e., $-N^+(CH_3)_3$, are preferred. Compounds having a group capable of being converted into a >N–CH_3 group in the culture medium, for example, N,N-methylenebisacrylamide, can also be used as the N-methyl compounds. The N-methyl compounds generally have molecular weights preferably in the range of from about 90 to 130. Water-soluble N-methyl compounds are used most advantageously. Examples of such N-methyl compounds are N-methyl acid amides, N-methylamino compounds, N-methylamines, N-methylammonium compounds, polymethylenediamines, N,N-methylenebisacrylamide, etc.

Satisfactory results can be obtained with N-methylammonium compounds, particularly with choline, betaine or tetramethylammonium. These N-methyl compounds can be used alone or as a mixture of two or more species. The process also includes adding a large amount of an N-methyl compound-containing substance, for example, beet molasses containing betaine, soybean meal containing lecithin or hen's egg containing both choline and lecithin, etc. to a culture medium so as to provide more than 3 mM of N-methyl compounds to the medium. The concentration is preferably 4 to 200 mM, more preferably 7 to 50 mM.

It is preferred to use N-methyl compounds in combination with naturally-occurring substances. The N-methyl compound is incorporated in the culture medium preferably before the culture medium is inoculated with the microorganism, but it can be added at any appropriate time during the cultivation.

The cultivation can be conducted by any of the conventional culture methods, but generally submerged aerobic culture is preferred. The cultivation temperature is preferably from 24° to 34°C.

From the culture broth obtained in the above manner, mildiomycin can be isolated, if necessary, by conventional procedures which are normally utilized for the recovery of basic water-soluble antibiotics.

Since the antibiotic is a basic substance, it is adsorbed well on a cation-exchange resin and can be eluted with a suitable acid, alkali or buffer solution. Such procedures can be suitably used selectively and in combination and thereafter the eluate is concentrated and crystallized to obtain mildiomycin. Mildiomycin thus obtained can be used safely as an active ingredient of an antibacterial and miticidal agent for plants as described in U.S. Patent 4,007,267 and British Patent 1,507,193.

GLYCOSIDE INHIBITORS

1-Desoxynojirimycin

It is known that a number of organisms of the family Actinomycetes, above all Actinoplanaceae, form inhibitors for glycoside hydrolases, preferably carbohydrate splitting enzymes of the digestive tract. Furthermore, it is known that nojirimycin, an antibiotic having a bacteriostatic action, derived from strains of organisms of the genus Streptomyces, inhibits certain microbial α-glucosidases.

In the course of studies of inhibitors for glycoside hydrolases, especially saccharase inhibitors, which are present in the digestive tract, *W. Frommer, L. Muller, D. Schmidt, W. Puls, H.-P. Krause and U. Heber; U.S. Patent 4,307,194; Dec. 22, 1981; assigned to Bayer AG, Germany* found that certain inhibitors could be formed by culturing organisms of the family Bacillaceae, particularly by strains of the genus Bacillus.

In addition, they found that certain strains of these organisms of the family Bacillaceae, produce the antibiotic known as 1-desoxynojirimycin. Particularly useful strains for this purpose were *B. subtilis* DSM 704 (ATCC 31324); *B. subtilis* var. *niger* DSM 675 (ATCC 9372); *B. amyloliquefaciens* DSM 7 (ATCC 23350); and *B. polymyxa* DSM 372 (ATCC 31322). (The DSM numbers refer to the deposit numbers in the German Collection of Microorganisms in Gottingen. The ATCC numbers are those given to the culture deposits at the American Type Culture Collection.) All four of these Bacillus strains had saccharase inhibiting action as well as producing 1-desoxynojirimycin in relatively good yields.

The method of producing 1-desoxynojirimycin therefore comprises culturing a 1-desoxynojirimycin-producing organism of the genus Bacillus in a nutrient solution of about 15° to 40°C for nonthermophilic Bacilli and 15° to 80°C for thermophilic Bacilli containing a source of protein, a source of carbon and a nutrient salt for a time sufficient to produce the 1-desoxynojirimycin, then separating it from the culture.

This process makes it possible for the first time to prepare desoxynojirimycin in good yields in one operation by direct microbiological synthesis, without using a roundabout route via nojirimycin which is relatively unstable and therefore difficult to handle.

Furthermore, individual strains of Bacillus, for example DSM 372, form saccharase inhibitors which, in a mixture, in addition to desoxynojirimycin and/or nojirimycin, also produce other components, having a saccharase inhibiting action, which can be clearly differentiated in a thin layer chromatogram.

It is known that in warm-blooded animals, after intake of carbohydrate-containing foodstuffs and beverages (for example, cereal starch, potato starch, fruit, fruit juices, beer and chocolate), hyperglycemias arise which are brought about as a result of a rapid degradation of the carbohydrates by glycoside hydrolases (for example, salivary and pancreatic amylases, maltases and saccharases).

In the case of diabetics, these hyperglycemias are particularly strong and of long-lasting pronounced character. In the case of adipose subjects, the alimentary hyperglycemia frequently leads to a particularly intense insulin secretion which, in turn, leads to increased fat synthesis and decreased fat degradation. Following such hyperglycemias, a hypoglycemia frequently occurs in the case of adipose persons of sound metabolism, as a result of the insulin secretion. It is known that both hypoglycemias and foodstuff sludge remaining in the stomach promote the production of gastric juice which, in turn, causes, or favors, the formation of a gastritis or a gastric or duodenal ulcer.

It is also known that carbohydrates, particularly sucrose, are split in the oral cavity by microorganisms and the formation of caries is thereby promoted.

Malabsorption of carbohydrates, for example, as a result of intestinal saccharase deficiency, causes diarrhea. Suitable doses of a glucosidase inhibitor effect a synthetic malabsorption and are thus suitable for counteracting constipation.

These inhibitors are thus suitable for use as therapeutic agents for the following indications: adiposity, hyperlipoproteinemia, atherosclerosis, diabetes, prediabetes, gastritis, constipation and caries.

In further work, *W. Frommer and D. Schmidt; U.S. Patent 4,282,320; August 4, 1981; also assigned to Bayer AG, Germany* describe a process for producing 1-desoxynojirimycin which involves culturing an organism of the Bacillaceae family, such as those described in the previous patent, in a nutrient solution in which sorbitol is used as the carbon source.

Preferably, the culturing is started at a temperature between 15° and 30°C and the temperature is increased to 30° to 50°C over a period covering the second half of the logarithmic growth phase and the first half of the stationary phase. Particularly preferably, the process is furthermore carried out by a procedure in which, during multiplication of the inoculum, which is optionally stepwise multiplication, the inoculum is heated to a temperature of 80° to 100°C and cooled again to the fermentation temperature at least once.

Particularly pure 1-desoxynojirimicin is obtained in very good yield by carrying out the isolation in the following manner. After separating off the cells, the culture solution is first adsorbed onto strongly acid ion exchangers (H^+ form) and desorbed with 0.5 to 2 N ammonia solution, the active fractions are adsorbed on a weakly acid ion exchanger (H^+ form) and desorbed with 0.02 to 0.1 N mineral acids, for example, hydrochloric acid, the active fractions are chromatographed over a strongly basic ion exchanger (OH^- form), the basic active eluate is evaporated to dryness, the residue is taken up in a polar solvent, preferably 60 to 100% strength aqueous, methanol, at elevated temperature and the product is crystallized by cooling.

N-Substituted Derivatives of 1-Desoxynojirimycin

G. Kinast and M. Schedel; U.S. Patent 4,266,025; May 5, 1981; assigned to Bayer AG, Germany have developed a process for producing compounds of the Formula (1) by microbiological reaction of compounds having Formula (2).

(1)

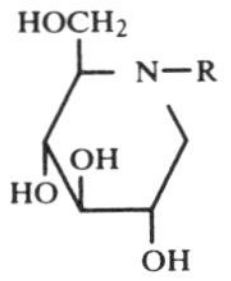

(2)

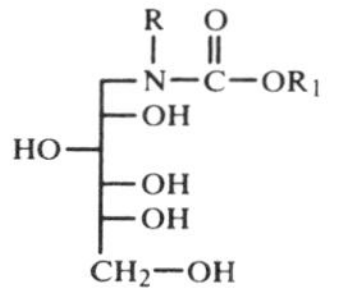

In the above formulas, R denotes a hydrogen atom or an optionally substituted alkyl, alkenyl, aralkyl or aryl group; and R_1 denotes an optically substituted benzyl radical or substituted β-alkenyl group.

The process consists of the microbiological reaction of a compound of Formula (2) with an aerobic microorganism or an extract of an aerobic microorganism capable, in a nutrient medium containing the compound of Formula (2), of accumulating a compound of the general Formula (3):

(3)

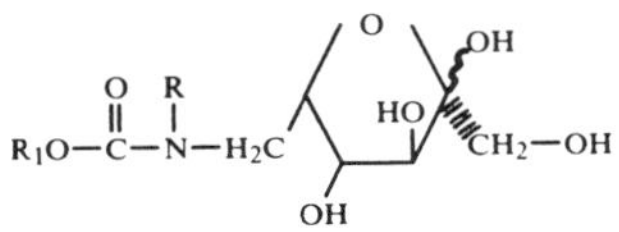

in which R and R_1 have the meaning given above, and subjecting the resulting compound to catalytic hydrogenation. The process has been found to give compounds of Formula (1) in excellent yields.

Microorganisms suitable for the process are, for example, bacteria of the order Pseudomonadales, and within this order, in particular representatives of the Pseudomonadaceae family, and here, above all, bacteria of the Gluconobacter genus. Furthermore, bacteria from the coryneform bacteria group, in particular those of the Corynebacterium genus, have also proved suitable. Finally, the process according to this method could also be carried out with fungi, thus, for example, with yeasts of the Endomycetales order, in particular with those of the Spermophthoraceae family, and here principally with representatives of the Metschnikowia genus.

Examples which may be mentioned are: *Gluconobacter oxydans* ssp. *suboxydans* (DSM 50 049), *Glucobacter oxydans* ssp. *suboxydans* (DSM 2003), *Corynebacterium betae* (DSM 20 141) and *Metschnikowia pulcherrima* (ATCC 20 515).

The DSM numbers give the numbers under which the microorganisms mentioned are stored in the German Collection of Microorganisms in Gottingen. *Metschnikowia pulcherrima* is stored in the American Type Culture Collection.

The microorganisms can be cultured in all nutrient media which are known to be used for culturing microorganisms of the abovementioned groups and which contain the compound of Formula (2) to be oxidized in the process.

The desired compounds are isolated from the culture solution and converted to compounds of Formula (1) by conventional catalytic hydrogenation methods.

Of particular interest are compounds of Formula (1) where R denotes a hydrogen atom or a C_1 to C_{10} (preferably C_1 to C_6) alkyl, hydroxyethyl or allyl group and R_1 denotes a benzyl or allyl group.

The compounds obtained are useful for the treatment of diabetes, hyperlipoproteinemia and adiposity.

Inhibitor of Pancreatic α-Amylase

The process of *V. Oeding, W. Pfaff, L. Vertesy and H.-L. Weidenmuller; U.S. Patent 4,282,318; August 4, 1981; assigned to Hoechst AG, Germany* relates to a new inhibitor of the glycoside hydrolases of the digestive tract, more particularly of the pancreatic α-amylase. Preparation of the inhibitor is accomplished by fermentation of the specific microorganism *Streptomyces tendae*, strain 4158, as well as the variants and mutants thereof.

The inhibitor of pancreatic α-amylase can be chemically classified among the peptides, is characterized by a molecular weight of 5,000 to 10,000, an absorption maximum in the ultraviolet light at 279 nm, and an isoelectric point of 4.4.

It is a characteristic feature of the inhibitor that a molecular proportion above average of the peptide is formed by the amino acids aspartic acid, glutamic acid, threonine, glycine, alanine and valine and that the pure substance does not contain methionine. The absence of methionine constitutes a good characteristic of identification of the inhibitor which may serve to determine the purity of the product, especially in concentration processes.

The inhibitor has a positive reaction to peptide reagents and a negative reaction to phenol-sulfuric acid.

The α-amylase inhibitor is free from sugar and in this respect and by its amino acid composition, its molecular weight and its isoelectric point it differs from all known α-amylase inhibitors. Pure preparations of the inhibitor have an activity of 2 to 3 times 10^6 AIU/g.

As compared to other proteins, the inhibitor is inactivated by proteolysis very slowly only by pepsin, trypsin or chymotrypsin; therefore, during the period of therapeutic action a noteworthy reduction of the activity in the digestive tract is not to be expected.

The inhibitor is distinguished by a high specificity of action. The inhibition of pancreatic α-amylase is extremely high while bacterial α-amylase, for example, those from *Bacillus subtilis* are not inhibited to a measurable degree. An effect on β-amylases has not been observed either. A small dose of the amylase inhibitor ensures a complete inhibition of the enzymatic activity of pancreatic amylase.

The strain *Streptomyces tendae* 4158 used to prepare the inhibitor is deposited at the American Type Culture Collection (ATCC) under the registration number 31210.

The fermentation can be carried out at 25° to 35°C, preferably 28° to 30°C, either immersed in a shake culture or in fermentation vessels of different dimensions, while stirring and aerating.

A culture medium having an optimum composition contains (% by weight in solution) 3 to 5% of soluble starch, 0.2 to 0.6% of corn steep, 0.5 to 1.5% of glucose, 0.5 to 1% of $(NH_4)_2HPO_4$, 0.3 to 0.6% of soy flour, 0.5 to 1.5% of casein peptone. The amylase inhibitor is also obtained in a good yield on other starch-containing culture media.

The inhibitor can be isolated from the culture liquid by processes known in protein and peptide chemistry, for example, by precipitation with organic solvents miscible with water such as acetone, isopropanol and other alcohols, or with salts, for example, ammonium sulfate.

The inhibitor has interesting properties as a therapeutic agent for the treatment of diabetes and prediabetes and adiposis and for the assistance of digestion.

PHARMACEUTICAL INTERMEDIATES

6-Amino-6-Deoxy-L-Sorbose

G. Kinast and M. Schedel; U.S. Patent 4,246,345; January 20, 1981; assigned to Bayer AG, Germany describe a process in which 6-amino-6-deoxy-L-sorbose is obtained when 1-amino-1-deoxy-D-glucitol or a salt thereof is oxidized with either an aerobic microorganism or with an extract obtained from an aerobic microorganism, which is capable of catalyzing the oxidation, in suitable media and under suitable conditions.

The desired product can be isolated as such or can be further reacted, for example, to 1-deoxynojirimycin by catalytic hydrogenation.

1-Deoxynojirimycin is identical with 2-hydroxymethyl-3,4,5-trihydroxypiperidine and is useful as an antidiabetic agent and as an inhibitor of increased lipid biosynthesis (U.S. Patent 4,065,562).

The starting material employed in the process, 1-amino-1-deoxy-D-glucitol, is known; it can be prepared, for example, from D-glucose by reductive amination with ammonia, hydrogen and nickel as the catalyst. The starting material can be employed as such or in the form of a salt thereof, for example, in the form of the chloride, sulfate, nitrate, acetate, oxalate or dihydrogen phosphate.

Microorganisms which are suitable for carrying out the process, or from which active extracts for carrying out the process can be obtained, can be bacteria or fungi, etc. It has been found that, for example, bacteria of the order Pseudomonadales, and within this order in particular representatives of the family Pseudomonadaceae and among these in particular bacteria of the genus Gluconobacter, are suitable microorganisms for the process. Bacteria from the group of the coryneform bacteria, especially those of the genus Corynebacterium, have also proved suitable. It has been found to be possible to carry out the process with fungi such as yeasts of the order Endomycetales, especially with those of the family Spermophthoraceae and among these in the main with representatives of the genus Metschnikowia.

Particularly preferred examples which may be mentioned are: *Gluconobacter oxidans* ssp. *suboxydans* (DSM 50 049), *Corynebacterium betae* (DSM 20 141) and *Metschnikowia pulcherrima* (ATCC 20 515). Such microorganisms may be used in the form of a growing culture, a concentrated cell suspension, a nonfractionated crude extract or a purified extract fraction.

If the process is carried out with intact microorganisms in a growing culture, solid, semisolid or liquid nutrient media can be used. Conventional aqueous-liquid nutrient media are preferable.

It has been found that the amount of 6-amino-6-deoxy-L-sorbose which accumulates in the culture broth generally reaches its maximum between 5 hours and 5 days after adding 1-amino-1-deoxy-D-glucitol.

It is also possible to carry out the oxidation reaction with concentrated cell suspensions of suitable microorganisms which has the advantage of shortening the reaction time of the process to a few hours.

D(–)-β-Hydroxyisobutyric Acid

The process of *J. Hasegawa, M. Ogura, S. Hamaguchi, M. Shimazaki, H. Kawaharada and K. Watanabe; U.S. Patent 4,310,635; January 12, 1982; assigned to Kanegafuchi Chemical Industry Co., Ltd., Japan* is for production of D(–)-β-hydroxyisobutyric acid (hereafter referred to as D-HIBA). More specifically, the process is one for advantageous production of D-HIBA utilizing the special ability of microorganisms to produce optically active compound by asymmetric synthesis. D-HIBA is a useful intermediate for synthesis of physiologically active substances having asymmetric carbon atoms such as medicines exemplified by 1-(D-3-mercapto-2-methyl propanoyl)-L-proline, an antihypertensive.

Chemical synthesis of β-hydroxyisobutyric acid is known; however, the product obtained by this method is the optically inactive DL(±) form. Methods of obtaining optically active forms of the compound have proved expensive and complicated.

Hasegawa et al have developed, therefore, this process for use of microorganisms to convert isobutyric acid (hereafter referred to as IBA) or methacrylic acid (hereafter referred to as MA) into D-HIBA.

In their process, a substrate selected from the group consisting of IBA, MA, and a mixture thereof, which is readily available at a low price and in large quantities, is subjected to the action of a microorganism having the ability to convert the substrate into D-HIBA. Conversion of the substrate into D-HIBA by catalysis of the microorganism is carried out in two ways. In one way, the microorganism is cultivated aerobically in an aqueous nutrient medium containing the substrate, whereby propagation of the microorganism and subjection of the substrate to the action of the microorganism are carried out simultaneously in one step.

In the other method, the microorganism is first cultivated in an aqueous nutrient medium, then the substrate is added to the resulting culture broth or to the suspension of cells obtained in the first step, and subsequently the mixture is incubated aerobically. Microorganisms employed convert the substrate into only the D(–) form of HIBA in high yield; therefore, D-HIBA can be produced by a simple process at a low cost.

Among microorganisms useful in the process are the following strains: *Candida rugosa* IFO 0750, *Candida rugosa* IFO 0591, *Candida parapsilosis* IFO 0708, *Candida utilis* IFO 0396, *Torulopis candida* IFO 0380, *Trygonopsis variabilis* IFO 0671, *Saccharomyces cerevisiae* IAM 4274, *Saccharomyces rouxii* IFO 0493, *Pichia membranaefaciens* IAM 4904, *Debaryomyces hansenii* IFO 0026, *Wingea robertsii* IFO 1277, *Rhodosporidium toruloides* IFO 0559, *Aspergillus niger* IAM 2532, *Choanephora circinanus* HUT 1324, and *Zygorhynchus moelleri* HUT 1305. (IFO: Institute for Fermentation, Osaka, Japan; IAM: Institute of Applied Microbiology, University of Tokyo; HUT: Faculty of Engineering, Hiroshima University.)

Preparation of Coproporphyrin III in Improved Yields

Coproporphyrin III has a porphyrin structure and is useful in a wide range of applications such as pharmaceuticals, intermediates for pharmaceuticals or red dyes for drinks and foods.

Japanese Laid-Open Patent Publication 7492/77 discloses a process which comprises cultivating a microorganism capable of producing coproporphyrin III of the following formula selected from microorganisms of the genera Arthrobacter and Brevibacterium, and recovering coproporphyrin III from the culture broth.

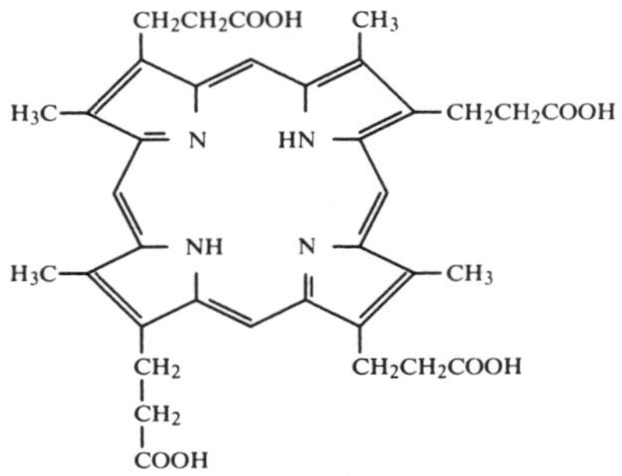

I. Kojima, K. Maruhashi and Y. Fujiwara; U.S. Patent 4,334,021; June 8, 1982; assigned to Nippon Oil Company, Ltd., Japan have found that coproporphyrin III can be obtained in markedly improved yields by performing the cultivation in a culture medium which contains a suitable amount, preferably at least 0.1 g/ℓ of the culture medium, of L-cystine. They also found that yields are improved by performing the cultivation in a culture medium containing a compound capable of producing magnesium ions, such as those from magnesium salts, particularly in an amount of at least 0.5 g/ℓ of the culture medium calculated as the magnesium ion.

Some specific examples of coproporphyrin-producing strains of Arthrobacter which can be used in the process, along with their accession numbers are the following (other strains are also listed in this patent): (1) *Arthrobacter hyalinus* (FERM-P 3125; ATCC 31263), (2) *A. globiformis* (ATCC 8010; IFO 12137), (3) *A. aurescens* (IFO 12136; ATCC 13344), (4) *A. pascens* (IFO 12139; ATCC 14358) and (5) *A. cremeus* (FERM-P 3126).

MISCELLANEOUS PHARMACEUTICALS

Coenzyme Q_{10} for Cardiac Insufficiency

Coenzyme Q_{10} plays an important function as an element of the electron transmission system in an organism. It is known to exhibit an excellent pharmaceutical effect against various diseases. In recent years, it has been clinically employed for curing of cardiac insufficiency by oral administration.

Coenzyme Q_{10} has the formula:

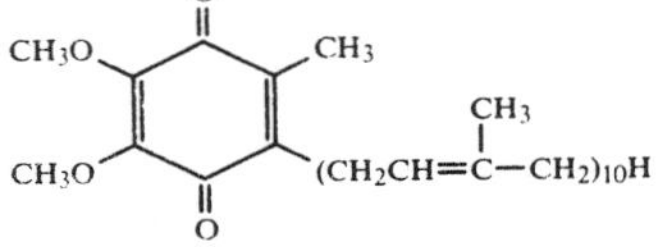

Coenzyme Q_{10} is industrially produced either by a semisynthetic process that uses a material originating in plants or by a fermentation process that extracts the compound from microbial cells. In the fermentation process, it is necessary to produce microbial cells containing coenzyme Q_{10} at a low cost in order to realize an industrially economical process because the amount of coenzyme Q_{10} accumulated in the individual microbial cells is small.

In processes utilizing microbial cells, it is usually of importance that the component of the culture medium serving as a primary source of carbon for growing the microbial cells be an inexpensive material. It is an objective of *K. Hata, K. Ohshima, I. Kano, M. Matsui and T. Sato; U.S. Patent 4,245,048; January 13, 1981; assigned to Jujo Paper Co., Ltd., Japan* to provide an economic fermentation process for producing coenzyme Q_{10} , and particularly one which uses an inexpensive carbon source as a component of the culture medium.

The process of producing sulfite pulp from wood produces a waste liquor from the cooking of wood referred to industrially and hereafter as SWL (sulfite waste liquor). Because of the inexpensive nature of SWL, Hata et al have endeavored to provide a process for producing coenzyme Q_{10} from a microorganism which can assimilate SWL.

Their process thus comprises cultivating a microorganism JY-155 which belongs to the genus Trichosporon (FERM-P 4650, ATCC 20566) in a culture medium containing SWL as the carbon source until coenzyme Q_{10} is formed and accumulated in a substantial amount in the culture, and recovering the coenzyme Q_{10} therefrom. The sulfite waste liquor is preferably one which contains from 0.5 to 4% by weight of sugar in terms of glucoside content. The culture medium preferably also contains a nitrogen source and inorganic salt.

Cultivation is performed by shaking aerobically or stirring with aeration. The cultivation temperature is preferably in the range of from 25° to 37°C. The pH value for cultivation is maintained in the range of from 4 to 8. The type of cultivation may be of either the batch type (10 hours to 5 days in cultivation period) or the continuous type. The latter is preferable for cultivation on an industrial scale.

Microbial cells are separated from the culture broth by methods such as centrifugating, filtration, etc. The separated cells contain coenzyme Q_{10} which has accumulated thereon. After being dried and optionally additionally processed, they can be used as nutrients, medicine and other applications.

Optically Pure N-Acyl-D-Phenylalanine Ester with Analgesic Activity

D-phenylalanine previously had little, if any, use. However, it has recently been found that D-phenylalanine is a simple nonaddictive analgesic compound that stimulates the body's own pain-fighting system, provides significant relief in chronic-pain patients and produces long term analgesia. Therefore, a process for producing N-acyl-D-phenylalanine esters and, particularly, N-C_{1-9} acyl-D-phenylalanine esters which can be converted easily to D-phenylalanine would be very desirable. An especially effective process would involve (1) readily available and inexpensive material, such as enzyme, and (2) produce high purity material in high yield at a rapid rate.

D.P. Bauer; U.S. Patent 4,262,092; April 14, 1981; assigned to Ethyl Corporation provides an especially effective process for obtaining N-C_{1-9} acyl-D-phenylalanine ester which can be subsequently converted to D-phenylalanine. In one aspect, there is provided a process for producing optically pure N-acyl-D-phenylalanine ester, comprising: (a) subjecting a mixture of N-acyl-D-phenylalanine ester and N-acyl-L-phenylalanine ester to the action of a proteolytic enzyme selected from the group consisting of microbially derived serine proteinases; and (b) separating the resulting N-acyl-L-phenylalanine from the unreacted N-acyl-D-phenylalanine ester.

In another aspect, the process for preparing optically pure N-acetyl-D-phenylalanine ester comprises (a) subjecting an aqueous solution of a mixture of N-acetyl-D-phenylalanine methyl ester and N-acetyl-L-phenylalanine methyl ester to the action of a proteolytic enzyme which is a member selected from the group consisting of the microbially derived serine proteinases, at a pH in the range of from 5 to 10; (b) separating the resulting N-acetyl-L-phenylalanine from the unreacted N-acyl-D-phenylalanine ester; and (c) recovering the N-acyl-D-phenylalanine ester.

It has been found that certain readily available proteinases, i.e., microbially derived serine proteinases exhibit high esterase activity for N-acyl-L-phenylalanine esters. Furthermore, it has been found that the high esterase activity exhibited for the L-isomer is not inhibited by the presence of the D-isomer. Subjecting N-acyl-D,L-phenylalanine ester to the action of such a proteinase provides a mixture of N-acyl-D-phenylalanine ester and N-acyl-L-phenylalanine. The N-acyl-L-phenylalanine can be readily separated from the mixure by conventional means, for example, by adjusting the pH of an aqueous mixture thereof and extracting with an organic solvent such as chloroform, ethyl acetate, butyl acetate, methylene chloride and the like.

A variety of specific N-acyl-D,L-phenylalanine ester mixtures can be employed in the process. Preferably, compounds will have the acyl and ester groups mentioned below.

Preferably, the acyl group is derived from fatty acids containing from 1 to 9 carbon atoms. More particularly, the N-acyl group will preferably be formyl, acetyl, propionoyl, butyroyl, valeroyl, caproyl, enanthoyl, capryryl, or pelargonoyl. The ester group can be derived from a variety of alcohols containing preferably from 1 to 6 carbon atoms. Especially suitable examples of ester groups are methyl, ethyl, n-propyl, isopropyl, n-butyl, sec-butyl and isobutyl. Especially suitable examples of acyl groups are formyl, acetyl and propionyl. Racemic N-acetyl-D,L-phenylalanine methyl ester is most preferred for use in the process.

Serine proteinases suitable for use in the process are derived from microorganisms such as bacteria, fungi and mold. These proteinases are relatively inexpensive and commercially available. An example of preferred serine proteinases are those derived from the bacterial organism *Bacillus subtilis* and termed subtilisins. A preferred subtilisin is the *Bacillus subtilis*-derived Carlsberg strain. Other suitable proteinases are derived from x-ray mutated *Bacillus subtilis* and serine proteinases derived from *Aspergillus oryzae.*

Because of the high selective esterase activity of the particular proteolytic enzymes employed, very small amounts of the proteolytic enzyme are required in order to rapidly produce N-acyl-L-phenylalanine and separate it from N-acyl-D-phenylalanine ester. For example, aqueous solutions containing preferably from about 0.005 to 0.5% by weight of pure crystalline enzymes are employed. The amount of N-acyl-D,L-phenylalanine ester employed will preferably be about 10% by weight of the aqueous solution.

The N-acyl-D-phenylalanine ester provided by this process can be converted to D-phenylalanine in a simple procedure. In one procedure, it can be treated at elevated temperatures with dilute acid; for example, it can be dissolved in 2 N HBr and this solution warmed for a time at 80° to 100°C. Other simple hydrolysis procedures can be employed to convert the N-acyl-D-phenylalanine esters into the desired D-phenylalanine.

D-phenylalanine has recently been determined to provide significant analgesic action. It is believed to work by inhibiting enzymes responsible for destroying naturally produced, short-acting pain killers. Specifically, D-phenylalanine has been found to be an inhibitor of carboxypeptidase A, and is useful in preventing such enzymes from breaking down enkephalins which are natural analgesics. Tests have shown marked long term analgesia using D-phenylalanine as an investigational new drug.

Cyclosporin Derivatives as Antiinflammatory Agents

R.P. Traber, M. Kuhn, H. Hofmann and E. Harri; U.S. Patent 4,289,851; Sept. 15, 1981; assigned to Sandoz Ltd., Switzerland have devised a microbiological procedure for making cyclosporin D and several derivatives of this antibiotic.

Cyclosporin D has the formula (1) where A is formula (2):

(1)

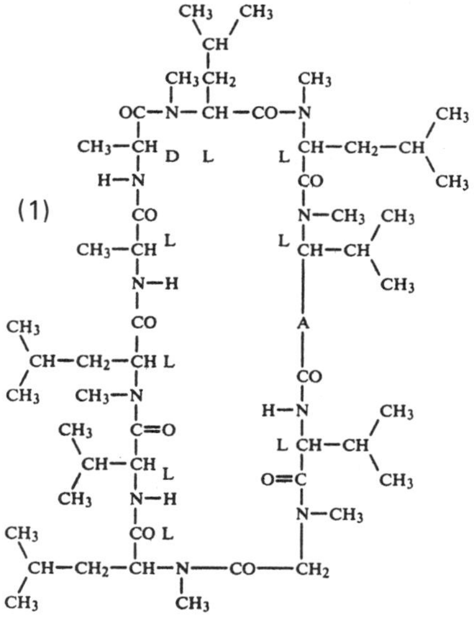

(2)

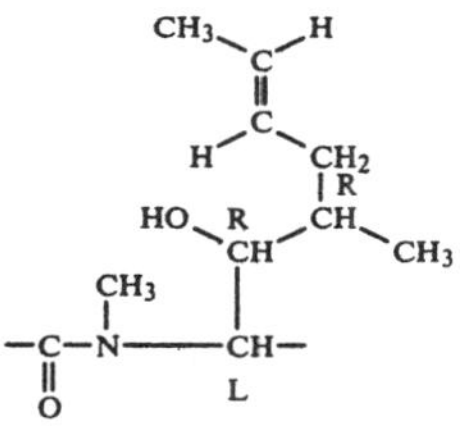

Two other derivatives of cyclosporin are also prepared by Traber et al: dihydrocyclosporin, which is produced by hydrogenation of the double bond present in formula (2); and isocyclosporin D, which is made by rearranging cyclosporin D under acidic conditions.

Cyclosporin D is made by a conventional fermentation process, preferably by using the strain NRRL 8044 of the species *Tolypocladium inflatum Gams*, a culture of which is available from USDA (Northern Research and Development Division). This strain was formerly described as a strain of the species *Tricho-*

derma polysporum (Link ex Pers.) and is described in the literature, e.g., in DOS 2,455,859.

Alternatively, cyclosporin D producing strains obtained by selection or mutation of NRRL 8044, or treatment of this strain by ultraviolet light or x-rays, or treatment of this strain with laboratory animals, may be used.

Cyclosporin D may be isolated in conventional manner from other natural products that may be produced in greater amounts, e.g., the somewhat more polar cyclosporin A (also known as S 7481/F-1), the more polar cyclosporin B (also known as S 7481/F-2) and the yet more polar cyclosporin C.

The hydrogenation of cyclosporin D may be effected in conventional manner, e.g., by catalytic hydrogenation. Suitable solvents include methanol, ethanol, isopropanol or ethyl acetate. The process is conveniently effected in a neutral medium at a temperature between 20° and 30°C and at atmospheric or slightly elevated pressure. A suitable catalyst is palladium on charcoal.

The acid treatment of cyclosporin D may be effected in conventional manner, e.g., with trifluoroacetic acid or preferably methanesulfonic acid or p-toluenesulfonic acid. The mol ratio of acid to cyclosporin D is preferably between 1:1 and 4:1. Suitable solvents include methanol, chloroform, and dioxane. Suitable temperatures are between 20° and 65°C, preferably 40° to 55°C.

These compounds exhibit pharmacological activity. In particular, the compounds exhibit antiinflammatory activity and antiarthritic activity as indicated by an inhibition of swellings in the Freunds adjuvant arthritis test in rats on p.o. administration of 3 to 100 mg/kg of the compounds.

The compounds are therefore useful for the treatment and prophylaxis of chronic inflammations, e.g., arthritis and rheumatic disorders.

Furthermore, cyclosporin D and dihydrocyclosporin D exhibit immunosuppressive activity, e.g., by their effect on hormonal and cellular immunity, as indicated in standard tests. Isocyclosporin D exhibits especially interesting activity.

Triazene Compounds as Smooth Muscle Relaxants and Hypotensives

The process of *K. Yoshida, H. Tanaka, M. Okamoto, E. Iguchi, M. Kohsaka, H. Aoki and H. Imanaka; U.S. Patent 4,297,096; October 27, 1981; assigned to Fujisawa Pharmaceutical Co. Ltd., Japan* is one for preparing new triazene compounds. These compounds have activity as smooth muscle relaxants and as hypotensives. These triazenes can be represented by Formulas (1) and (2) below, which are tautomers.

RCH=N–N=N–OH (1)   RCH=N–NH–N=O (2)

where R is alkyl, alkenyl or aryl and its pharmaceutically acceptable salt. R is more particularly a lower alkyl (C_1 to C_6); alkenyl is unsaturated hydrocarbon residue and preferably includes higher (C_6 to C_{18}) alkatrienyl and aryl includes phenyl, tolyl, naphthyl and the like.

The pharmaceutically acceptable salts of the triazene compound may include a salt with a base such as an inorganic base (e.g., sodium hydroxide, potassium hydroxide, calcium hydroxide, ammonia, etc.), and an organic base (e.g., ethanolamine, triethylamine, dicyclohexylamine, etc.).

The triazene compound (1) can be prepared by a synthetic process and/or a fermentation process.

The synthetic process comprises reacting an aldehyde RCHO (where R is as previously described) with hydrazine; reacting the resultant hydrazone derivative with a silylating agent taken from the class consisting of trialkylhalosilane, trialkylsilylamide, dialkyldihalosilane, alkyltrihalosilane, dialkylarylhalosilane, triarylhalosilane, dialkylaralkylhalosilane, dialkoxydihalosilane, and trialkoxyhalosilane; and reacting the resultant silylated compound with an n-nitrosating agent.

Two compounds of particular interest may be prepared by fermentation. They are: FR-900184 = 1-hydroxy-3-(trans, trans, trans-2,4,6-undecatrienylidene)triazene and FR-900190 = 1-hydroxy-3-(trans, trans, trans-2,4,7-undecatrienylidene) triazene. These compounds are prepared by a microorganism of the Streptomyces genus, such as *Streptomyces aureofaciens* ATCC 31442 (FERM-P 4671). The two compounds are produced simultaneously by the fermentation of this microorganism. Submerged aerobic cultivation in a conventional nutrient medium is preferred, at a temperature about 30°C, for 50 to 100 hours. Separation of the two compounds may be made in the conventional manner, such as high pressure liquid chromatography.

Antihemolytic M-9337

Toxins discharged from streptococci, staphylococci, tetanus and the like cause a hemolytic reaction and bring about various diseases. *S. Miyamura; U.S. Patent 4,334,025; June 8, 1982; assigned to SS Pharmaceutical Co., Ltd., Japan* has found a new strain of the genus Streptomyces which produces a substance arbitrarily named M-9337 which is capable of inhibiting the hemolytic reaction.

This strain belongs to the genus Streptomyces and resembles *St. robefuscus*, *St. albaduncus* and *St. naganishii*. However, this strain differs from these species in that with *St. robefuscus*, the vegetative hyphae are a tint or shade of brown in color, that with *St. albaduncus*, the utilization of sucrose is (±), and its spore silhouette is spiny, and that *St. naganishii* makes no use of sucrose and its spore bearing aerial hyphae are not wavy or spiral. This strain is a new species and has been called *Streptomyces antihaemolyticus*. Cultures of this strain have been deposited at the Fermentation and Research Institute in Japan as FERM-P 4651, and in the American Type Culture Collection under Accession No. 31801.

The substance M-9337 has the following characteristics: (1) white or yellow powder with a neutral pH; (2) elementary analysis: C = 52 to 53%; H = 7 to 8%; 0 = 39 to 41%; (3) soluble in dimethylsulfoxide and mixed chloroform and methanol, slightly soluble in water and methanol; insoluble in ethyl acetate and chloroform; and (4) melting point 170° to 175°C.

The ultraviolet and infrared absorption spectrum are shown in figures of the patent.

The cultivation of *St. antihaemolyticus* is feasible by any of methods ordinarily employed for the species belonging to the genus Streptomyces and is preferably carried out by a shaking culture using a liquid medium or a submerged culture.

Anthelmintic-Acaricidal Compound B-41D

Y. Takiguchi, H. Mishima, S. Yamamoto and M. Terao; U.S. Patent 4,346,171; August 24, 1982; assigned to Sankyo Company, Limited, Japan have provided a compound, designated Compound B-41D, which has valuable acaracidal and anthelmintic activity. The compound has the formula:

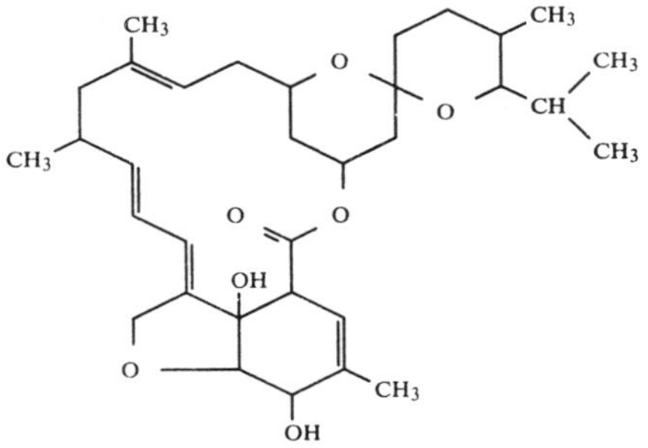

and is prepared by cultivating a microorganism of the genus Streptomyces in a culture medium. The microorganism used is Streptomyces strain B-41-146 (FERM 1438). The cultivation medium used is a conventional one. Surprisingly good yields of Compound B-41D may be obtained by using a culture medium containing 6 to 8% w/v glucose, 0.5 to 2% w/v of lactose and/or maltose and/or cornstarch, 0.5 to 1.0% w/v soybean meal, and 1.0 to 2.0% w/v skimmed milk. Although any known method of culturing the Streptomyces strain B-41-146 may be employed, liquid culture, and particularly submerged culture, is the most suitable method. The culture is preferably conducted aerobically and at a temperature which may vary over a wide range, preferably 28°C. The production reaches a maximum after 5 to 10 days culture, either by shaking or in a tank.

Although the Compound B-41D may be separated and purified for use, it is also possible to discontinue the purification procedure at any desired stage and to use the crude product, which contains a mixture of milbemycins as well as Compound B-41D. If a mixture containing two or more such compounds is used without complete separation, it is sufficient that it should be so purified as to obtain a 100% acaricidal effect at a concentration of 5 ppm. The content of Compound B-41D in the crude mixture in this case is preferably at least 25% by weight, and more preferably about 50% by weight, the remainder being impurities.

The compound has superior acaricidal activity against adults and eggs of such mites as the two-spotted spider mite, the European red mite, the citrus red mite and rust mites, which are parasitic on fruits, vegetables and flowers. It is also active against mites of the genera Ixodidae, Dermanyssidae and Sarcoptidae, which are parasitic on animals. It is also highly active against such ectoparasites as Oestrus, Lucilia, Hypoderma, Gastrophylus, fleas, lice, etc., as well as insects of sanitary importance (cockroaches or flies) and other agriculturally damaging insects, such as aphids and larvae of insects of the order Lepidoptera. Furthermore, it is active against nematodes, such as those of the genus Meloidogyne, and bulb mites such as those of the genus Rhizoglyphus. It is also highly active as a parasiticide for humans.

STEROIDS AND VARIOUS BIOLOGICALS

STEROIDS AND STEROID INTERMEDIATES

By Employment of Culture Media Containing Hydrocarbons in Place of Carbohydrates

The microbiological conversions of steroids have long been used for the production of derivatives having particular pharmacological activities or for the preparation of intermediates useful in the production of such derivatives. These transformations are normally performed by cultivating appropriate cultural microbial strains which contain carbohydrates and/or other complex substrates as a source of carbon and of energy. Under these conditions the microbiological transformations of the steroidal substrates have no direct relationship with the metabolism of the microbial strains which perform them and it seems that they could rather be attributed to detoxification mechanisms.

P. Zaffaroni, V. Vitobello and A.M. Gamalerio; U.S. Patent 4,273,872; June 16, 1981; assigned to Snamprogetti SpA, Italy have found that microbiological transformations on steroids of preeminent industrial importance (hydroxylation, dehydrogenation, side chain demolition) can be encouraged by using hydrocarbons rather than carbohydrates in the culture media. The microbial strains used in the practice of the process have been isolated from samples of different origin (soil, sewage waters) usually drawn in the neighborhood of refineries or of petrochemical installations and were used for inoculating enrichment cultures containing mixtures of hydrocarbons as the only source of carbon and energy. The strains were purified according to conventional microbiological procedures. The isolated strains were maintained on agar slants.

Inoculated cultures were grown in a nutrient mixture containing a mixture of normal chain C_{15-20} paraffins as the single source of carbon.

When the cultures showed a satisfactory growth, they were supplemented with the steroidal substances to be transformed in suspension, in water or in the mixture of paraffins, micronized with supersonic vibrations, or as a solution in an organic solvent (ethanol, acetone, dimethylformamide). After an appropriate period of time, the cultures were extracted with a solvent immiscible with water, such as chloroform or ethyl acetate. The extracts were evaporated and examined by thin-layer chromatography (TLC) and/or gas chromatography.

By using this method, it is possible to carry out very selective oxidation with very high yields, such as the oxidation of cholesterol to Δ^4-cholestene-3-one, thus obtaining from cheap sources more appreciable products (production of $\Delta^{1,4}$-androstadiene-3,17-dione from cholesterol, production of Δ^4-androstene-3,17-dione and $\Delta^{1,4}$-androstadiene-3,17-dione from progesterone).

The reactions which have been observed under conditions of cooxidation with hydrocarbons are:

(1) demolition of the steroid side-chain such as in cholesterol or progesterone, to 17-ketosteroids;

(2) oxidation of the –OH group in the 3-position of several steroids to a group =CO in the 3-position;

(3) Δ^1-dehydrogenation of several steroids.

The bacterial strains useful in carrying out these steroid transformations are SP2T (NRRL B-11,112); SP3T bis (NRRL B-11,113), SP5T bis (NRRL B-11,114) and SP7T/1 (NRRL B-11,115).

Preparation of 19-Hydroxysteroids of the Androstane and Pregnane Series

19-Hydroxysteroids are, as is known, important intermediates for the partial synthesis of pharmacologically active 19-nor steroids. The chemical synthesis of these compounds from the corresponding 10-methyl steroids comprises many stages and is very expensive. Investigations have also been conducted on hydroxylation of 10-methyl steroids in the 19-position by means of microorganism cultures. However, the previously known methods yield such low quantities of 19-hydroxysteroids that they are useless from a technical viewpoint.

K. Petzoldt; U.S. Patent 4,284,720; August 18, 1981; assigned to Schering, AG, Germany have found that it is possible to convert 10-methyl steroids into the corresponding 19-hydroxysteroids in high yields.

This process comprises fermenting a 10-methyl steroid of the androstane or pregnane series with a fungal culture of the genus Nigrospora. One useful species is *Nigrospora sphaerica* (ATCC 12,772).

The 19-hydroxysteroid (1) is prepared from the 10-methyl steroid (2) by the fermentation.

(1)

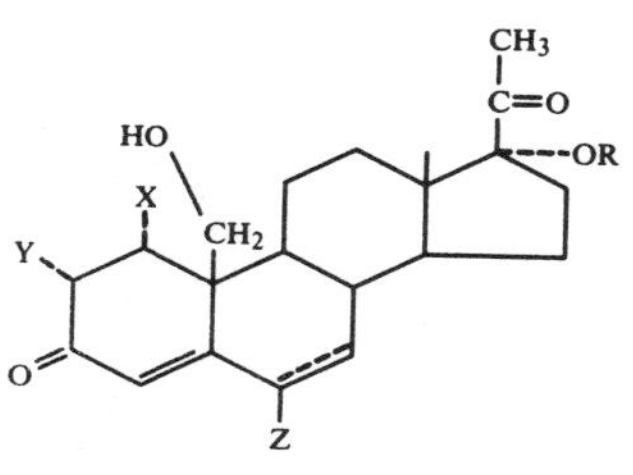

(2)

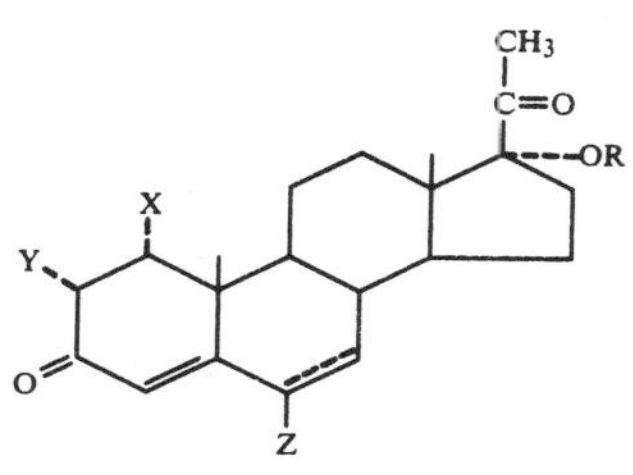

In both formulas: ------ is a single bond or a double bond, X and Y each is hydrogen or together represent a carbon-carbon bond or a methylene group, Z is hydrogen, fluorine, chlorine or methyl, and R is hydrogen or an acyloxy group of 1 to 8 carbon atoms.

The process is accomplished under the conditions customarily employed in the microbiological hydroxylation of steroids with fungal cultures.

Preferred 10-methyl steroids are those of general formula (2) which carry as the substituent R, for example, a hydrogen atom, a formyloxy group, a propionyloxy group, or a butyryloxy group.

Androst-4-Ene-3,17-Dione

The transformation of steroids by microorganisms has been widely studied and documented. Apparently, the earliest such work was by Mamoli and Vercellone in 1937 (*Ber.* 70, 470 and *Ber.* 70, 2079).

Marsheck et al in U.S. Patent 3,759,791 show the selective microbiological preparation of androst-4-ene-3,17-dione by fermenting a steroid of the cholestane or stigmastane series containing at least 8 carbons in the 17-alkyl side chain with Mycobacterium sp. NRRL B-3805, which has been characterized as *Mycobacterium vaccae.*

Three patents–*U.S. Patents 4,293,644, Oct. 6, 1981; 4,345,030 and 4,345,033, August 17, 1982–the first and second by M.G. Wovcha and C.B. Biggs and the third by Wovcha and Biggs plus T.R. Pyke; all assigned to The Upjohn Company,* describe a mutant which is characterized by its ability to selectively degrade steroids having 17-alkyl side chains of from 2 to 10 carbon atoms, inclusive, and accumulate androst-4-ene-3,17-dione, hereinafter referred to as AD, in the fermentation beer. This mutant can be obtained from microorganisms of the following genera by using the mutation procedures discussed here or other mutation procedures: Arthrobacter, Bacillus, Brevibacterium, Corynebacterium, Microbacterium, Mycobacterium, Nocardia, Protaminobacter, Serratia, and Streptomyces. A preferred genus is Mycobacterium. Exemplary species of this genus are *M. phlei, M. smegmatis, M. rhodochrous, M. mucosum, M. fortuitum,* and *M. butyricum.* Specifically exemplified herein is the mutant microorganism, *Myobacterium fortuitum,* NRRL B-11045. Examples of suitable steroid substrates are sitosterols, cholesterol, stigmasterol, campesterol, and the like. These steroid substrates can be in either the pure or crude form.

Mutation of *M. fortuitum* ATCC 6842, accomplished by the use of nitrosoguanidine, has resulted in the production of a mutant which selectively degrades steroids having 17-alkyl side chains of from 2 to 10 carbon atoms, inclusive, to produce AD and small amounts of androsta-1,4-diene-3,17-dione (ADD). This mutant microorganism has been given the accession number NRRL B-11045 by the Northern Regional Research Laboratory, U.S.D.A., where it has been deposited in the permanent collection. This microorganism has been distinguished from the Mycobacterium species NRRL B-3805 of U.S. Patent 3,759,791, which is discussed above. NRRL B-3805 has the general characteristics of *Mycobacterium vaccae* which is a distinctly different species than the *M. fortuitum* of this process. See Bergey's *Manual of Determinative Bacteriology,* 8th edition, p 695 for a comparison of these microorganisms.

The morphology and drug sensitivities of *M. fortuitum* NRRL B-11045 are indistinguishable from that of the parent *M. fortuitum,* ATCC 6842. *M. fortuitum* is an acid-fast nonmotile, nonspore-forming bacilli belonging to the family Mycobacteriaceae of the order Actinomycetales.

The selective transformation of the steroid can be effected in a growing culture of *M. fortuitum* NRRL B-11045 by either adding the selected steroid substrate to the culture during the incubation period, or incorporating it in the nutrient medium prior to inoculation. The steroid can be added singly or in combination with another steroid. The preferred, but not limiting, range of concentration of the steroid in the culture is about 0.1 to about 100 g/ℓ. The culture is grown in a conventional nutrient medium containing a carbon source, and a nitrogen source. Trace metals need not be added to the fermentation medium since tap water and unpurified ingredients are used as components of the medium prior to its sterilization.

The transformation process can range from about 72 hours to 15 days or more. The incubation temperature can range from about 25° to about 37°C, with 30°C being preferred for NRRL B-11045. The contents are aerated with sterilized air and agitated to facilitate growth of the microorganism, and, thus, enhance the effectiveness of the transformation process.

Upon completion of the transformation process, as evidenced by thin layer chromatography using silica gel plates (E. Merck, Darmstadt) and a solvent system consisting of 2:3 (by volume) ethyl acetate-cyclohexane, the desired transformed steroid is received by means well known in the art. For example, the fermentation (transformation) reaction mixture, including the fermentation liquor and cells, can be extracted with a water-immiscible organic solvent for steroids. A preferred solvent is dichloromethane.

The desired product of the transformation process is the known steroid intermediate AD. This compound is useful as an intermediate in the synthesis of useful steroidal hormones such as testosterone.

Three further patents use two other microorganisms of the genus Mycobacterium to effect this same steroid transformation: *M.G. Wovcha and C.B. Biggs; U.S. Patent 4,345,029; August 17, 1982; Wovcha, Biggs and T.R. Pyke; U.S. Patents 4,293,645; October 6, 1981 and 4,328,315; May 4, 1982; all assigned to The Upjohn Company.*

Specifically exemplified in these patents are the mutant microorganisms, *Mycobacterium fortuitum,* NRRL B-8153, and *Mycobacterium phlei,* NRRL B-8145, which are used to selectively degrade steroids having 17-alkyl chains of from 2 to 10 carbon atoms, inclusive, to ADD and AD. Examples of suitable steroid substrates are sitosterols, cholesterol, stigmasterol, campesterol, and like steroids with 17-alkyl side chains of from 2 to 10 carbon atoms, inclusive. These steroid substrates can be in either the pure or crude form.

Mutation of *M. fortuitum,* ATCC 6842, and *M. phlei,* UC 3533 (UC denotes The Upjohn Company Culture Collection), using nitrosoguanidine has resulted in the production of these mutants, which have been given the accession numbers NRRL B-8153 and NRRL B-8154, respectively, by the NRRL, U.S.D.A. where they have been deposited in the permanent collection.

M. fortuitum NRRL B-8153 or *M. phlei* NRRL B-8154 are cultivated, therefore, in a conventional aqueous nutrient medium under aerobic conditions and containing a steroid such as sitosterol, cholesterol, stigmasterol or campesterol or mixtures of two or more of these sterols and from the fermentation beer. The steroids androst-4-ene-3,17-dione (AD) and androsta-1,4-diene-3,17-dione (ADD) are recovered by conventional methods.

ADD and AD can be separated from the silica gel by elution with the solvent system ethyl acetate-chloroform (15:85). The compounds then can be isolated as separate entities by evaporation of the solvent and recrystallization from hexane.

The products of this transformation process are the known steroid intermediates, ADD and AD, which are useful as intermediates in the synthesis of useful steroidal hormones. For example, ADD can be used to make estrone according to the process disclosed in U.S. Patent 3,274,183. Also, AD can be used to make testosterone according to processes disclosed in U.S. Patents 2,143,453; 2,253,798; 2,264,888 and 2,356,154.

In two other patents *M.G. Wovcha and K.E. Brooks; U.S. Patents 4,293,646; October 6, 1981 and 4,345,034; August 17, 1982; assigned to The Upjohn Co.,* discuss *Mycobacterium fortuitum* mutants which give improved yields of AD in the microbiological steroid transformation process.

The mutant of the process, *Mycobacterium fortuitum* NRRL B-11359, is obtained from *M. fortuitum,* NRRL B-11358, which is an adaptive mutant from *Mycobacterium fortuitum* NRRL B-11045, previously described.

M. fortuitum NRRL B-11359 is characterized by its ability to selectively transform steroids having 17-alkyl side chains of from 2 to 10 carbon atoms, inclusive, and accumulate AD as essentially the sole transformed product in the fermentation beer. The previously described process for preparing AD accumulates about 4 times more ADD in the fermentation beer than this process. A smaller amount of ADD in the fermentation beer facilitates the isolation of the desired product AD.

The mutant can be obtained by growing *M. fortuitum* NRRL B-11045 on a suitable medium containing 9α-hydroxyandrostene dione (9α–OH AD). A rapid growth colony, identified as an adaptive mutant, is selected. This adaptive mutant, *M. fortuitum* NRRL B-11358, is then subjected to nitrosoguanidine (NTG) mutagenesis. The mutant is then selected on agar plates using a 9α–OH AD medium, as above. The desired mutant *M. fortuitum* NRRL B-11359 will not grow on a medium containing AD or 9α–OH AD as the sole carbon source.

Sterol-Degrading Microorganism *Mycobacterium fortuitum* **NRRL B-8128**

J.C. Knight and M.G. Wovcha; U.S. Patent 4,304,860; December 8, 1981; assigned to The Upjohn Company have developed other mutants which are characterized by their ability to selectively degrade steroids with or without 17-alkyl side chains of from 2 to 10 carbon atoms, inclusive.

These microorganisms accumulate predominantly the following compounds in the fermentation beer:

(1) 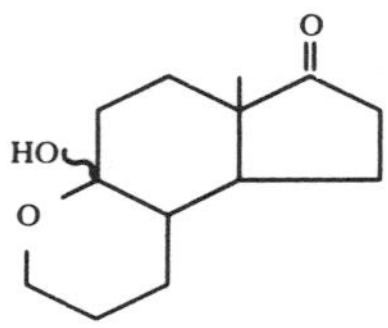 (2) 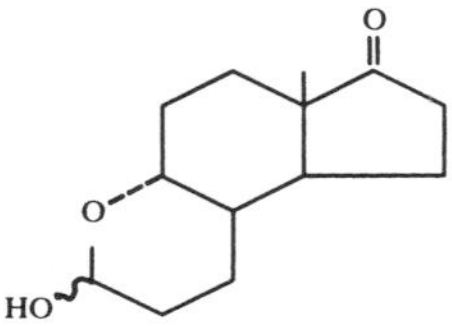

(3) 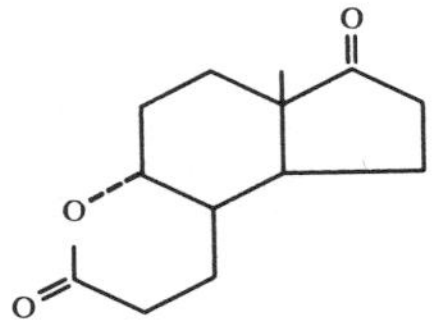 (4) 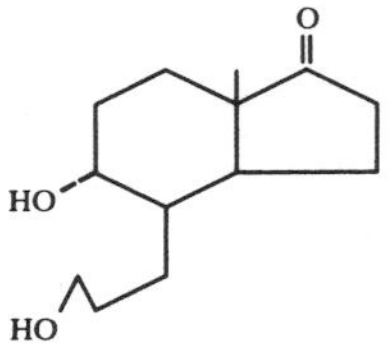

The formula (1) shows Compound I: 3aα-H-4α-[3'-propanol]-7aβ-methylhexahydro-1,5-indandione hemiketal; formula (2) shows Compound II: 3aα-H-4α-[3'-propanol]-5α-hydroxy-7aβ-methylhexahydro-1-indanone hemiacetal; Compound III shown in formula (3) is: 3aα-H-4α[3'-propionic acid]-5α-hydroxy-7aβ-methylhexahydro-1-indanone-δ-lactone; Compound IV of formula (4) is: 3aα-H-4α-[3'-propanol]-5α-hydroxy-7aβ-methylhexahydro-1-indanone.

The mutant used to selectively degrade steroids to Compounds I, II, III and IV was *Mycobacterium fortuitum* NRRL B-8128. Examples of suitable steroids on which to practice the degradation are sitosterols, cholesterol, stigmasterol, campesterol, and like steroids, and androst-4-ene-3,17-dione, androsta-1,4-diene-3,17-dione, dehydroepiandrosterone, and testosterone.

The mutant named above was prepared from *M. fortuitum* ATCC 6842, using nitrosoguanidine.

7β-Hydroxylated Steroids of the Cholane Group

The patent by *I. Macdonald; U.S. Patent 4,303,754; December 1, 1981* relates to the production of steroids having the general structure of formula (1) from those having the B-ring structure of formula (2):

(1) 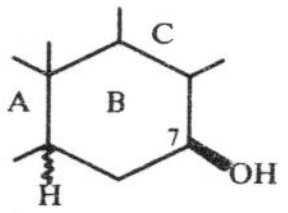 (2) 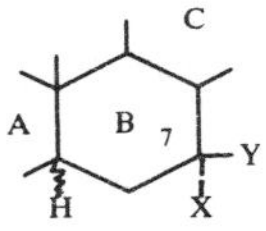

wherein X is OH, Y is H and when taken together X and Y is oxo (O=).

This steroid transformation is accomplished by a microorganism of the genus Clostridium. The final products which may be produced by this process are 7β-OH substituted steroids of the cholane group, and include such specific steroid compounds as 3α, 7β-dihydroxy-5β-cholanoic acid; 3α, 7β, 12α-trihydroxy-5β-cholanoic acid, etc.

It has been found that satisfactory results may be obtained when the steroid starting materials possess the above identified B-ring structure and are also cholane derivatives. Most preferable results are obtained when the starting materials are cholanoic acid derivatives including such compounds as 3α, 7α, 12α-trihydroxycholanoic acid, 3α,7α-dihydroxycholanoic acid, 3α-hydroxy-7-oxo-cholanoic acid, 3α, 12α-dihydroxy-7-oxo-cholanoic acid, and other such compounds.

The curved line employed in the chemical structures in the linkage of atoms, is meant to denote that the linked atom or substituent may be either in the alpha or beta stereochemical position on the molecule as the case may be.

In order to produce the final products of this process, the 7-oxygenated steroid starting materials are subjected to the action of enzymes of the microorganisms to be described or directly to the action of the microorganisms themselves under the proper conditions and in the necessary medium in which the microorganisms can be propagated in the presence of the desired starting steroid compounds.

Among the microorganisms which may be employed are those enzyme-producing microorganisms of the genus Clostridium. More particularly, are included such microorganisms as *Clostridium absonum.* Even more particularly, most satisfactory results have been obtained where the specific microorganism employed has been *Clostridium absonum* ATCC 27555, ATCC 27635, ATCC 27636, ATCC 27637, or *Clostridium absonum* Nakamura's strain, KZ1214, KZ1215, KZ1216, or KZ1218. These microorganisms have been deposited in various culture collections; those designated with ATCC accession numbers with the American Type Culture Collection; and those designated with KZ accession numbers have been deposited with Kanazawa University, Japan; and specimens thereof may be obtained in the usual manner as is known to the worker skilled in the art.

In general, the conditions of culturing the microorganisms which may be utilized are, except for the inclusion of the 7-substituted steroid starting material, the same as those employed in the propagation of like organisms for such purpose, i.e., the microorganism is grown either aerobically or anaerobically (depending on the microorganism employed, for example, Clostridium are generally propagated under anaerobic conditions) in contrast with (in or on) a suitable fermentation medium.

It is also possible to obtain equivalent results by employing cell free solutions of the enzyme materials produced by the propagation of the desired Clostridia organisms. These cell free compositions may be obtained in any manner known to the skilled worker, for example, ultrasonic cell disruption techniques when employed in conjunction with proper filtering and collection methods will provide satisfactory results. However, when cell free preparations are employed in the practice of this process, it has been found necessary to employ an NADP-coenzyme, which may be prepared and employed in accordance with procedures which are well known and accepted by the skilled worker in the art.

Production of 17-C-Steroid-Alpha-Propionic Acid Compounds

Due to the potential importance of natural sterol compounds having 17-C side chain of plant (phytosterols) or animal (cholesterol) origin as starting material for high quality pharmaceutic compositions having steroid basic structure, a great number of attempts have been made to achieve selective side chain degradation by the microbiological route.

The prior art proposals fall within the following three groups: Transformation of the ring structure of the sterol starting compound in such a manner that the mechanism of ring cleavage is prevented so that the selective side chain cleavage becomes possible; concomitant use of inhibitors to inhibit the steroid ring degradation and/or the growth of the microorganism; and, finally, the search for microorganism mutants which lead to as extensive a selective side chain degradation as is possible.

According to these processes, the final products of the microbial degradation are 17-ketosteroid compound which, while valuable as intermediates for the technical production of hormones of the estrane-androstane- and spirostane-series, are less suitable for the preparation of steroid hormones of the pregnane series, e.g., progesterone, hydrocortisone, prednisone, prednisolone, triamcinolone and the like. In the latter cases, it is necessary to incorporate again chemically stereospecific side chains in the 17-position.

It is an object of the process developed by *F. Hill, W. Preuss, J. Schindler, R. Schmid and A. Struve; U.S. Patent 4,320,195; March 16, 1982; assigned to Henkel Kommanditgesellschaft auf Aktien, Germany* to permit the production of 17-C-steroid-α-propionic acid compounds in commercially useful and improved yields by selective microbial degradation of 1-C-side chain steroid substrata. This is achieved with the use of defective block mutant microorganisms which have been prepared and selected in a directive manner and previously recovered from suitable wild strains. In particular, the selected defective block mutants are used in the absence of inhibitors which inhibit the steroid ring degradation and/or the growth of microorganisms.

In a narrower sense, a particular object of the process is the microbial production of 3-oxo-pregna-4-ene-20-carboxylic acid (Δ-4 BNC) and/or 3-oxo-pregna-1,4-diene-20-carboxylic acid (Δ-1,4 BNC) by the method to be described. The starting materials for the microbiological production of these compounds are natural and/or vegetable sterols which, as waste products, have hardly gained any practical importance.

The process is carried out by

(1) isolating and cultivating a microorganism wild strain capable of aerobic growth on sterol compounds as the sole carbon source while giving preference to 17-C-side chain degradation over ring degradation,

(2) subjecting the wild strain to a mutation treatment known per se,

(3) cultivating the mutant population on a separating medium (accumulation medium for the mutants desired) on which the mutants producing 17-C-steroid-α-propionic acid compounds do not or substanially do not grow while the undesirable accompanying mutant strains grow and are killed off thereby or during their growth, and

(4) cultivating the remaining fraction of the mutant strains on 17-C-side chain sterol compounds while isolating the strains having optimum production of the 17-C-steroid-α-propionic acid compounds.

A new block mutant, Chol. spec. 73-MII, having the accession numbers CBS 437.77 and ATCC 31385 was found to be suitable for the production of the Δ-4 BNC and Δ-1,4 BNC. It was recovered according to the process from the wild strain Chol. spec. 73, (CBS 660.77 ATCC 3184) and belongs to the group of Coryneform bacteria.

7-Alpha-Hydroxylated Steroids of the Pregnane or Androstane Series

The process developed by *M. Fujiwara, A. Fujiwara and C. Miyamoto; U.S. Patent 4,336,332; June 22, 1982; assigned to Hoffmann-La Roche, Inc.* is one for the manufacture of 7α-hydroxylated steroids by fermentation of 7-unsubstituted steroids of the pregnane or androstane series with microorganisms of the genus Botryodiplodia or by reaction with enzyme extracts thereof. The process produces steroid compounds which are intermediates for the manufacture of pharmacologically valuable substances and which themselves exhibit pharmacological (e.g. hormonal) activity.

More particularly, in a preferred embodiment, the steroids to be hydroxylated (starting compounds) have been selected from dehydroepiandrosterone, pregnenolone or steroids of the formula (1) to produce compounds of formula (2). In both formulas, X represents one of the following groups: >CHOH, >C=O, or $>COHCOCH_2OH$.

(1)

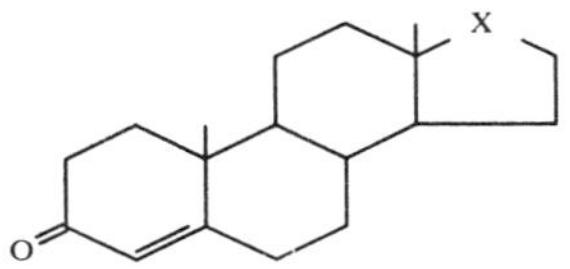

(2)

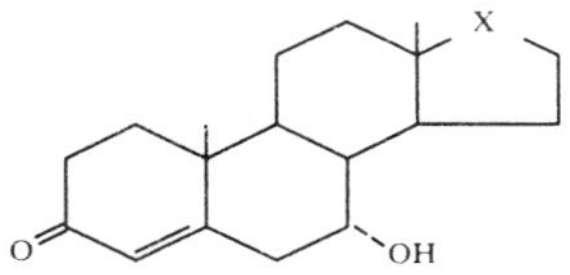

In the case of the fermentation of dehydroepiandrosterone by the process there are obtained 7α-hydroxy-4-androstene-3,17-dione and 7α,17β-dihydroxy-4-androstene-3-one; in the case of the fermentation of pregnenolone there is obtained the 7α-hydroxy-progesterone.

Any strain of the microorganisms of the genus Botryodiplodia capable of the 7α-hydroxylation of steroids, especially such steroids of the formula (1) along with pregnenolone and dehydroepiandrosterone, as well as variants thereof, can be used in the process. Preferred strains are, for example, IFO 6469 and *Botryodiplodia malorum* CBS 134.50.

With the use of 4-androstene-3,17-dione as the starting material there is obtained a mixture of 7α-hydroxy-4-androstene-3,17-dione and 7α,17β-dihydroxy-4-androstene-3-one. This mixture can be readily separated by chromatography, whereby one varies the polarity of the elution agent.

The hydroxylating microorganisms used include all strains belonging to the genus Botryodiplodia which are capable of hydroxylation as well as mutants and variants thereof. Particularly preferred strains are IFO 6469 and *Botryodiplodia malorum* CBS 134.50. A subculture of IFO 6469 has been deposited at Northern Regional Research Laboratory of the U.S.D.A. under NRRL No. 11174. Cultures of *B. malorum* CBS 134.50 were obtained from Centraal-Bureau voor Schimmelcultures, The Netherlands.

The hydroxylated steroids produced, in particular the 7α-hydroxylated steroids, can be employed as intermediates by any art-recognized procedure for producing synthetic hormones, cholic acid and various pharmaceuticals. The following example illustrates the process. Temperatures are given in degrees centigrade.

Example: A fermentation medium containing 1% lactose, 3% Bacto-liver (Difco), 0.1% KH_2PO_4 and 0.05% KCl was adjusted to pH 6.3 and sterilized at 120°C for 15 minutes. The nutrient medium was inoculated in ten 100 ml portions with the mycelium of a two weeks old malt extract-agar slant culture of IFO 6469. The cultures were then shaken on the rotary machine at 26.5°C with 180 movements per minute. After 22 hours, there were added in each case 50 mg of 17α,21-dihydroxy-4-pregnene-3,20-dione (previously emulsified by exposure to ultrasonic wave for 10 minutes in 1 ml of 0.1% Tween 80), so that the concentration of steroid amounted to 0.5 mg/ml of fermentation solution. The cultures were then incubated for a further 92 hours; thereafter pooled, filtered and washed with water; so that from 1,000 ml of nutrient solution there was obtained a total volume of 1,100 ml.

The thus-obtained 1,100 ml of culture solution were extracted with ethyl acetate and concentrated to a small volume under reduced pressure. The residue was chromatographed on silicic acid with the use of chloroform-acetone as the elution agent. The homogeneous fractions were pooled and crystallized from ethyl acetate. There were obtained 188 mg of 7α,17α,21-trihydroxy-4-pregnene-3,20-dione, melting point 221.5° to 223.5°C.

Preparation of 1α, 2α-Methylene Steroids

The process of *K. Petzoldt, H. Steinbeck and W. Elger; U.S. Patent 4,337,311; June 29, 1982; assigned to Schering AG, Germany* relates to pharmaceutically active 1α,2α-methylene steroids. It is their object to provide new steroids possessing high antiandrogenic activity with minor progestational activity.

These objects have been attained by providing hydroxylated 1α,2α-methylene steroids of formula (1) by fermentation of compounds of formula (2).

(1)

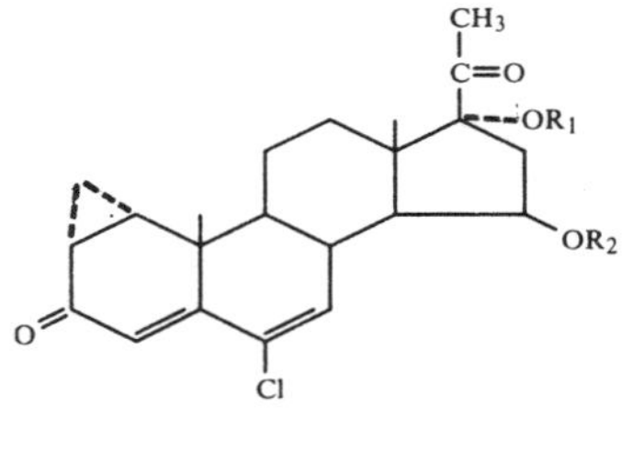

(2)

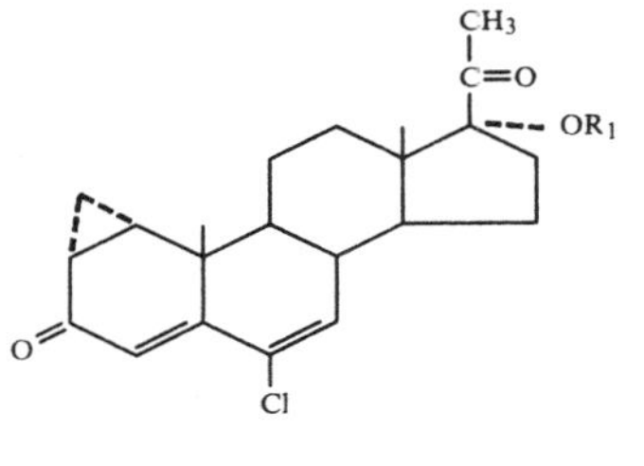

In the above formulas R_1 and R_2 are identical or different and each is hydrogen or the residue of an organic or inorganic acid. Esters of formula (1) may also be formed in cases where R_1 and R_2 are not hydrogen. Especially suitable acids for making these esters are alkanoic and alkane sulfonic acids having 1 to 12 carbon atoms. Physiologically suitable salts may also be used.

The 15β-hydroxylation is accomplished using a bacterial culture of the genus Bacillus or with a fungal culture of the genus Mucor. Suitable for this fermentation are the bacterial strain, for example, *Bacillus megaterium* (ATCC 13 368) and the fungal strain, for example, *Mucor griseocyanus* (ATCC 1207 b). The fermentation is conducted under the same conditions employed in conventional fermentative conversions of steroids using bacterial and fungal cultures.

Free hydroxy groups in the 17- and/or 15-positions are then esterified when necessary to produce the desired R_1 and R_2 moieties.

When R_1 = H, formula (2) represents the known steroid, cyproterone. The starting material cyproterone esters of formula (2) can be prepared in accordance with the esterification procedures discussed below, if desired, or the corresponding esterification can be performed after the 15β-hydroxylation.

For the esterification step, conventional steroid chemistry processes can be utilized as disclosed for example in U.S. Patent 4,011,314. Since the secondary hydroxy group in the 15-position and the tertiary hydroxy group in the 17-position exhibit different reactivities, the hydroxy groups can also be esterified in stages. Moreover, it is frequently useful to start with compounds already esterified in the 17α-position, particularly if relatively drastic conditions are required for the esterification of the 15-hydroxy group, such as, for example, in the esterification with inorganic acids or when producing the aminoacylates from the corresponding chloroacylates. The partial esterification in the 15-position is preferably conducted in the presence of an alkaline catalyst at room temperature. Suitable catalytic bases are preferably tertiary amines, e.g. triethylamine, pyridine, collidine, optionally with the addition of 4-dimethylaminopyridine. Suitable esterifying agents are reactive acid derivatives, such as acid halogenides or anhydrides of monocarboxylic acids.

The esterification of the hydroxy group in the 17-position with a monocarboxylic acid may be conducted by utilizing fully conventional methods.

The esterification of the free hydroxy groups with dicarboxylic acids is conducted, according to a preferred embodiment, with a mixture of trifluoroacetic anhydride and the desired dicarboxylic acid. The esterification is advantageously carried out at 10° to 50°C in the presence of a solvent inert to the reactants. Suitable solvents include water-miscible solvents, such as tetrahydrofuran or dioxane, as well as water-immiscible solvents, such as benzene.

The water solubility of suitable salts of the compounds esterified with one or two dicarboxylic acids is of special significance. For purposes of such salt formation, the acidic monoesters are dissolved, for example, in alcohol and neutralized with the base, especially sodium hydroxide solution. A compound monoacylated in the 17-position with a carboxylic acid can be converted into the 15-sulfonic acid ester by conventionally reacting the 17-monoacyl compound in the presence of a tertiary amine with a sulfonic acid halogenide at room temperature. If a 15,17-dihydroxy compound is reacted with a sulfonic acid halogenide in the presence of a tertiary amine at room temperature, the 15,17-disulfonic acid ester is obtained.

The hydroxylated 1α,2α -methylene steroids of formula (1) are pharmacologically active compounds. For example, these compounds possess a strong antiandrogenic activity with minor progestational activity.

Preparation of 11β-Hydroxy Steroids

As is known, 11β-hydroxy steroids having antiinflammatory activity (such as, for example, the corticoids; hydrocortisone, prednisolone, dexamethasone, betamethasone, prednylidene and flurandrenolone) are produced from naturally occurring steroids (such as diosgenin) by means of a very expensive, multistage partial synthesis. Within the multistage synthesis of these compounds, the microbiological introduction of the 11β-hydroxy group into the steroid skeleton is normally the most expensive synthesis step and also the one wherein most of the losses are incurred due to the formation of by-products.

K. Petzoldt, K. Annen, H. Laurent and R. Wiechert; U.S. Patent 4,353,985; October 12, 1982; assigned to Schering AG, Germany have provided a process for preparing 11β-hydroxy steroids in a simple and inexpensive way with high yield. This process is less expensive, produces higher yields, and has shorter reaction times than earlier ones. Moreover, it does not require pure forms of the 11-deoxy steroids as starting materials.

The process produces 11β-hydroxy steroids of the general formula (1) below by fermentation of 11-deoxy steroids of general formula (2) below.

(1)

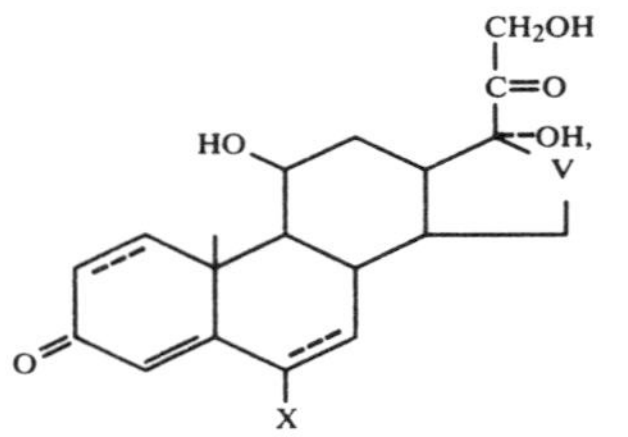

(2)

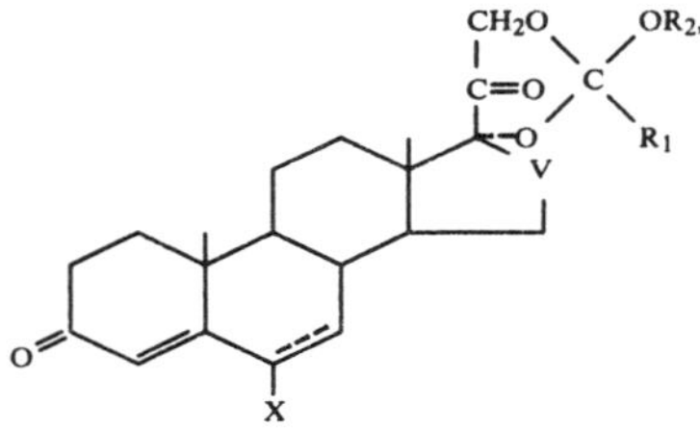

In the above formulas, - - - represents a single bond or a double bond; X is hydrogen, fluorine, chlorine or methyl; and V is a methylene, ethylene, ethylidene or vinylidene; R_1 is hydrogen or alkyl of 1-6 carbon atoms; and R_2 is alkyl of 1-6 carbon atoms. An especially preferred alkyl group R_1 is methyl. Particularly preferred alkyl groups R_2 are methyl and ethyl.

In the 11-deoxy steroids saturated in the 6,7-position of formula (2), the substituent X is preferably in the α-position.

In a suitable nutrient medium, submerged cultures are grown under aeration using the culturing conditions customarily employed for these microorganisms. Then the substrate (dissolved in a suitable solvent or preferably in emulsified form) is added to the cultures. The latter are fermented until maximum substrate conversion has been attained.

The fungal culture used for the hydroxylation is of the species *Curvularia falcuta* Qu-102 H, *Curvularia genticulata* IFO (6284), *Curvularia lunata* NRRL 2380, NRRL 2434, NRRL 2178, ATCC 12071 or IFO (6286), or *Curvularia maculans* IFO (6292).

Chenodeoxycholic Acid from β-Sitosterol

Chenodeoxycholic acid has exhibited valuable therapeutic activity, particularly as an agent for dissolution of cholesterol gallstones. This compound is obtained from cholic acid. However, since cholic acid is extracted from cow bile or with some additional processing from chicken bile, these natural sources even if utilized to maximum potential can provide only a minor portion of the demand. Thus, an efficient synthesis from highly abundant starting materials is an important factor in determining whether chenodeoxycholic acid achieves its potential role in medicine.

C. Despreaux, T.A. Narwid, N.J. Palleroni and M.R. Uskokovic; U.S. Patent 4,301,246; November 17, 1981; assigned to Hoffmann-La Roche Inc. have developed an efficient synthesis of chenodeoxycholic acid starting from 3-keto-bisnorcholenol (also named 22-hydroxy-23,24-bisnorchol-4-en-3-one which is readily obtained by the microbiological degradation of the commercially available plant sterol β-sitosterol by procedures well known to the art.

In the key step of the process 3-keto-bisnorcholenol is microbiologically hydroxylated in the 7-position to produce 7-alpha-hydroxy-3-ketobisnorcholenol (7α,22-dihydroxy-23,24-bisnorchol-4-en-3-one.

Of the 152 cultures examined for 7-alpha hydroxylation, it has been found that only nine very closely related cultures, *Botryodiplodia theobromae* IFO 6469, ATCC 28570, DSM 62-678, DSM 62-679; *Botryosphaeria ribis* ATCC 22802, *B. berengeriana* ATCC 12557, *B. rhodina* CBS 374.54, CBS 287.47 and CBS 306.58, are capable of carrying out the desired 7-alpha hydroxylation in this sterol substrate.

The microorganism may be used in the form of the culture broth, the mycelia or an enzyme extract thereof. The culture broth may be prepared by inoculating the organism into a suitable medium. The culture medium can contain carbon sources, nitrogen sources, inorganic salts and other nutrients suitable for the growth of the microorganisms. Since aerobic conditions are required for organisms used in this process, it is preferable to cultivate under conditions which promote aeration.

It has further been found that a number of procedures can be employed to optimize the yield of desired product from the fermentation, for example, addition of a chelating agent such as 2,2'-dipyridyl in a final concentration ranging from 0.5×10^{-4} M to 0.75×10^{-3} M, addition with the substrate of either glucose or sucrose in a final concentration of about 5%, lowering the temperature of the incubation fermentation to about 24°C, after adding the substrate and using a suspension of substrate at a concentration of 5% in 0.1% Tween 80. An even greater increase in yield is obtainable by the addition of adsorbants to the fermentation medium. For example yield improvement is obtained when polymeric resin absorbents, such as Amberlite XAD7 (Rohm & Haas Co.), a polymer of the methyl ester of acrylic acid is added at a concentration 0.3 to 0.6 wt % to a fermentation medium where the substrate is present in a concentration of up to about 1 g/ℓ. Best yield improvement was obtained at about the 0.6 wt % concentration level for the adsorbent.

The 7-alpha-hydroxy-3-keto-bisnorcholenol may be converted to chenodeoxycholic acid by a series of steps which are detailed in the patent.

INSULIN

Semisynthesis of B30-Threonine-Insulin

Insulin is an indispensable medicament for treatment of diabetes. Bovine and porcine insulins are commercially available at present but differ from human insulin in a few amino acid components, which cause formation of an antibody in some patients which is known to decrease the effectiveness of the insulin treatment. Therefore, human insulin is favored and its commercial availability is much desired. Human insulin is different from other animal insulins in amino acid components at the positions 8, 9, and 10 of the A chain and at the position 30 of the B chain.

K. Morihara, T. Oka and H. Tsuzuki; U.S. Patent 4,320,196; March 16, 1982; assigned to Shionogi & Co., Ltd., Japan provide a process to substitute the amino acid at the 30 position of the B chain of an animal insulin with L-threonine. Accordingly, human insulin can be easily prepared by this process using porcine insulin as starting material.

The process for preparing the B30-threonine-insulin comprises reacting a porcine or bovine des-B30-insulin with a threonine derivative of the formula $Thr(R^1)(R^2)$ in which Thr is an L-threonine residue, R^1 is hydrogen or a hydroxy-protecting group and R^2 is a carboxyl-protecting group, in a molar ratio of the threonine derivative to the des-B30-insulin of 5:1 to 1000:1, and in the presence of an enzyme specifically acting on the basic-amino-acid carbonyl in peptide bondings. The enzyme is selected from the group consisting of trypsin, a trypsin-like enzyme produced by *Streptomyces fradiae* ATCC 3535, and a trypsin-like enzyme produced by *Streptomyces erythreus.* The resultant product is subjected to one or more reactions to remove all of the protecting groups.

In the process, the molar ratio of the threonine derivative to the des-B30-insulin is about 20:1 to 100:1. It is effected in a medium containing one or more water-miscible organic solvents, such as ethanol or dimethylformamide.

The threonine derivative may be, for example, L-threonine-tert-butyl ester. Preferable pH for the reaction is 6 to 7, and preferred temperatures are 20° to 40°C.

In a similar patent *K. Morihara, T. Oka, M. Soejima and T. Masaki; U.S. Patent 4,320,197; March 16, 1982; assigned to Shionogi & Co., Ltd., Japan* describe a similar process for producing human insulin from various animal insulins. In this case the threonine derivative is the same as described in the previous patent. The reaction conditions are also the same, the difference in the processes being that the enzyme specifically acts on the lysine carbonyl in the peptide bondings. The enzyme used for this process is *Achromobacter lyticus* M 497-1.

The removal of the protective groups in both patents, if the tert-butyl group is the protective group, may be carried out by treatment with trifluoroacetic acid in the presence of a cation-trapping agent such as anisole.

Production of Insulin by Living Cell Systems

The therapeutic value of insulin and, in particular, human insulin in treating diabetic patients has long been appreciated. The primary commercial source of insulin for therapeutic application, however, has been that prepared from bovine or porcine pancreas. Because of such factors as the immunological differences between human and animal insulin which may result in the formation of antiinsulin antibodies after prolonged use in humans and the often uncertain availability of bovine and porcine pancreas, efforts to produce significant quantities of insulin from human pancreas have increased.

The primary objective of *R.A. Rosenberg; U.S. Patent 4,332,893; June 1, 1982;* is to provide a process for producing a line of insulin-producing pancreatic cells which may be utilized either in vitro or in vivo as a source of insulin.

The process comprises the following steps

(a) selectively isolating insulin-producing beta cells from a human pancreas to produce a suspension of single cells rich in insulin producing beta cells, pelleting those cells by low speed centrifugation, suspending the pellet thus formed in nutrient medium and plating the cells;

(b) growing a culture of a mutant strain of Rous sarcoma virus selected from the strains ts-sarc-SR-RSV-D mutant FU-19 and ts-sarc-Bryan-RSV including a temperature sensitive lesion in its transforming gene which causes the gene to be expressed at a permissive environmental temperature and to be inactivated at a nonpermissive environmental temperature by infecting chick embryo fibroblasts with the mutant strain and incubating the infected fibroblasts in a nutrient medium in a manner to produce a supernatant portion which contains the virion of the mutant strain of Rous sarcoma virus for a time sufficient to confluently transform the culture;

(c) harvesting the supernatant portion of the culture and adjusting the pH to 7.2 to 7.8;

(d) infecting the pancreatic cells obtained in step (a) with the virion-containing supernatant obtained in step (c);

(e) culturing the infected pancreatic cells at the permissive temperature for the expression of the temperature sensitive transforming gene in insulin-producing pancreatic beta cells for a time sufficient to cause insulin-producing pancreatic cells to be conditionally transformed by the expression of the temperature sensitive transforming gene;

(f) culturing in a culture medium the conditionally transformed insulin-producing pancreatic cells at the permissive temperature for a time sufficient to produce a desirably large population; and

(g) replacing the culture medium containing the conditionally transformed insulin-producing pancreatic cells with a culture medium containing glucose, incubating the cells in the glucose medium at the nonpermissive temperature for the inactivation of the temperature sensitive transforming gene of the mutant virus strain for a time sufficient to cause them to revert to a nontransformed state and to produce insulin, and harvesting the insulin produced by the cells from the glucose medium.

The problems which could arise from the oncogenic potential of the Rous sarcoma virus strain ts-sarc-SR-RSV-D could be diminished by using infected pancreatic cells which have reverted to the nontransformed phenotype at the permissive temperature. To isolate complete or partial revertants, conditionally transformed pancreatic cells produced as described above and shown to be hormonally active are dispersed through the use of an enzyme such as trypsin or collagenase and diluted for seeding in the wells of standard culture plates in a nutrient medium, such as Eagle's medium without serum. Each well should preferably contain only one cell. The plated cells are then incubated at the permissive temperature (33° to 34°C) for at least about 24 hours and examined to identify cells which appear morphologically normal and evidence no morphological transformation during growth. These revertant insulin-producing pancreatic cells are then used in place of the cells exhibiting transforming activity at the permissive temperature as described above. The preferred use of a population of revertant cells would be in the proposed in vivo insulin producing system.

It should be noted, however, that the revertant cells still contain both viral genetic information and the ts-sarc gene product, and, therefore, the oncogenic biohazard potential must be determined by a safety evaluation procedure prior to their in vivo application.

This process will find its primary application in the production of an insulin-producing system which could be utilized either as an in vitro source of insulin which is then administered to diabetic patients by conventional means, or implanted in vivo to provide a continuous endogenous supply of insulin.

Conversion of Porcine Insulin into a B30-Threonine Ester of Human Insulin

The process developed by *J. Markussen; U.S. Patent 4,343,898; August 10, 1982; assigned to Novo Industri A/S, Denmark* is one for converting crude porcine insulin into a threonine B30 ester of human insulin in high yields.

The process is based upon the discovery described in U.S. Patent 3,276,961 that the amino acid or peptide chain bound to the carbonyl group of Lys^{B29} in the insulin compound can be interchanged with a threonine ester. The interchange is referred to as a transpeptidation.

The term "insulin compounds" as used herein encompasses insulins and insulin-like compounds containing the human des(Thr^{B30}) insulin moiety, the B30 amino acid of the insulin being alanine (in insulin from, e.g., hog, dog, and fin and sperm whale) or serine (rabbit). The term "insulin-like compounds" as used herein encompasses proinsulin derived from any of the above species and primates, together with intermediates from the conversion of proinsulin into insulin.

This process comprises transpeptidizing an insulin compound or a salt or complex thereof with an excess of an L-threonine ester or a salt thereof in a mixture of water and a water miscible organic solvent in the presence of trypsin.

The reaction medium in which the threonine ester and insulin compounds are dissolved comprises water and a water miscible organic solvent, the water content of the reaction mixture being less than 50%, preferably 10 to 40% v/v. The reaction temperature range is from the freezing point of the reaction mixture to 50°C, but preferably is below ambient temperature. The preferred range is above 0°C. The reaction may require several days. Optionally present in the reaction mixture is an acid, preferably an organic acid, in up to 10 equivalents per equivalent of the threonine ester.

A substantial excess of threonine ester is present in the reaction mixture solution, with the molar ratio of threonine ester to insulin compound preferably exceeding 5:1. The threonine ester concentration of the reaction mixture should preferably exceed 0.1 molar, with the upper concentration of the threonine ester being the solubility thereof.

To obtain the 60% transpeptidation yield considered herein as an important aspect to practice of this process the reaction temperature, water content and acid content are interrelated within the described ranges. The attainment of high yields (60%) is positively correlated to low water concentration and presence of acid.

A preferred material for practice of this process is crude porcine insulin. Under properly selected reaction conditions a yield exceeding 95% may be obtained from the transpeptidation reaction. If crude porcine insulin containing PLI (proinsulin-like immunoreactive compounds) is transpeptidized the PLI may be reduced by a factor of about 100, being advantageously converted into Thr^{B30} ester of human insulin. A preferred procedure for preparing human insulin is as follows:

(1) The starting material used for the transpeptidation is crude porcine insulin, e.g., crystalline insulin obtained by the use of a citrate buffer.

(2) If there is any trypsin activity left after the transpeptidation, it is preferred to remove it, e.g., under conditions where trypsin is inactive, e.g., in acid medium below pH 3. Trypsin can be removed by separation according to molecular weight, e.g., by gel filtration on Sephadex G-50 or Bio-gel P-30 in 1 M acetic acid.

(3) Other impurities such as unreacted porcine insulin may be removed by the use of anion and/or cation exchange chromatography.

(4) Thereafter, the $threonine^{B30}$ ester of human insulin is deblocked and human insulin is isolated, e.g., crystallized, in a manner known per se.

By this process human insulin of an acceptable pharmaceutical purity can be obtained. It may be further purified, if desired.

INTERFERON

From Permanent Cell Lines of Lymphoblast Origin or Human Fibroblast Cells

P. Swetly, G.R. Adolf, G. Bodo, S.J. Lindner-Frimmel, P. Meindl and H. Tuppy;

U.S. Patent 4,266,024; May 5, 1981; assigned to Boehringer Ingelheim GmbH, Germany have found that the addition of a stimulator substance to cell culture medium significantly increases the interferon production by a factor of from about 4 to 60, as compared to production from cultures not exposed to stimulator substances. According to their process, interferon is produced by incubating permanent cell lines of lymphoblast origin or human fibroblast cells with interferon inducers and stimulator substances and recovering the interferon produced.

The cells in which interferon is produced are human cells, either permanent cell lines of lymphoblast origin like Namalwa cells or human fibroblast cells which can be diploid with a restricted life span or heteroploid with an infinite capacity for division.

The interferon inducers can be any known ones. Inducing viruses are usually paramyxo viruses such as hemagglutinating virus Japan (HVJ = Sendai virus) or Newcastle disease virus; rhabdoviruses like vesicular stomatitis virus; measle viruses; or Reogroup viruses like reo virus or blue tongue virus. Nuclei acid inducers are usually double stranded ribonucleic acids like poly (I:C), either alone or in combination with metabolic inhibitors and/or polycations.

Useful stimulator substances include substances which, upon addition to the cell culture medium, increase the proportion of interferon-producing cells. More specifically, they are defined as substances which upon addition to the cell culture medium, result in the cells supported by this medium at optimal nutritional conditions, transgressing from an exponential growth phase to a resting stage.

Especially potent stimulator substances include propionic acid, n-butyric acid, n-valeric acid and their alkali salts, such as sodium salts, hexamethylenebisacetamide, dimethylsulfoxide, dexamethasone, triamcinolone, hydrocortisone, and prednisolone. The addition of stimulator substances is preferably prior to the interferon induction with interferon inducers; however, addition prior to and during the induction period can be applied. The increase in yield of interferon is due both to an increase of the percentage of interferon producing cells and to an increase in interferon yield on a per cell basis. The process is preferably carried out as follows:

Namalwa cells are cultivated in a growth medium, preferably the medium RPMI 1640 (Gibco Co., USA, or Flow Co., Great Britain) which contains the following substitutions: Tryptose phosphate broth (10 to 25% by volume, preferably however, 20% by volume); and partially purified human serum (1 to 10% by volume, preferably, however, 2 to 6% by volume). As an alternative to human serum, serum or serum fractions of other species can be applied such as fetal bovine serum at a concentration of from about 5 to 15% by volume.

To such a medium, which supports cell proliferation in suspension at a density of from 0.2 to 5 x 10^6 cells/ml, preferably, however, at a density of from 0.5 to 1 x 10^6 cells/ml, the stimulator substance (i.e., the sodium salt of n-butyric acid) is added at a concentration of from 0.1 μmol to 250 mmol per liter. Cells are further incubated at these conditions at a preferred temperature of from 35° to 37°C for from 6 to 72 hours, preferably, however, for from 24 to 48 hours. After the stimulation period, cells are harvested, preferably by centrifugation, and resuspended at a density of approximately 50 x 10^6 cells/ml medium, with or without stimulator substance. To this suspension the interferon inducer, such as HVJ (Sendai virus) is added and incubated for approximately 1 to 3 hours.

The thus induced cells are collected by centrifugation, resuspended at a density of from 5 to 10 x 10^6 cells/ml in a serum-free cell culture medium (Gibco Co., U.S.A., or Flow Co., Great Britain) such as in Eagle's Minimal Essential Medium and further incubated at from 35° to 39°C at a pH value of from 6.7 to 8.0, preferably 7.3, for a time span of from 15 to 24 hours, preferably from 18 to 20 hours, in the absence or in the presence of the stimulator substance. The suspension containing the crude lymphoblastoid interferon is separated from cells and cell debris by centrifugation. To inactivate residual inducer virus the supernatant is adjusted to a pH of 2.0 by addition of a mineral acid such as hydrochloric acid and incubated at from 0° to 4°C for from 3 to 5 days.

Steps by which this crude human lymphoblastoid interferon may be recovered, purified, concentrated and stabilized are given in detail in the patent.

Enhanced Production in Media Containing High Sugar Concentrations

J.H. Robinson; U.S. Patent 4,289,850; September 15, 1981; assigned to Beecham Group Limited, England has found that enhanced interferon production may be obtained by inducing interferon in media containing high sugar concentrations. It is thought that the enhancement is produced by a hypertonic effect and would therefore be generally applicable regardless of the interferon-inducer used. Therefore, a method is provided for preparing interferon which comprises contacting interferon-producing cells in vitro with an interferon-inducing medium comprising a nontoxic quantity of an interferon-inducing double-stranded polynucleotide, diethylamino ethyl dextran (DEAE-Dextran) and at least 100 mmol/ℓ of a water-soluble sugar in a physiologically acceptable aqueous vehicle at physiological conditions of temperature and pH, and thereafter separating the interferon so formed.

Preferably, the sugar is a low molecular weight sugar, which is used herein to mean a pentose or hexose monomer or oligomer. Examples of preferred monomeric sugars (monosaccharides) which may be used in the method are glucose, mannose, galactose and fructose. Examples of oligosaccharides which may be used in the method include the disaccharides, maltose, lactose and sucrose. Preferred sugars are glucose, galactose and sucrose. A sugar concentration of 300 to 400 mM, is preferred.

An interferon-inducing double-stranded polynucleotide here means synthetic or naturally occurring double-stranded ribonucleic acid or a double-stranded derivative thereof which is able to induce interferon.

Naturally occurring ds-RNAs are most suitable for the purpose of this process and the ds-RNA isolated from the fungus *P. chrysogenum* is preferred.

The quantity of polynucleotide is not particularly critical to the process, and appropriate quantities may be determined by trial and error.

Suitable concentrations of the preferred ds-RNA lie in the range 0.1 to 50 mg/ℓ inclusive, a concentration of 1 mg/ℓ being convenient.

DEAE-Dextran is a polymer of DEAE-glucose which has a molecular weight between 10^3 to 10^8. A number of DEAE-Dextrans of different molecular weight are available commercially. DEAE-Dextran of molecular weight 5 x 10^5 as supplied by Pharmacia is particularly convenient. The concentration of DEAE-Dextran included in the induction medium will not exceed 1,000 mg/ℓ, 100 to 300 mg/ℓ being particularly convenient.

In order to put this process into practice it is generally most convenient to culture an appropriate quantity of cells by any of the standard techniques, and thereafter to treat the cell population with the inducer in the presence of the sugar and DEAE-Dextran.

For the purposes of in vitro propagation, cells may be considered to fall into two categories depending upon whether they may be grown in suspension or as sheets in a monolayer. This process may be applied equally to cells which grow in suspension and cells which grow in monolayers.

The process may be put into effect with either normal or transformed human or nonhuman cells. Examples of suitable cells which grow in suspension are leucocytes, particularly peripheral blood leucocytes and examples of suitable sheet forming cells include epithelial cells and fibroblasts. Preferred cells for clinical grade material are MRC-5 human diploid fibroblasts.

Examples of suitable nutrient media include any standard one such as Eagle's, Fischer's, Ham's, Liebovitz's, McCoy's, etc.

When, in the case of cells cultured by a suspension technique, a suitable population of cells is obtained, or, in the case where cells are cultured by a monolayer technique, a confluent cell sheet is obtained, the cells are contacted with the polynucleotide and DEAE-Dextran in the presence of sugar. This may be done in two ways: either an appropriate quantity of sugar followed by polynucleotide and DEAE-Dextran may be added so that the desired concentrations of sugar and polynucleotide are achieved in the nutrient medium, or the nutrient medium used for culturing the cells may be drained and replaced by an inducing medium which comprises a nutrient medium containing the appropriate levels of sugar, polynucleotide and DEAE-Dextran.

The time for which the induction should be carried out will of course vary. A period of 2 hours is generally preferred when using preferred quantities of the preferred inducer DEAE-Dextran and cell line.

The induction step is carried out under physiological pH conditions, i.e. pH 6.5 to 8.0, and at a temperature in the range 2° to 40°C, 34° to 38°C being most apt. When the induction step is completed, the inducing medium is drained, optionally washed by contacting with a suitable wash medium, which may be simply phosphate buffered saline or a nutrient medium as previously described, and thereafter contacted with a harvest medium. The harvest medium is left in contact with the cells until interferon production has ceased. Suitable harvest media include phosphate buffered saline, and nutrient media as previously discussed. The length of time for which the cells are left in harvest medium depends on the length of time for which the cell continues to produce interferon. This will of course vary depending upon the type of cell which is employed, but may be determined by titrating aliquots at given time intervals.

When all interferon production has ceased, the harvest medium is drained from the cells, and the interferon recovered by standard techniques.

Production of Cell Cultures on a Large Scale

The process devised by *M. Iizuka; U.S. Patent 4,307,193; December 22, 1981; assigned to Toray Industries, Inc., Japan* is one for growing animal cells, and in particular, one for growing cell cultures on a large scale. Such large scale growth systems are desirable in view of the large quantities of cells required in the development and production of biological products such as viral vaccines, interferon, enzymes such as urokinase, and hormones.

Therefore, it is a principal object of this process to provide a large surface area for cell growth in a small total volume wherein a sufficient oxygen supply is provided without movement of the cells and culture media and thus eliminating any diffusion of microexudate.

The apparatus which was developed for use in the process for propagating tissue culture cells comprises a plurality of spaced-apart, substantially flat plates on which cells, liquid and gas phases can exist, and means for maintaining these plates substantially parallel to each other. The plates are enclosed by frames to form dishes, which are easily separable from each other and are easily separable from the other components of the apparatus.

Example: A strain of human diploid cells (2×10^7 cells) derived from human fetal lung, cultivated in four Roux bottles of 150 cm^2/bottle was trypsinized and suspended in a nutrient medium comprising 1750 ml of Eagle's basal medium, 32 ml of 7.5% sodium bicarbonate, 20 ml of 3.0% glutamine and 200 ml calf serum.

The cell suspension was placed in the reservoir and poured into an apparatus having overflow passages and comprising a polycarbonate container and 10 glass dishes 20 cm x 30 cm in size. The apparatus was inclined to the degree necessary for 200 ml of liquid to be contained on each dish, and then returned to the horizontal position. The cells attached themselves to the submerged bottom of each dish and grew to form confluent monolayers on each dish after 5 days of cultivation at 37°C. The old medium was removed and the sheets were replenished with a new medium comprising 1910 ml of Eagle's basal medium, 32 ml of 7.5% sodium bicarbonate, 20 ml of 3.0% glutamine and 40 ml calf serum.

After cultivation at 37°C for one day, the medium was again removed. Five hundred milliliters of solution of a potent interferon inducer, polyriboinosinic polyribocytidylic acid complex (Poly IC 10 μg/ml) were added and dispensed evenly (50 ml each) to all the cell sheets and remained in contact with the cell sheets for one hour at 37°C. After removal of nonabsorbed poly IC, the front panel of the container was detached and each dish sequentially removed and irradiated with ultraviolet light (70,000 ergs/cm^2) under aseptic conditions. The irradiated dishes were immediately returned to the container and the apparatus reconstructed. Serum-free medium (2 ℓ) was then fed onto the poly IC-stimulated and UV-irradiated cell sheets on each dish. Incubation was continued for an additional 24 hours at 37°C, after which time the supernatant fluid (about 2 ℓ) was harvested and found to contain 10,000 international units/ml of human interferon.

Enhancing Interferon Production by Fibroblast Cells

D.J. Giard and R.J. Fleischaker, Jr.; U.S. Patent 4,357,422; November 2, 1982; assigned to Massachusetts Institute of Technology have devised a method of enhancing interferon production, particularly interferon production by fibroblast cells. In its most general sense, their process relates to the establishment of a distinct interferon-production phase in which the temperature is initially elevated for a short period of time and then subsequently reduced to a selected interferon-production temperature capable of sustaining interferon production for prolonged periods. The method can be practiced by forming a cell culture of interferon-producing cells and introducing an interferon-inducer into the culture in an amount sufficient to induce interferon production. In a particularly preferred embodiment, the fibroblast cells are cultured on microcarriers suspended in cell culture medium. One method by which the distinct production phase can be established is by introducing one or more antimetabolites to effectively block interferon production during the induction period and subsequently removing the antimetabolites to initiate the distinct interferon-production phase.

The cells employed in the experimental work were human diploid foreskin cells designated as FS-4 cells. These are known to be good producers of F-interferon and probably also produce some L-interferon. Generally, it is believed that any mammalian cell strain or line which produces interferon can be employed.

The preferred culturing techniques involve the use of microcarrier cultures. Microcarrier cultures offer many potential advantages including ease of innovative cellular manipulation and potential high density cell growth in very limited volumes. A specific example of microcarriers found suitable are those cell culture microcarriers called Superbead (Flow Laboratories). An extensive description of such microcarriers and their use in cell cultures is presented in U.S. Patent 4,189,534, issued to D.W. Levine et al, the teachings of which are hereby incorporated by reference.

Such microcarrier cultures employ standard cell culture media, such as Dulbecco's Modified Eagle Medium (DMEM), RPMI-1640 medium, Weymouth's 199, etc. Serum supplement is usually added, but it was found that the growth of FS-4 cells in microcarrier culture required less serum than generally employed with other culturing systems, such as roller bottles.

It was found that good growth of FS-4 cells could be routinely achieved if careful attention was paid to the preparation of the inoculum and to control of the cell density at inoculation, despite previous findings that normal human cells were sometimes more difficult to grow on microcarriers than most cell lines. One example of a technique which was found to be consistently successful employed an inoculum grown in Corning 490 plastic roller bottles using DMEM supplemented with 10% fetal bovine serum (FBS). Confluent FS-4 cells in roller bottles, not older than 4 weeks, were split 1:2 and grown in roller bottles and then harvested. These cells were then used to inoculate microcarrier cultures. In order to obtain uniform growth of cells on the microcarriers, it was found that the cells should be well dispersed and that it was desirable to exceed a certain minimum requirement of cells per surface area. As an example, it was found preferable, at a microcarrier concentration of 5 g/ℓ, to have an inoculum density of 3 x 10^5 cells/ml or more.

Also, as previously mentioned, the FBS concentration was reduced from 10% to 5% in order to promote good attachment of cells to the microcarriers during the stationary growth phase and during the interferon induction and production phases.

Priming, which is the addition of a small amount of interferon prior to induction, was found to be useful with the FS-4 cells grown in microcarrier cultures. Induction of the FS-4 cells was achieved using the known interferon inducer, polyinosinic-polycytidylic acid (poly I-poly C). Other synthetic inducers could also be employed. Additionally, viral inducers including Sendai Virus, New Castle Disease Virus (NDV) and others, could be employed, live or inactivated (such as by irradiation with UV light). These inducers are all known to those skilled in the art.

A superinduction technique using both cycloheximide (at a concentration of 10 μg/ml) and actinomycin-D (at a concentration of 1 μg/ml) was used. A study of the time of exposure to the antimetabolites showed that the yield of interferon was optimum in a 5-hour or greater exposure to cycloheximide and a 3-hour exposure to the actinomycin-D.

The crux of this process, however, is the temperature switch made during the distinct production phase. Production is started at an initial temperature, which is elevated in terms of the interferon production temperature at which production will be sustained. As an example, it was found with the FS-4 microcarrier cultures that an initial temperature of 37°C for one hour in the production phase followed by sustained interferon production at 30°C produced outstanding results.

In general, the initial temperature and period should be sufficient to produce the increased rate at the later and lower production temperature. Preferably, the initial temperature is at least about 3°C above the final production temperature.

In terms of time, studies were made involving the use of an elevated temperature for 1, 2, 3 and 4 hours. The results indicated that one hour was sufficient and that ultimate production fell off if this period were extended. Of course, these temperatures and times will vary with the specific cells and other culture parameters involved.

MISCELLANEOUS BIOLOGICALS

Mutants of *Vibrio cholerae* for Making Cholera Vaccine

It is widely known that the profuse watery diarrhea of cholera results from the action of cholera toxin on the epithelial cell lining of the small intestine. Cholera toxin is elaborated from toxinogenic strains of *V. cholerae.* Toxinogenic strains may be either the classical or el tor biotype and the Oqawa or Inaba serotype; however, in all instances so far examined the toxin produced by these different types appears to be the same.

Over the past decade a number of different cholera vaccines have been developed and tested by field trial. In general, these vaccines have been composed of either killed whole cell preparations, cell wall preparations, or most recently a toxoid prepared from the formalin or glutaraldehyde inactivation of purified toxin. The vaccines have been administered parenterally and the results from field trials have been disappointing in that the protection against disease has been short lived for only a fraction of the population immunized.

Many mutant strains of *V. cholerae* 569B that produce low levels of biologically active toxin have been isolated, but most all the mutants that have been isolated by single-step methods have produced either low or elevated levels of holocholera toxin. As a class these mutant strains have been regarded as cholera toxin regulatory mutants, rather than mutants with an alteration in the structural gene for either Subunit A or Subunit B of toxin. Most all of the mutants that have been described have also been found to have reverted to "wild-type" toxinogenicity at a measurable frequency either in vitro and/or in vivo.

J.R. Murphy; U.S. Patent 4,264,737; April 28, 1981; assigned to President and Fellows of Harvard College has found that genetic stability of mutant strains of *V. cholerae* can be ensured by a two-step process, one of which is induced by incubating a strain at elevated temperature to produce a variant having both hypotoxinogenicity and genetic stability, and the second is mutagenesis and selection of strains with decreased toxicity. Preferably, the selected mutants would also be nonpathogenic.

The heat-induced variant is preferably first formed by incubating a standard parent strain of *V. cholerae,* such as the widely available *V. cholerae* 569B, to produce a strain which is hypotoxinogenic as assayed in S49 mouse lymphosarcoma cells, the toxicity being reduced by a factor of more than 750, preferably by a factor of about 1000, and which is genetically stable. The incubation temperature is from 40° to 42°C; at 43°C, there is no growth of the cells. The time of incubation is not critical, a matter of several hours usually being required for adequate growth, preferably from 15 to 20 hours. A hypotoxinogenic strain thus induced is then subjected to a second mutation by any conventional mutagenesis, either by a chemical mutagen or by irradiation, preferably the former.

Mutant strains thus produced are found to retain the biotype and antigens of the original *V. cholerae* strain (e.g., 569B) and the heat-induced variant, but differ from them in that their toxicity is further reduced. These mutant strains display a further reduction in toxicity, being decreased by a factor of at least 7500, usually from 7500 to 15,000, with respect to the original or parent strain, and also display genetic stability both in vitro and in vivo, being capable of colonizing the gastrointestinal system. Further selection of nonpathogenic mutants may be made following challenge of rodent and porcine animal model systems used for the determination of choleragenicity. These nonpathogenic mutants would be consequently useful as live oral vaccines for immunization against cholera.

The isolation of genetically stable mutants of heat-induced variants offers the advantage of a two-step genetic alteration. If the frequency of reversion is on the order of 10^{-8} for the heat induction step and 10^{-8} for the mutational step, then the rate of reversion of mutants to wild-type should be 10^{-16}. This offers a high degree of confidence in the safety of such a live oral vaccine strain. The importance of genetic stability for such a strain cannot be overemphasized for scientific, ethical, and political reasons.

Large-Scale Production of Human Growth Hormone

Many nutrient culture mediums have been developed and are, to a large extent, specifically designed to permit prolonged and sustained growth of specific types of cells, such as those derived from skin, from tumors, and from various other tissues in mammalian species. These are grown by primary culture systems as explants, monolayers or suspension cultures or secondary cell lines. However, there is no culture system or nutrient media available which will permit prolonged growth and proliferation of human anterior pituitary growth hormone producing cells in sufficient number to provide a source of large amounts of growth hormone. This lack of success is related to many factors, among which are (1) limited knowledge concerning the specific nutrients for the specific cells, (2) the general trend for the pituitary growth hormone-producing cells to lose their function of producing growth hormone, (3) death of cells after a few days to a few weeks, and (4) transformation of cells to nonfunctioning fibroblastic or fibroblastoid type cells.

M.J. Narasimhan and J.A. Anderson; U.S. Patent 4,288,546; September 8, 1981; and Reissue 30,753; September 29, 1981; both assigned to The Regents of the University of Minnesota, describe a new culture system and new culture media for growing in vitro normal human pituitary cells for producing large amounts of human growth hormone. More specifically, the process for producing human growth hormone is one in which cells of the human anterior pituitary gland are dispersed in an amino acid-rich nutrient medium supplemented with liver extract, insulin and antibiotic and antifungal agents and incubated under open-aeration cell growth conditions. The resulting culture is serially subcultured several times until the optimum desired cell growth level is achieved. The human growth hormone (hGH) is extracted using conventional techniques.

The preferred nutrient composition is as follows:

(1) Medium 199-1X with Earle's Modified Salts, GIBCO (Grand Island Biological Co.), 1,000 ml

(2) Liver extract containing B vitamins (Lexavite Injectable, Eli Lilly & Co.), 10 ml

(3) Insulin, crystalline, 10 International Units

(4) Crystalline sodium penicillin G, 500,000 units

(5) Streptomycin sulfate, 500 mg

(6) Nystatin, a polyene antifungal antibiotic, 50,000 units.

The insulin should not be added to the medium until time of use because insulin will degrade in the medium between 4 and 8 days, particularly when kept at room temperature or above.

To achieve sustained growth and proliferation of human anterior pituitary cells and obtain a large cell mass to extract large amounts of human growth hormone, a modified system of culturing is used. This involves the growing of dispersed cells of the human anterior pituitary in containers containing this medium with an open-aeration system for a period of several days, preferably about 8 days, followed by serial subculturing about every 4 to 8 days. The cultures may be grown as still-suspension cultures or as rotary suspension cultures. The technique involves greater degrees of aeration than is employed in culturing of human and mammalian cells. This is achieved by having open air vent caps on the containers which are placed in a 37°C incubator with water troughs for moisture, provided with an air-flow system of about 8 to 16 ℓ of sterile air per minute. Addition of 5% CO_2 to the air-flow which is usually employed in available tissue culture techniques is avoided since it has been found that human pituitary cells do not grow well in the presence of 5% CO_2.

Following about 8 days growth the cultures are subcultured serially, utilizing an appropriate dilution range of 1 volume of medium containing cells added to 1 volume of fresh medium. The original culture may be divided into 2, 3 or more parts, depending upon the quantitative cell count found at each subculture time. If cell numbers are large by microscopic examination, greater subculture, dilution transfers should be employed. If cells are fewer in number, less dilution transfers of medium containing cells to fresh medium should be used. Cultures should be protected from bacterial fungal and viral contamination at all times.

Example: Using the preferred nutrient composition described above, an initial tissue inoculum (about ⅓ of an anterior pituitary gland) was grown in serial culture with ½ volume transfers every 8 days for 32 days. The following average concentration of hGH in ng/ml found in the medium were: at 8 days 21,166; at 16 days 18,666; at 24 days 21,750; and at 32 days 20,000. Each culture flask contained 30 ml and hence because of one-half dilutions and volume reconstitution at subculture times, the total amount of hGH produced at each subculture time was as follows: 8 days 5.01 mg in 6 flasks; at 16 days 6.7 mg in 12 flasks; at 24 days 15.6 mg in 24 flasks; and, by 32 days 28 mg in 48 flasks. Cultures have been grown for as long as two months. Preliminary characterization studies utilizing a Sephadex gel filtration procedure indicate that 80 to 90% of the immune reacting hGH is a monomer. Cells were epithelial type, 10 to 14 μ in diameter, contained chromatin granules and had a large highly chromatic nucleus. Mitosis was present.

By this technique of serial subculture, one is able to obtain a large mass of human pituitary cells for producing large amounts of human growth hormone, in relatively short periods of time.

Solubilization of Gonococcal Antigens

It is essential that gonococcal antigens, especially those isolated from the cell surface of *Neisseria gonorrheae,* be solubilized in a physiologically acceptable medium if they are to be used as a vaccine against *N. gonorrheae* in humans. *Y.D. Karkhanis; U.S. Patent 4,330,623; May 18, 1982; assigned to Merck & Co., Inc.* has devised such a solubilization process involving trypsin digestion of the antigen. The following example illustrates the process.

Example: Lyophilized G_c antigen (37.8 mg) is suspended in 8.0 ml of 0.001 N aqueous sodium hydroxide with a final pH of 7.8. The resulting suspension is prewarmed to 37°C in a heater-block before 100 μg of TPCK-trypsin (Worthington Biochemical Corp.) in 100 ml of water are added. The mixture is incubated at 37°C for 50 min followed by two additions of 130 μg of trypsin at 50 min intervals. The total incubation time is about 150 minutes.

The trypsin digest is applied directly to a 2 x 4 cm column packed with Ultrogel AcA34 coupled with soybean trypsin inhibitor and preequilibrated with water. It is eluted with about 25 ml of water. The eluate (25 ml) is mixed with 25 ml of 0.3 N aq. sodium chloride and filtered to give a solution of about 0.76 mg/ml of the solubilized antigen.

Polysaccharide Vaccine or Antigen Against Group B Streptococci

D.L. Kasper, U.S. Patents 4,324,887; April 13, 1982 and 4,356,263; October 26, 1982; both assigned to The President and Fellows of Harvard College, has found that polysaccharide antigens or vaccines against *Streptococcus agalactiae* having a large molecular size (0.8-6 x 10^6 daltons) of increased immunogenicity and distinctly different in chemical composition from those previously prepared can be made by culturing or propagating a Group B streptococcus in a nutrient broth containing a high concentration of glucose, preferably 10-15 g/ℓ or more, and by maintaining the pH of the broth at pH 6-8, preferably 6.8-7.6, throughout the culturing. The term "molecular size" as used herein means molecular weight as determined by molecular exclusion chromatography. These high molecular size antigens are immunoreactive to Type III-specific antibodies. It has further been found that still greater improvement can be achieved and higher yields of polysaccharide antigen obtained by culturing the organisms in a broth free from constituents having a molecular size above 10^5 d, separating or harvesting the organisms from the broth after culturing, then recovering from the residual broth polysaccharides including a fraction having molecular sizes above 0.8 x 10^6 d.

An additional but smaller quantity of the large molecular size polysaccharides can be obtained by extraction from the organisms as well. The same procedures, it has been found, can be used to prepare large molecular size antigens which are immunoreactive to Type Ia- and Type Ic-specific antibodies, and antigens having a molecular size from about 0.2 x 10^6 to about 6 x 10^6 daltons which are immunoreactive to Type II specific antibodies.

Example: Todd-Hewitt broth from a commercial supplier was subjected to dialysis against distilled water through a suitable membrane to separate from it all ingredients having molecular size above 100,000 d. It was then sterilized by autoclaving and there were dissolved in it an additional 14 g/ℓ of sterile glucose.

The contents of a starter flask containing Type III Group B streptococci were inoculated into the broth and allowed to grow at room temperature while maintaining the pH of the broth continually within the range pH 6.8-7.6 by means of a pH titrator. When the amount of organisms was greater than 3 x 10^8 organisms/ml, the bacteria were harvested by centrifugation at 4°C.

The residual or supernatant broth was concentrated to 1/10 volume by ultrafiltration while retaining all material of molecular size greater than 10^5 d, then dialyzed exhaustively against distilled water to remove small molecular size components. Cold absolute alcohol was added in an amount of 30% by volume and the precipitate or flocculate containing some extraneous nucleic acids separated by centrifugation. To the supernatant there was added 80% by volume of cold absolute alcohol and the precipitate containing the polysaccharides was separated by centrifugation. The solid pellet was dried, then suspended in a buffered solution containing 0.01 M Tris (hydroxymethyl aminomethane) hydrochloride, 0.001 M calcium chloride, and 0.001 M magnesium chloride at pH 7.3. This solution was then twice digested for 3 hours at 37°C with ribonuclease and deoxyribonuclease at concentrations of 0.5 g/ℓ, then twice with Pronase B at 1.0 g/ℓ to destroy unwanted nucleic acids and proteins.

The enzyme-treated solution was then fractionated by chromatography on a column of Sepharose 4B equilibrated in 0.01 M Tris hydrochloride buffer at pH 7.3 to separate the fraction containing substances of small molecular size (below 10^5). The large molecular size fractions (8-11 x 10^5) were eluted, combined, and mixed with 4 volumes of absolute ethanol to precipitate the large molecular size polysaccharide which was separated by centrifugation, and lyophilized to provide a purified dry solid product consisting of repeating units of Type III antigen of the structure defined above. In general, each 1 ℓ lot of Type III organisms yielded about 5-15 mg of polysaccharide antigen (MW 0.8 to 5×10^6) by this procedure.

Smaller amounts of the same molecular weight antigen were obtainable by extraction of the harvested bacteria using the same procedures as employed in the prior art for lower molecular weight polysaccharide Type III antigens. In general each 1 ℓ of organisms yielded about 1-4 mg of the antigen by such procedure. Indeed, it has been found that lower molecular size fractions, as low as about 0.2×10^6 daltons, of the same Type III antigen having the chemical structure described above can be made by the same procedure and possesses the same immunoreactivity, which is different from that of the Type III polysaccharide antigen described in the prior art.

For use as a vaccine, suspensions were prepared containing 100 μ/ml of the dried polysaccharide Types Ia and Ic, II, and III antigens, respectively, in a normal saline solution, and there was injected subcutaneously into each volunteer 0.5 ml of one such suspension. Both Type III polysaccharide antigen of large molecular size prepared in accordance with this process and that of smaller molecular size ($5\text{-}6 \times 10^5$) exhibited high immunogenicity, developing from 0.1 to 2000 μg/ml antiserum of specific antibody in selected vaccines as determined by radioimmunoassay of the antiserum using as tracer polysaccharide antigens intrinsically labelled by the introduction of tritiated acetate into the nutrient broth.

Microbiological Reduction of 15-Ketoprostaglandin Intermediates

The objective of *K. Kieslich, B. Raduchel, W. Skubalia, H. Vorbruggen and H. Dahl; U.S. Patent 4,247,635; January 27, 1981; assigned to Schering AG, Germany* was to provide a process for stereoselectively reducing the 15-keto function of bicyclic prostaglandins in relatively good yields and without using the basidiomycetes already suggested for such a process, since these fungi exhibit only slow growth and are difficult to handle. These objectives have been attained by preparation of a 15-α-hydroxy-prostaglandin intermediate having the formula (1) below by the reduction of the corresponding 15-ketone of formula (2) using a strain of Kloeckera, Saccharomyces or Hansenula.

(1)

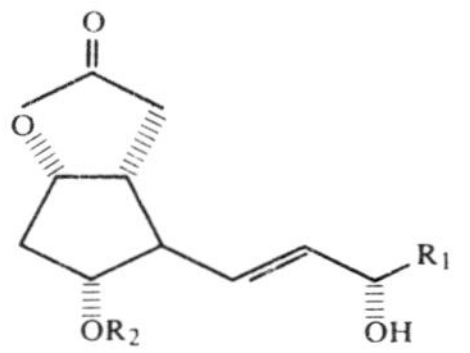

(2)

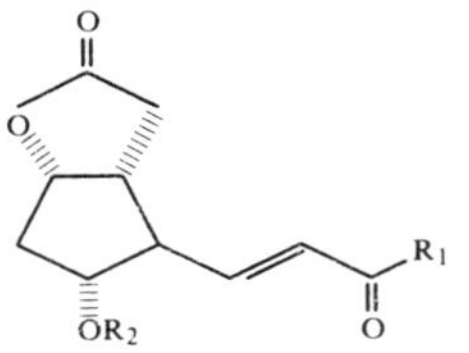

In both of the above formulas, R_1 is phenoxymethyl, phenoxymethyl substituted on the phenyl moiety by halogen or trifluromethyl, or alkyl of 1-5 carbon atoms, and R_2 is hydrogen, acetyl, benzoyl or p-phenylbenzoyl.

Suitable strains of the aforementioned microbiological reduction agents include *Kloeckera magna* (ATCC 20,109), *Kloeckera jensenii* (ATCC 20,110), *Saccharomyces carlsbergensis* (CBS 1506), *Saccharomyces carlsbergensis* (CBS 1513) or

Hansenula anomala (NRRL-Y-366). The strain *Kloeckera jensenii* (ATCC 20,110) has proven to be especially suitable.

Suitable halogens in formula (1) include fluorine, chlorine and bromine, fluorine and chlorine being preferred. If the phenyl ring in the phenoxymethyl residue of R_1 is substituted by halogen or trifluoromethyl, the 3- or 4-position is suitable for this purpose. The 3-position is preferred. Suitable substituted phenoxymethyl residues include: 3-chlorophenoxymethyl, 4-chlorophenoxymethyl, 3-fluorophenoxymethyl, 4-fluorophenoxymethyl, 3-trifluoromethylphenoxymethyl, 4-trifluoromethylphenoxymethyl and the like.

The R_1 C_{1-5} alkyl groups may be straight-chained or branched. Straight-chain alkyl groups of 3-5 carbon atoms are preferred.

The process works especially advantageously when R_2 in the compounds of formula (2) is benzoyl or p-phenylbenzoyl. Depending on the R_2 moiety desired in the final product, the benzoyl or p-phenylbenzoyl blocking groups can be split off according to conventional methods.

To carry out the microbiological reduction conventional methodology is employed. Submerged cultures are incubated under the conventional culturing conditions for each of the aforementioned microorganisms, in a suitable nutrient medium with aeration. The substrate (dissolved in a suitable solvent or preferably in emulsified form) is then added to the culture, and the latter is fermented until maximum substrate conversion has been achieved.

It is possible to produce pharmacologically active prostaglandins from the compounds of formula (1) prepared by this process using methods which maintain the center of asymmetry in the 15-position (prostaglandin nomenclature). For example, starting with (1S,5R,6R,7R)-6-[(E)-3-oxo-4-phenoxy-1-butenyl]-7-benzoyloxy-2-oxabicyclo[3.3.0]-octan-3-one, a multistage synthesis yields the active agent Sulproston (described in DOS No. 2 355 540).

Production of Angiogenic Factor by Cell Culture

Angiogenesis, or the ability to stimulate blood vessel growth, is important to burn and wound healing and certain inflammatory reactions as well as to tumor growth. The substance that is released by tumors and provides vascularization has been named tumor antiogenesis factor (TAF) by Dr. Judah Folkman of the Harvard Medical School. Provision for the availability of TAF is particularly useful as an aid to the search for ways to inhibit neovascularization. While TAF also finds use in the development of tests such as an angiogenic assay or a diagnostic screening test for neoplasia, for use in patient treatment an angiogenic material derived from normal rather than tumor cells would be much preferred from the standpoint of safety.

W.R. Tolbert, M.-J. Kuo and J. Feder; U.S. Patent 4,268,629; May 19, 1981; assigned to Monsanto Company have investigated numerous normal human cell lines for the production of angiogenic factor by cell culture but most of them have been eliminated as unsuitable candidates in view of their relatively poor angiogenic activity production or poor growth characteristics as above-defined.

A cell line that has unexpectedly been found by Tolbert et al to have good growth characteristics in cell culture and to be able to elaborate the desired angiogenic factor in suitable quantities is the human diploid cell line IMR-90. This cell line was derived from lung tissue of a normal human female fetus and was obtained from the American Type Culture Collection under the code designation ATCC CCL 186.

The process for the production of human angiogenic factor in vitro comprised growing the human diploid cell line IMR-90 on support surface in nutrient culture medium at about 35° to 38°C for a sufficient time to elaborate the angiogenic factor.

The nutrient culture medium used in the process was Dulbecco's modification of Eagle's minimum essential medium fortified with fetal bovine serum. The angiogenic factor was isolated by extraction from the cells and concentration by carboxymethyl Sephadex chromatography. The following examples will further illustrate the method.

Example 1: A sample of the IMR-90 (CCL 186) cell line was obtained from the ATCC, at passage 7. The cell line was maintained at 37°C as a monolayer in a series of T-75 cm^2 tissue culture flasks (Falcon Plastics) containing 50 ml Dulbecco's MEM supplemented with 4.5 mg/ml glucose and 20% fetal bovine serum (KC Biologicals) without addition of any antibiotics. The flasks were charged with fresh medium every 1 to 2 days. When confluency was reached after 6 to 8 days, the cells were subcultured 1:3 in the following manner:

The spent medium was poured off the monolayer and discarded. The monolayer then was rinsed twice with 10 ml phosphate buffered saline (PBS) with 0.02% ethylenediaminetetraacetic acid (EDTA), at pH 7.4.

The PBS was prepared by dissolving 80 g NaCl, 2 g KCl, 2 g KH_2PO_4 and 21.6 g $Na_2HPO_4{\cdot}7H_2O$ in 10 ℓ distilled water. The cells were released from the surface of the flask by adding 5 ml 0.05% trypsin in PBS with 0.02% EDTA and allowing the thus treated cells to stand 5 minutes at room temperature (ca. 22° to 25°C). The suspension was divided equally into 3 fresh T-flasks containing 50 ml medium each.

Example 2: IMR-90 cells grown to confluency in T-flasks as in Example 1 were then subcultured in roller bottles through 13 passages. To inoculate a roller bottle, three T-flasks were washed with PBS with EDTA and then trypsinized as in Example 1. Two-thirds of the suspension from three T-flasks were added to a CO_2 gassed roller bottle containing 100 ml medium as in Example 1. The remainder of the suspension was added to three fresh T-flasks. Confluent roller bottles were split 1:3 by the same procedure as the T-flasks in Example 1, except that 25-30 ml PBS with EDTA were used for each rinse and 10 ml of the trypsin solution was used to suspend the cells. Roller bottles were incubated 5 minutes at 37°C on a roller deck during the enzyme treatment.

After 18 roller bottles at passage 13 had been grown to confluency (750 cm^2 available surface/bottle), each bottle was washed twice with 25 ml PBS without EDTA. Twenty-five ml of sterile water was then added to each bottle, and the cells were incubated for 30 minutes at 37°C. The cells were then lysed by mechanical disruption. The lysate plus a 25 ml water rinse of each bottle was stored at -20°C for further use or fractionation.

A portion of the above stored lysate was fractionated as follows: The lysate was unfrozen and clarified by centrifugation (one 5-minute period at 5000 x *g*, and the supernatant again at 5000 x *g* for 30 minutes). The clarified lysate was concentrated to about 200 ml and dialyzed into low salt phosphate buffer (0.1 M NaH_2PO_4, 0.02% NaN_3, pH 6.1) in a cassette concentration system. The concentrated lysate was added to 1.5 g of carboxymethyl (CM) Sephadex (particle size 40 to 120 μ), which had been previously swollen in low salt phosphate buffer. After stirring for 30 minutes at room temperature, the mixture was filtered and the filtrate was retained as the CM-I fraction. The CM-Sephadex gel was rinsed with low salt phosphate buffer and then poured into a glass column (2 cm diameter). A high salt phosphate buffer (1 M NaCl, 0.1 M NaH_2PO_4, 0.02% NaN_3, pH 6.1) was applied to the column to release the protein from the CM-Sephadex. The column was set up so that the eluate flowed through a LKB Unicord II monitor set at 280 nm which recorded any protein eluted from the gel. The eluate was collected in fractions, with retention of that fraction (CM-II) which corresponded to the protein peak on the monitor.

Both the foregoing CM-I (35.3 mg protein) and CM-II (1.9 mg protein) fractions were dialyzed against distilled water and lyophilized. The lyophilized fractions were tested for angiogenic factor in CAM (chorioallantoic membrane) assay with the results:

CM-I	6 positive out of 6 eggs
CM-II	4 positive out of 5 eggs
Control sample from Walker 256 carcinosarcoma cells	4 positive out of 5 eggs

The investigators *W.R. Tolbert, M.M. Hitt and J. Feder; U.S. Patent 4,273,871; June 16, 1981; assigned to Monsanto Company* also found that the cell lines of human foreskin fibroblasts have good growth characteristics and are able to elaborate the desired angiogenic factor in suitable quantities.

These cell lines are readily available to the public for cell culture research and other such purposes from cell culture repositories such as the American Type Culture Collection and the Human Genetic Mutant Cell Repository of the Institute for Medical Research. Suitable such cell lines also can arise from subculturing of primary cell cultures which are started from tissues or cells obtained directly from surgical procedures. Illustrative examples of publically available human foreskin fibroblast cell lines are the cell lines available under the code designations AG 1518, AG 1519 and AG 1523 from the Institute for Medical Research and the cell line HR 218 which is available from HEM Research Inc.

The process for producing the angiogenic factor in vitro is the same as that described in the previous patent, using human foreskin fibroblast cell lives instead of the human diploid cell line IMR-90 used in that patent.

ORGANIC CHEMICALS PRODUCED BY MICROORGANISM FERMENTATION

AMINO ACIDS

Preparation of L-Tryptophan Using a Strain of Serratia

L-Tryptophan is useful as an essential amino acid, and an economical process for producing it on an industrial scale is in demand. Conventionally, L-tryptophan may be prepared by fermentation on a medium containing a precursor (e.g., indole, anthranilic acid or serine), or by the direct fermentation method which does not use such a precursor. The direct fermentation method is considered advantageous for production of the tryptophan on an industrial scale.

H. Yukawa, K. Osumi, T. Nara, Y. Takayama; U.S. Patent 4,271,267; June 2, 1981; assigned to Mitsubishi Petrochemical Co., Ltd., Japan have found that a significant amount of L-tryptophan is produced when a microorganism of the genus Serratia that utilizes ethanol and has the ability to produce L-tryptophan is cultivated on a medium containing ethanol as the main carbon source. Tryptophan has been recovered and purified.

An illustrative example of a useful microorganism for this process is *Serratia marcescens* MT-5 which was derived by induction from *Serratia marcescens* variety MAY-110 (FERM-P No. 3521) as a strain resistant to 5-methyl-DL-tryptophan. The MT-5 strain has been deposited with the Fermentation Research Institute, Japan, and accepted as FERM-P No. 4735.

The mutated strains resistant to 5-methyl-DL-tryptophan, particularly *S. marcescens* MT-5, are obtained in the conventional manner by subjecting the MAY-110 strain, to ultraviolet ray irradiation treatment or treating it with chemicals such as N-methyl-N'-nitro-N-nitrosoguanidine. The concentration of N-methyl-N'-nitro-N-nitrosoguanidine used is 100 μg/ml (tris-maleate buffer solution of pH 7.0 is used), and the treatment time is 15 minutes at 30°C. The specific process is as follows.

Ethanol is used as the carbon source of the medium, and its initial concentration is properly selected from the range of about 1 to 5% v/v depending on the strain used. Any loss in ethanol due to digestion is to be compensated with care so that its concentration is not detrimental to the growth of the culture or the formation of L-tryptophan. Ammonium sulfate, ammonium nitrate, ammonium chloride, ammonium phosphate, urea or other suitable substances are used as a nitrogen source depending on the strain's ability to utilize nitrogen in an amount of about 0.5 to 5% w/v. The medium may contain required amounts (about 0.01 to 5% w/v) of other organic nutrients such as amino acids, corn steep liquor, seasoned liquid and yeast extract, inorganic salts, and vitamins. The cultivating temperature ranges preferably from 25° to 35°C, and the pH preferably from 6 to 8, with optimum conditions being selected depending on the strain used. The incubation generally requires 2 to 7 days. After incubation, the resulting L-tryptophan is recovered from the fermentation liquor using an ion exchange resin, activated charcoal, concentration and precipitation, or any other conventional method.

Preparation of L-Lysine or L-Valine Using a Strain of Acinetobacter

L-valine and L-lysine are essential amino acids and they may be used for pharmaceutical purposes, incorporated into feeds, etc. It would be desirable to achieve a process for producing them which used a starting material which is readily available and inexpensive. Processes using hydrocarbons as starting material have been investigated but have the drawback that most are insoluble in water, which restricts their commercial application and their yields.

The process developed by *H. Yukawa, K. Osumi, T. Nara, Y. Takayama; U.S. Patent 4,276,380; June 30, 1981; assigned to Mitsubishi Petrochemical Co., Ltd., Japan* is one for producing L-valine or L-lysine by aerobically culturing bacteria belonging to the genus Acinetobacter in a culture medium in which ethanol is the main carbon source.

An example of the bacteria which produces and accumulates L-valine or L-lysine from ethanol is *Acinetobacter calcoaceticum* YK-1011 which has been deposited with the Fermentation Research Institute as FERM-P No. 4818. This bacterium is derived and isolated from *Acinetobacter calcoaceticum* ATCC 19606 as the 5-methyl-DL-tryptophane resistant strain, and has the same characteristics as *Acinetobacter calcoaceticum* ATCC 19606 except that it is resistant to 5-methyl-DL-tryptophane.

The growth of the *Acinetobacter calcoaceticum* ATCC 19606 was almost completely inhibited in a plate culture medium prepared by adding 200 μg/ml of 5-methyl-DL-tryptophane to a conventional artificial culture medium. The derivation of its mutant was carried out by subjecting it to the nitrosoguanidine treatment in a known manner, culturing it on the plate culture medium containing 200 μg/ml of 5-methyl-DL-tryptophane as above at 30°C for 3 to 5 days and isolating the colony produced. Needless to say the derivation of the mutant can be effected by ultraviolet irradiation or treatment with other chemicals.

The preferred embodiment for the production of the amino acids is as follows: Ethanol is used as the carbon source in the culture medium and the initial concentration is suitably chosen in a range of about 1 to 5% v/v depending on the particular strain used. With its consumption, ethanol is intermittently supplemented to give an optimum concentration (about 0.01 to 5% v/v and more preferably about 0.01 to 2% v/v) which does not inhibit the growth of the strain or the production of L-valine or L-lysine.

A normal nitrogen source is provided, plus conventional organic nutrient sources, vitamins, etc. Temperatures are preferably 25° to 35°C and the pH is advantageously 6 to 8. After cultivation L-valine and L-lysine may be recovered by methods well known in the art.

From Cellulose-Containing Agricultural Wastes

N.V. Gluschenko, V.N. Bukin, M.E. Beker, L.V. Dmitrenko, V.A. Utenkova, M.A. Kuzmina, L.S. Kutseva, N.M. Bazdyreva, G.K. Liepinsh, E.B. Trusle and T.A. Pavlova; U.S. Patent 4,286,060; August 25, 1981 describe a process for producing amino acids in good yield from economical and readily-available agricultural materials. The process uses, as a source of carbon, a purified mixture of hexose and pentose monosaccharides obtained by percolation hydrolysis of cellulose-containing plant raw materials.

In the USSR plant raw material is processed by the method of percolation hydrolysis with diluted sulfuric acid. The percolation hydrolysis enables processing of all kinds of cellulose-containing plant raw material—rejects of agricultural production (rice and cotton husk, maize stump, sunflower seed husk and the like), rejects of woodworking and sawmilling (slabs, edgings, sawdust), and fuel wood, regardless of the species.

The purification of this mixture of pentose and hexose monosaccharides to remove furfural is conducted by filtering at a pH of 3.3 to 3.7, evaporating in a vacuum at 60° to 80°C to bring the content of the monosaccharides to 6 to 11% by weight, oxidating the mixture by an oxygen-containing gas in the presence of an absorbent such as activated charcoal or an alkali, and sterilizing the resultant mixture.

Microorganisms are then used to produce the amino acid from a culture medium of the purified monosaccharides described above, a nitrogen source, and mineral salts in an aerobic fermentation. As microorganisms suitable for the fermentation, there are suggested strains of Brevibacterium 22 to produce L-lysine and *Brevibacterium flavum* ATSS 14067 or *Micrococcus glutamicus* to prepare L-glutamic acid.

L-Aspartic Acid Using a Strain of Brevibacterium

L-Aspartic acid is one of the important amino acids, and is used in medicines and as a food additive.

H. Yukawa, T. Nara and Y. Takayama; U.S. Patent 4,326,029; April 20, 1982; assigned to Mitsubishi Petrochemical Co., Ltd., Japan have produced a process for producing L-aspartic acid from fumaric acid or a salt thereof and ammonia or an ammonium salt using a cultured product obtained by aerobically culturing a microorganism belonging to the Genus Brevibacterium and having resistance to α-amino-n-butyric acid.

It has been found that impartation of resistance to α-amino-n-butyric acid to microorganisms belonging to the genus Brevibacterium greatly enhances their aspartase activity, i.e., their ability to biosynthesize L-aspartic acid from fumaric acid and ammonia.

A representative strain of microorganism useful in the process is *Brevibacterium flavum* MJ-233-AB-41 (FERM-P 3812). This strain may be derived from *Brevibacterium flavum* MJ-233 (FERM-P 3068) as an α-amino-n-butyric acid resistant strain by a typical procedure–mutation using N-methyl-N'-nitrosoguanidine, for example, and culturing the organism so obtained in a plate culture containing the α-amino-β-butyric acid, and harvesting the colony to obtain the resistant variant.

The biosynthesis of L-aspartic acid from fumaric acid or one of its salts and ammonia or an ammonium salt is then carried out in a conventional manner with the *B. flavum* FERM-P 3812.

Preparation of L-Isoleucine

Production of L-isoleucine, L-valine and other amino acids via fermentation has been the subject of considerable research. Numerous genera of microorganisms have been employed along with various analogs of L-isoleucine, threonine, valine, etc.

The process developed by *M.H. Updike and G.J. Calton; U.S. Patent 4,329,427; May 11, 1982; assigned to W.R. Grace & Co.* for preparing L-isoleucine comprises cultivating under aerobic conditions a mutant strain of *Brevibacterium thiogenitalis* resistant to an analogue of L-isoleucine. Cultivation, i.e., fermentation, is carried out in the presence of a post-threonine biosynthetic precursor of L-isoleucine to accumulate L-isoleucine in the fermentation broth.

Wild strains of *Brevibacterium thiogenitalis* (e.g., ATCC 19240) selected for mutation are characterized by overproduction of glutamic acid. The mutant strains useful in the process do not require precursors for growth but do require the precursor for production of L-isoleucine. In the absence of the precursor, production is shifted to L-valine.

The term "post-threonine precursors" is intended to include precursors of L-isoleucine subsequent to threonine and compounds similar thereto, e.g., α-hydroxy butyric acid and α-amino-n-butyric acid. Generally the precursors can be employed in the form of acids or water-soluble salts thereof, e.g., alkali metal salts with the sodium salt being preferred. The mutants described above are characterized in that the threonine conversion to α-ketobutyrate is hindered; i.e., to produce L-isoleucine rather than L-valine, the post-threonine L-isoleucine precursor must be present.

Certain analogues of the naturally occurring amino acids are suitable for isolating the mutant strains of this microorganism. These analogues are toxic to strains which do not overproduce L-isoleucine. Such analogues include α-amino-β-hydroxyvaleric acid; methylglycine; gamma-dehydroisoleucine; 3-cyclopentene-1-glycine; 2-cyclopentene-1-glycine; o-methylthreonine; and β-hydroxyleucine.

The isoleucine analogue resistant mutant may be obtained by ultraviolet irradiation of a wild type strain of *Brevibacterium thiogenitalis* or by treating the wild strain with a mutagen, e.g., ethyl methane sulfonate, N-methyl-N_1-nitro-N-nitrosoguanidine, etc. Thereafter the strain can be cultured in the presence of the analogue to isolate the colonies which overproduce L-isoleucine. For example, the unrelated strain can be cultured at 30°C for 2 to 7 days on agar plates of the following composition: gelatin hydrolyzate peptone, 5.0 g/ℓ; beef extractives, 3.0 g/ℓ; agar, 15 g/ℓ; 2 hydroxy-β-valeric acid sodium salt, 25 g/ℓ.

A viable culture of an L-isoleucine-producing mutant strain of *Brevibacterium thiogenitalis* resistant to α-amino-β-hydroxyvaleric acid has been deposited with the American Type Culture Collection, ATCC 31723.

Fermentation of the isolated mutant strains of *Brevibacterium thiogenitalis* can be accomplished by shaking cultivation or submerged fermentation under aerobic conditions. The fermentation is carried out at 20° to 45°C and at a pH of 5 to 9. Calcium carbonate and ammonia may be employed for ajustment of the pH of the medium. The fermentation medium contains a source of carbon, a source of nitrogen and other elements. Suitable sources of carbon for the fermentation include fermentable sugars, protein hydrolyzates and proteins. Examples of suitable sources of nitrogen are urea, ammonium salts of organic acids (e.g., ammonium acetate and ammonium oxalate) and ammonium salts of inorganic acids (e.g., ammonium sulfate, ammonium nitrate or ammonium chloride). The amounts of the carbon and nitrogen sources in the medium are from 0.001 to 20 w/v percent. Also, organic nutrients (e.g., corn steep liquor, peptone, yeast extracts) and/or inorganic elements (e.g., potassium phosphate, magnesium sulfate, vitamins such as biotin and thiamine, and amino acids, e.g., isoleucine and valine) may be added to the medium. The amount of the L-isoleucine precursor is from 0.001 to 20 w/v percent of the medium. The fermentation is accomplished in 16 to 176 hr, and L-isoleucine is accumulated in the fermentation broth.

After the fermentation is completed, i.e., from 0.1 to 6 w/v percent of L-isoleucine is accumulated in the broth, cells and other solid culture components are removed from the fermentation broth by conventional procedures such as heating followed by filtration or centrifugation. Known procedures may be employed in the recovery and/or purification of L-isoleucine from the filtrate or the supernatant solution. For instance, the filtered fermentation broth is treated with a strong cation exchange resin. Then the resin is eluted with a dilute alkaline solution such as aqueous ammonia. The eluates containing L-isoleucine are combined and concentrated. An alkanol such as methanol or ethanol is added to the concentrated solution. The precipitated crystals can be recrystallized from an aqueous alkanol such as aqueous methanol and aqueous ethanol to yield pure crystals of L-isoleucine.

L-Arginine Produced by Mutants of Brevibacterium or Corynebacterium

L-arginine is produced by a fermentation process, in which mutants of the genus Brevibacterium or Corynebacterium resistant to a sulfa drug or arginine antagonist are used. *K. Akashi, Y. Nakamura, T. Tsuchida, H. Yoshii and S. Ikeda; U.S. Patent 4,346,169; Aug. 24, 1982; assigned to Ajinomoto Co., Inc., Japan* have found productivity of L-arginine is significantly increased when a resistance to ketomalonic acid, fluoro-malonic acid, monofluoro acetic acid, or aspartate-antagonist is given to the known mutants which belong to the genus Brevibacterium or Corynebacterium and are capable of producing L-arginine. These mutants may be induced from parent strains of the genus Brevibacterium or Corynebacterium by conventional mutation means, e.g., irradiation with UV-ray or exposure to N-methyl-N'-nitro-N-nitrosoguanidine, and thereafter picking up the colonies formed on the nutrient agar-medium containing the amount of the chemical agents inhibitive to the growth of the parent strain.

When the wild strains are used, L-arginine productivity is produced in the wild strains before giving them resistance to the above chemicals or, after the wild strains have been given the chemical resistance, L-arginine productivity is produced in them by making them resistant to arginine-antagonists such as 2-thiazole alanine, arginine hydroxamate or sulfa drugs.

The arginine antagonists are such chemicals as those which inhibit the growth of the microorganisms of the genus Brevibacterium and Corynebacterium and the inhibition is suppressed when L-arginine coexists in the medium.

The preferred wild strains belonging to the genus Brevibacterium or Corynebacterium are, for example: *B. divaricatum* ATCC 14020, *B. flavum* ATCC 14067, *B. lactofermentum* ATCC 13869, *B. saccharolyticum* ATCC 14066, *B. roseum* ATCC 13825, *C. acetoacidophilum* ATCC 13870, *C. lilium* ATCC 15990 and *C. glutamicum* ATCC 13032.

Aspartate-antagonists inhibit the growth of the microorganisms of genera Brevibacterium and Corynebacterium and the inhibition is suppressed partly or completely when L-aspartate coexists in the medium, and are, for instance, β-aspartylhydrazide, diamino-succinic acid and hadacidin.

Several mutants of the process are: *B. flavum* FERM-P 4940 (NRRL B-12235), resistant to sulfadiazine and aspartylhydrazine; *C. acetoacidophilum* FERM-P 4945 (NRRL B-12237), resistant to sulfadiazine and diaminosuccinic acid; and *C. flavum* FERM-P 5638 (NRRL B-12243), resistant to sulfadiazine and fluoromalonic acid.

L-Lysine Produced by Mutants of Brevibacterium or Corynebacterium

In similar work *O. Tosaka, E. Ono, M. Ishihara, H. Morioka and K. Takinami; U.S. Patent 4,275,157; June 23, 1981; assigned to Ajinomoto Company, Inc., Japan* have produced L-lysine, an essential amino acid used in feeds, by fermentation using mutants belonging either to the Corynebacterium or Brevibacterium genus. These microorganisms must have the characteristics to be known necessary for the production of L-lysine such as homoserine-requirement and resistance to S-(2-aminoethyl)-L-cysteine (hereinafter referred to as AEC), and further have sensitivity to fluoropyruvic acid.

The mutants can be induced from the parent strains by conventional mutation means such as exposure to UV-rays or to N-methyl-N'-nitro-N-nitrosoguanidine.

The parent strains are, for example, *Brevibacterium divaricatum* ATCC 14020, *Brevibacterium flavum* ATCC 14067, *Brevibacterium lactofermentum* ATCC 13869, *Brevibacterium roseum* 13825, *Corynebacterium acetoacidophilum* ATCC 13870, and *Corynebacterium lilium* ATCC 15990. Those parent strains have the common characteristics that their mutant strains can produce L-lysine.

The mutants sensitive to fluoropyruvic acid grow more poorly than their parent strains in the medium which contains fluoropyruvic acid, and thus the mutants can be separated from their parents by the replication-method.

The microorganisms used to produce L-lysine may be selected from the following: *Brevibacterium lactofermentum* NRRL B-11471, *Corynebacterium acetoglutamicum* NRRL B-11473, and *Brevibacterium flavum* NRRL B-11475.

L-Glutamic Acid Produced by Mutants of Brevibacterium or Corynebacterium

L-Glutamic acid has been produced by fermentation using a microorganism of the genus Brevibacterium or Corynebacterium. Various attempts have been done to improve the productivity of the known glutamic acid producing strains by artificial mutation techniques.

M. Yoshimura, Y. Takenaka, S. Ikeda and H. Yoshii; U.S. Patent 4,347,317; August 31, 1982; assigned to Ajinomoto Company, Incorporated, Japan, have found that mutants of the genus Brevibacterium or Corynebacterium resistant to a respiratory inhibitor or ADP (adenosine diphosphate) phosphorylation inhibitor produce L-glutamic acid in an improved yield. These respiratory inhibitors inhibit, as is well known, respiration of the microorganisms, and then their growth. Examples of respiratory inhibitors are malonic acid, potassium cyanide, sodium azide, sodium arsenite.

ADP phosphorylation inhibitors inhibit the phosphorylation of ADP of microorganisms and production of ATP (adenosine triphosphate), and then their growth. ADP phosphorylation inhibitors include uncouplers, which inhibit coupling reaction in respiratory chain, such as 2,4-dinitrophenol, hydroxylamine and arsenic, and energy transfer inhibitors, which inhibit energy transferring reactions, such as guanidine.

Satisfactory mutants were produced from the same parent strains of Brevibacterium and Corynebacterium as those listed in the previous patents– ATCC 14067, ATCC 13869, ATCC 13032, etc.–by exposing them to 250 μg/ml of N-nitro-N'-methyl-N-nitrosoguanidine at 30°C for 20 minutes.

The mutants for producing the L-glutamic acid by aerobic fermentation in a conventional aqueous medium were as follows:

C. glutamicum NRRL B-12210	*B. flavum* NRRL B-12205
B. lactofermentum NRRL B-12212	*B. lactofermentum* NRRL B-12216
B. lactofermentum NRRL B-12213	*C. glutamicum* NRRL B-12208
B. flavum NRRL B-12204	*B. lactofermentum* NRRL B-12217
C. glutamicum NRRL B-12207	*B. lactofermentum* NRRL B-12218
B. lactofermentum NRRL B-12214	*B. flavum* NRRL B-12206
B. lactofermentum NRRL B-12215	*C. glutamicum* NRRL B-12209

In order to obtain the best yield of L-glutamic acid in known fermentation processes, the amount of L-biotin in the fermentation medium should be controlled in a very narrow range. However, beet or cane molasses which is available as the carbon source at a reasonable price, contains a very high amount of biotin, and then the fermentation medium prepared with such molasses inevitably contains an excessive amount of biotin. In order to avoid the undesirable effects of excessive amounts of biotin on L-glutamic acid production, surfactants, or antibiotics are added to the fermentation medium when an excessive amount of biotin is contained therein.

H. Nakazawa, I. Yamane and E. Akutsu; U.S. Patent 4,334,020; June 8, 1982; assigned to Ajinomoto Co., Inc., Japan have found that mutants of the genus Brevibacterium or Corynebacterium which are resistant to a compound having vitamin-P activity (hereinafter referred to as vitamin-P compound) produce L-glutamic acid in a remarkably high yield, even when they are cultured wihout addition of surfactants or antibiotics in the fermentation medium containing the excessive amount of biotin.

Their method for producing L-glutamic acid comprises culturing in an aqueous culture medium a mutant of the genus Brevibacterium or Corynebacterium which is resistant to a vitamin-P compound and recovering L-glutamic acid accumulated in the resultant culture liquid.

Representative of parent strains used for mutation are those listed in the previous three patents. Mutations were achieved in the conventional manner.

Examples of the vitamin-P compounds are acenocoumarin, apiin, coumetarol, cyclocumarol, dicumarol, diosmetin, esculetin, esculin, ethyl-bis-coumacetate, 3,3'-ethylidene-bis-4-hydroxy-coumarin, heseperetin, hesperidin, morin, naringenin, phenprocoumon quercetin, quercimeritrin, robinin, rutin, scoparone, skimmin, umbelliferone, and warfarin.

A mutant resistant to one of the vitamin-P compounds is usually resistant to another vitamin-P compound. The mutant resistant to the vitamin-P compounds is grown on an agar medium which contains the amount of the vitamin-P compound inhibitive to the growth of its parent strain.

Mutants used in the process and their NRRL numbers are as follows: *Brevibacterium flavum* NRRL B-12128, NRRL B-12129, and NRRL B-12130; *B. lactofermentum* NRRL B-12133, NRRL B-12134 and NRRL B-12135; *Corynebacterium glutamicum* NRRL B-12138, NRRL B-12139, and NRRL B-12140.

Preparation of L-Tryptophan Using an Enterobacter Strain

L-Tryptophan is one of the essential amino acids and is important as a medicine, nutrient, and as an additive for animal feed. Some L-tryptophan derivatives work as an antagonist against the metabolism of L-tryptophan and contain physiologically active substances that may be used to prepare pharmaceuticals that affect the central nervous system. This L-tryptophan and its derivatives can be produced by known methods of synthesis, biological process and many other methods. Known methods for producing L-tryptophan using microorganisms include: (1) direct fermentation using sugars to accumulate L-tryptophan in a culture; and (2) adding indole or anthranilic acid simultaneously with sugar to a culture, and permitting L-tryptophan to accumulate in the culture.

L-Tryptophan can also be produced from indole and serine or from indole, pyruvic acid and ammonium ion by using a microorganism-produced tryptophanase (enzyme). This method of using tryptophanase has the advantage that by varying the type of indole compound, various corresponding L-tryptophan derivatives can be produced, and that hence, a reaction that best suits a particular purpose can be selected.

A. Mimura, Y. Takahashi, K. Yuasa and M. Shibukawa; U.S. Patent 4,349,627; September 14, 1982; assigned to Asahi Kasei Kogyo KK, Japan have found a microorganism in the soil of the forest in Fuji, Shizuoka, Japan that produces L-tryptophan and derivatives thereof in high yield.

In the process, an indole compound is reacted with serine or with pyruvic acid and/or its salt and ammonium ion in the presence of a culture or treated culture of genus Enterobacter sp. AST 49-4 having the designation FERM-P 5543. The microorganism can be cultured on a common synthetic or natural medium.

Tryptophanase produced by Enterobacter sp. AST 49-4 is considered an adaptive enzyme, and L-tryptophan is preferably added to the medium for preparing a culture of the microorganism in an amount of from about 0.1 to 0.7% by weight. The microorganism is incubated in the resulting medium at from 25° to 37°C for from 16 to 96 hours.

The thus prepared culture of the microorganism has an enzyme system that produces L-tryptophan or its derivatives from an indole compound and serine, or from an indole compound, pyruvic acid and/or its salt, and ammonium ion. The culture may be immediately used in the desired enzymatic reaction, or it may be used as immobilized cells or enzyme obtained by polymerizing them with an acrylic acid amide monomer or the like.

The thus prepared tryptophanase-containing cells, treated cells and immobilized cells, as well as the purified tryptophanase and the immobilized tryptophanase are used as a catalyst for enzymatic reaction in a reaction liquor comprising an indole compound and serine, or an indole compound, pyruvic acid and/or its salt, and ammonium ion for production of L-tryptophan or its derivatives. In addition to the indole compounds serine, pyruvic acid and/or its salt, and ammonium ion used as substrate, the reaction liquor preferably contains ethylenediaminetetraacetic acid and pyridoxal phosphate for achieving a higher yield of L-tryptophan or derivatives thereof.

The L-tryptophan or its derivatives produced in the reaction liquor can be isolated by conventional methods including adsorption with ion exchange resin, activated carbon, etc. The L-tryptophan and derivatives thereof produced can be verified and quantified by high-pressure liquid chromatography or silica gel thin-layer chromatography.

ASCORBIC ACID AND ITS PRECURSORS

2-Keto-L-Gulonic Acid Using a Citrobacter

2-keto-L-gulonic acid is a valuable intermediate in the preparation of ascorbic acid (vitamin C), an essential vitamin for human nutrition. Processes for converting 2-keto-L-gulonic acid to ascorbic acid, for example by heating in the presence of a base, are well known in the art. Only a limited number of microorganisms are known in the art to be effective for the reduction of 2,5-diketo-D-gluconic acid to form 2-keto-L-gulonic acid.

D.A. Kita and K.E. Hall; U.S. Patent 4,245,049; January 13, 1981; assigned to Pfizer Inc. have found that a process for the preparation of 2-keto-L-gulonic acid or salts thereof in good yields is provided by cultivating a 2-keto-L-gulonic acid-producing microorganism of the genus Citrobacter or mutants thereof in an aqueous nutrient medium in the presence of 2,5-diketo-D-gluconic acid or a salt thereof. The fermentation is preferably conducted at a temperature of about 25° to about 35°C, most preferably at about 25° to 30°C and at a pH in the range from about 5.5 to 7.5. Preferred 2,5-diketo-D-gluconic acid salts are sodium 2,5-

diketo-D-gluconate and calcium 2,5-diketo-D-gluconate. The microorganisms include *Citrobacter freundii* and *Citrobacter diversus* and preferred strains of microorganism for use in this process are *Citrobacter freundii* ATCC 6750 and 10787.

Example: The following inoculum medium having a pH of 7.0 was prepared (all measurements in g/ℓ): sorbitol = 3, yeast extract = 2, peptone = 2, KH_2PO_4 = 1, $MgSO_4 \cdot 7H_2O$ = 0.2.

A 300 ml flask containing 100 ml of this inoculum medium was inoculated with cells of *Citrobacter freundii* ATCC 6750 from a nutrient agar slant (1 ml of a 10 ml sterile aqueous suspension). The cells were cultivated for 24 hours at a temperature of 28°C while the flask was on a rotary shaker. 100 ml of the fermentation medium containing the indicated number of g/ℓ: cerelose = 2, $(NH_4)_2HPO_4$ = 1, KH_2PO_4 = 1, $MgSO_4 \cdot 7H_2O$ = 0.5, beet molasses = 2, and glycine = 0.2, having a pH of 6.7 was placed in a 300 ml flask and inoculated with 5% (v/v) of the 24 hour inoculum culture described above and shaken for 22 hours on a rotary shaker, when the pH was approximately 5. 15 ml of fermentation broth containing 15 to 20% (wt/vol) sodium 2,5-diketo-D-gluconate and prepared by the fermentation of *Acetobacter cerinus* strain IFO 3263 in a glucose containing medium at 28°C for 50 hours, was added to the *Citrobacter freundii* culture and the pH was adjusted to 6.5 by addition of sodium hydroxide. The fermentation was continued at 28°C, readjusting the pH to 6.5 every 24 hours. After 52 hours the 2,5-diketogluconate was essentially consumed, to give sodium 2-keto-L-gulonate in a yield of about 30%, based on the 2,5-diketo-D-gluconate starting material.

Production of 2,5-Diketogluconic Acid

Heretofore, 2,5-diketogluconic acid has been produced by several different varieties of bacteria such as *Acetobacter melanogenum, Acetobacter aurantium, Gluconoacetobacter rubiginosus, Gluconoacetobacter liquifaciens* and *Pseudomonas sesami.* The use of these microorganisms, however, is not satisfactory from an industrial point of view because of relatively low yields of 2,5-diketogluconic acid, relatively long fermentation times and because of the production of large amounts of brown or yellow-brown pigments as by-products of cultivation, thereby decreasing the purity of the desired 2,5-diketogluconic acid.

2,5-Diketogluconic acid is useful as an intermediate for the preparation of ascorbic acid. A solution of 2,5-diketogluconic acid may be selectively reduced to 2-ketogulonic acid, which may be converted to ascorbic acid. The reduction of 2,5-diketogluconic acid may be effected by a fermentative reduction as described, for example, in the previous patent.

2,5-Diketogluconic acid is also useful an an intermediate for the preparation of comenic acid by heating in the presence of an acid, as described, for example, in U.S. Patent 3,654,316.

The process by *D.A. Kita and K.E. Hall; U.S. Patent 4,263,402; April 21, 1981; assigned to Pfizer Inc.* relates to the preparation of 2,5-diketogluconic acid by a process comprising aerobically propagating the microorganism *Acetobacter cerinus* in a glucose-containing fermentation medium. The glucose concentration in the fermentation medium is preferably between 10 and 15% (wt/vol), most preferably between 11 and 13%. The propagation is preferably conducted at 25° to 30°C and at a pH between 5 and 6. Particularly preferred microorganisms are

Acetobacter cerinus strains IFO 3263 and IFO 3266. The 2,5-diketogluconic acid produced may be isolated from the fermentation medium or may be converted to 2-ketogulonic acid or to comenic acid.

Further work by *D.A. Kita and D.M. Fenton; U.S. Patent 4,316,960; February 23, 1982; assigned to Pfizer Inc.* has shown that while total amounts of glucose from about 2.5 to about 20% (wt/vol) can be utilized, initial glucose concentrations in the medium greater than about 15% cannot be tolerated by the microorganisms. Accordingly, total amounts of glucose greater than about 15% (wt/vol) can only be utilized by conducting the fermentation at an initial glucose concentration of about 10 to 15% (wt/vol) and thereafter adding further increments of glucose to the fermentation medium during the course of the fermentation, the concentration of glucose in the medium not exceeding about 15% (wt/vol) at any given time.

Accordingly, heretofore the overall concentrations, or production capacities, of 2,5-diketogluconic acid have been limited by the relatively low initial glucose concentration that can be employed in the fermentation medium. The productivity of processes employing other microorganisms for the preparation of 2,5-diketogluconic acid is also limited by the necessity of using relatively low initial glucose concentration in the fermentation medium.

It will be readily apparent that a process wherein initial glucose concentrations higher than about 15% (wt/vol), especially levels above 20%, can be tolerated and utilized by the microorganisms for the preparation of 2,5-diketogluconic acid will provide a substantial increase in production capacity and will result in substantial economies of operation. Such a process also avoids the possibility of contamination of the fermentation medium that may occur when production capacities are increased by adding further increments of glucose during the course of the fermentation.

Kita and Fenton have found that initial glucose concentrations above 20% and up to about 30% (wt/vol) in a fermentation medium can be utilized by the microorganism *Acetobacter cerinus* for the production of 2,5-diketogluconic acid when at least about 0.04 wt % of choline, based on the amount of D-glucose in the medium, is added to the fermentation medium. They, therefor, provide a process for the production of 2,5-diketogluconic acid in high concentrations in the fermentation medium by aerobically propagating *Acetobacter cerinus* in a fermentation medium containing D-glucose in an initial concentration above about 20% and up to about 30% (wt/vol) and choline in an amount of at least about 0.04 wt % based on the amount of D-glucose in the medium.

The initial glucose concentration in the fermentation medium is preferably about 25 to 30% (wt/vol). The propagation is preferably conducted at a temperature of 25 to 30°C, preferably at a pH from about 5 to 6. Preferred strains are *Acetobacter cerinus* IFO 3263 and IFO 3266.

Conversion of L-Galactonate to 2-Keto-L-Galactonic Acid

The synthesis of vitamin C from pectin as the starting material involves a reaction sequence long known to the art, which offers the advantage of commencing with rather inexpensive widely available agriculture by-products (e.g., beet pulp or citrus pulp). In commercial practice, however, vitamin C is synthesized from glucose.

R.P. Lanzilotta and M.K. Weibel; U.S. Patent 4,246,348; January 20, 1981; assigned to Novo Laboratories, Inc. have reinvestigated the conversion of pectin substances into vitamin C and have concluded that one major step of the sequence, namely, conversion of L-galactonic acid into 2-keto-L-galactonic acid is difficult to carry out with high yield by chemical procedures, but might be carried out microbiologically. The microbiologic reaction offers considerable promise for reducing conversion cost substantially.

The process they have developed involves subjecting L-galactonic acid itself or an equivalent thereof such as L-galactono-1,4-lactone or nontoxic water-soluble salts, notably sodium L-galactonate, all being hereinafter referred to as L-galactonate, to microbiological oxidation by cultivating a microorganism that metabolically converts L-galactonate into 2-keto-L-galactonate in an L-galactonate containing medium, then after the 2-keto-L-galactonate has been elaborated in the medium, separating the medium from the microorganism. The metabolic product may be 2-keto-L-galactonic acid, or a salt thereof, all of which are herein referred to as 2-keto-L-galactonate. The 2-keto-L-galactonate is then recovered from the separated medium. The product 2-keto-L-galactonic acid may be recovered from cell free (filtered) broth by standard procedures including solvent precipitation (e.g. with methanol), insoluble salt precipitation (e.g. as calcium salt) or adsorption (e.g. on an anion exchange resin).

Examples of subcultures of various microorganisms found to be suitable for converting L-galactonate into 2-keto-L-galactonate are as follows (with their NRRL accession numbers):

Citrobacter amalonaticus	B-11365
Erwinia herbicola	B-11366 and B-11367
Pseudomonas cepacia	B-11369
Pseudomonas fluorescens	B-11370

2-Keto-L-Gulonic Acid by Microbial Conversion

2-Keto-L-gulonic acid, a useful intermediate for the production of ascorbic acid, can be produced commercially by Reichstein's method, one which comprises a number of steps and does not have very good yields. Various microbial conversion methods have been suggested for the production but these have not improved yields to any significant extent.

According to the process provided by *T. Sonoyama, B. Kageyama, and T. Honjo; U.S. Patent Reissue 30,872; February 23, 1982; assigned to Shionogi & Co., Ltd., Japan*, 2-keto-L-gulonic acid is made from 2,5-diketo-D-gluconic acid or its salts by treatment with one of the following microorganisms: *Brevibacterium ketosoreductum* ATCC 21914; *Bacillus megaterium* ATCC 21916; or *Staphylococcus aureus* ATCC 21915.

CITRIC ACID PRODUCTION

High Yields Using a Candida Strain

Several processes are described in the technical literature for the preparation of citric acid by fermentation of substrates such as linear paraffins and carbohydrates. For various reasons the current trend is to use carbohydrates (molasses) as substrates, and the strains commonly used belong to the genus Candida, and more particularly to the species *C. lipolitica* and *C. tropicalis.* The process basically comprises the following operations:

(1) preparing the culture broth in which molasses is diluted with water, reducing the impurity content down to acceptable values and adding measured quantities of nutrient elements;

(2) sterilizing the apparatus and the fluids;

(3) introducing the inoculum prepared in a separate culture or taken from a preceding fermentation cycle;

(4) carrying out the fermentation, which comprises an initial period of multiplication of the inoculated yeast, followed by a period of production of citric acid until the fermentable sugar present in the medium is exhausted.

As known, the yeasts of the genus Candida require the use of a medium with pH values of from 4 to 7 and the pH is controlled within this range by continuously feeding into the medium a neutralizing agent such as an alkali metal, alkaline earth metal, or ammonium hydroxide. As a result, the citric acid is in salified form in the fermentation medium.

The processes in which citric acid is neutralized in the form of a soluble citrate have drawbacks deriving from the inhibition of the fementation exerted by the salt, which inhibition increases as the concentration of the salt is increased. In other words there is a limit for the output of citric acid, expressed as the quantity of acid produced per useful volume of the reactor. On the other hand, the precipitation of citric acid in the form of an insoluble salt, with removal of the latter from the fermentation medium, has not been considered until now as a valid solution for the problem under discussion, since a non-negligible fraction of the cells is removed from the fermentation medium, together with the insoluble salt, so that the final balance is unfavorable as regards the conversion products of the sugars, and more particularly as regards the ratio between citric acid and yeast cells produced.

C. Rottigni and G. Cardini; U.S. Patent 4,278,764; July 14, 1981; assigned to Euteco Impianti SpA, Italy, however, have developed a process which is an improvement on previous fermentations for producing citric acid. In their process molasses, which has previously been hydrolyzed to break down the saccharose present into glucose and fructose, is submitted to a submerged fermentation, under aerobic conditions, with a strain of Candida, preferably *C. lipolitica.* The fermentation is operated at a controlled temperature and the pH of the fermentation is maintained at a value of 5 to 7 by adding calcium hydroxide in measured quantities during the course of the fermentation.

The citric acid is thus salified, as it forms, with calcium hydroxide, with consequent formation of an insoluble precipitate in the form of granules with a size of the order of 5 to 50 microns, comprising mainly calcium citrate, in addition to cells. According to the process, the fermentation broth is first submitted to centrifuging by using a low revolution number to separate the calcium citrate, and the supernatant is submitted to a further centrifuging by using a higher revolution number, thereby to separate the yeast cells from the exhausted broth.

The cells thus recovered, upon possible washing, are recycled or used for another fermentation. It has been ascertained that the centrifuging operations do not lead to a substantial pollution of the treated medium. Moreover, by operating in this way, the quantity of sugar converted into cells is reduced to a minimum, and the overall conversions of sugar into citric acid are higher than those achieved by using conventional processes. Another advantage consists in the higher output of citric acid, expressed as the quantity of acid produced per useful volume of the reactor. Finally, by operating as described above, there is obtained an acid of high purity, free or substantially free from isocitric acid.

By operating under these conditions, yields in citric acid typically of the order of 60 to 70% with respect to the sugar fed in are achieved, best results being achieved at about neutral pH values which permit practically theoretical yields in citric acid and yeast cells to be obtained with respect to the converted sugar. The output of citric acid is typically of the order of 130 g/ℓ when using a sugar concentration of 200 g/ℓ.

From Mutant Yeast Strains

Production of citric acid by means of yeast has suffered from the fact that isocitric acid is produced by the fermentation, and the separation of the two compounds is complicated. Furthermore, the presence of iron accelerates the isocitric acid production in the fermentation, which means that expensive stainless steel or glass-lined equipment must be used if the process is not to be impractical.

Accordingly, *K. Takayama, T. Adachi, M. Kohata, K. Hattori and T. Tomiyama; U.S. Patent 4,322,498; March 30, 1982; assigned to Kyowa Hakko Kogyo Co., Ltd., Japan* are concerned with suppressing the amount of isocitric acid produced in a fermentation process so that the yield of citric acid is high and the product does not need purification. They have accomplished these aims with the use of mutant yeast strains which produce good yields of citric acid and which have a nutritional requirement for iron.

The mutants useful in the process usually exhibit a growth comparable to that of the parent strains in the presence of 0.1 mg/ℓ, and preferably, 0.2 mg/ℓ or more of iron. Particularly preferred mutants are *Candida zeylanoides* ATCC 20391, ATCC 20392, ATCC 20393 and ATCC 20367, all of which are derived from *Candida zeylanoides* ATCC 20347. The aforementioned strains require a higher amount of iron for growth than the parent strain. *Candida zeylanoides* ATCC 20367 requires glycerin for growth in addition to an iron requirement.

In obtaining mutants requiring a higher amount of iron, any of the conventional methods for inducing mutation to obtain a strain having a requirement property may be employed. For example, such artificial mutation means as X-ray irradiation, ultraviolet ray irradiation, nitrogen-mustard treatment, nitrosoguanidine treatment, etc. are appropriate. As an example, microbial cells of a yeast strain are suspended in trismaleate buffer solution having a pH of 9.0 containing 200 γ/ml of N-methyl-N'-nitro-N-nitrosoguanidine at a concentration of 10^8 cells per 1 ml. The suspension is allowed to stand for 15 minutes. The cells are then collected by centrifugation, washed with sterile physiological sodium chloride solution, placed on an agar plate and incubated.

The resulting colonies are isolated into pure cultures by any of the well known methods. Each of the pure cultures is then tested against the parent strain, and those strains which have been mutated to acquire the desired iron requirement are selected as applicable for the process.

Any culture medium normally used for the culturing of yeasts is suitable for the process as long as it contains an assimilable carbon source, a nitrogen source, inorganic materials and other growth promoting factors which may be required by the specific yeast strain used.

Although ferrous sulfate and ferric chloride are preferred sources of iron ion in the culture medium, any soluble iron salt which does not prove toxic to the microorganism is appropriate.

Culturing is carried out under aerobic conditions at 20° to 40°C, and at a slightly acidic to neutral pH of about 3 to 7 for 2 to 5 days, at which time a considerable amount of citric acid is formed in the culture liquor. The pH may be adjusted with calcium carbonate, sodium hydroxide or an aqueous ammonia.

After the completion of culturing, the microbial cells are removed from the culture liquor by, for example, filtration and the filtrate is concentrated. By adding calcium hydroxide to the filtrate, citric acid is readily recovered as calcium citrate. The calcium citrate is converted to citric acid by the addition of sulfuric acid, thus precipitating out calcium sulfate. The citric acid may also be recovered by any other of the usually used purification techniques.

From Whey in a Two-Step Process

R.M. El-Sayed; U.S. Patent 4,326,030; April 20, 1982; assigned to Diamond Shamrock Corporation has found that citric acid can be produced in a two-step process by cultivating suitable microorganisms while using lactose as a carbon source. To be able to use lactose as a carbon source the lactose has to be transformed to pyruvic acid, which later can be transformed to citric acid. The problem is to get pyruvic acid to concentrate in the culture solution in such a way that it can be transformed to citric acid by influence of microorganisms. This is accomplished by the process of cultivation of *Escherichia coli* KG 93, F^- (DSM 1392) in a first step for 15-24 hours at a temperature of 20° to 37°C and a pH of 5.0 to 7.5 on a substrate consisting of whey permeate to which has been added phosphates in a content of 0.8 to 1.6 g/ℓ and nitrates in a content of 0.8 to 1.2 g/ℓ or a corresponding quantity of urea. *Hansenula wickerhamii* CBS 4308 (DSM 1380) in a second step is cultivated for 20 to 26 hours at a temperature of 15° to 35°C and a pH of 4.5 to 6.5 on the culture from the first step, whereupon citric acid is obtained.

Example: A whey permeate containing the following percentages by weight was prepared: fat, 0.20%; protein 0.05%, lactose 5.0%, lactic acid 0.1% and the conventional salts (Ca, P, K, N, NaCl) 0.6%. To this was added 1.2 g/ℓ phosphates and 1.0 g/ℓ nitrates.

The whey permeate was pumped into a continuous fermentor (Chemap), which was provided with a turbine agitator. The solution was inoculated with *E. coli* KG 93, F^-. The cultivation was made at a temperature of 37°C and at pH of about 7. The pH was kept at a stable value by means of ammonia gas. Air was supplied from a compressor to the cultivation.

After approximately 12 hr of cultivation the air supply was stopped completely to achieve a concentration of pyruvic acid. After approximately another 6 hr of cultivation without air supply the liquid from the fermentor was pumped continuously to another fermentor of the same type and there it was inoculated with *H. wickerhamii* CBS 4308. This second step of the cultivation was made at a temperature of 30°C and the pH of about 5. The cultivation in the second step went on for about 24 hr. Also in the second step the pH was kept at a stable value by means of ammonia gas. At the cultivation in the second step 0.3 g/ℓ of ferrocyanide was added.

The yield of pyruvic acid in step one was 32 g/ℓ. The yield of citric acid after the whole process was 43 g/ℓ.

OTHER ACIDS

Long-Chain Dicarboxylic Acids Using *Debaryomyces vanriji*

Monocarboxylic acids having straight and long carbon chains are useful raw materials for surfactants, detergents, stabilizers, and the like. However, their use has been limited since natural fats, such as beef fat and palm oil, have been mostly employed for the preparation of the above mentioned chemicals.

Dicarboxylic acids having straight and long carbon chains are useful raw materials for the preparation of plasticizers, synthetic resins, synthetic lubricants, oils, perfumes, and the like. The establishment of a method of the manufacture of dicarboxylic acids with varied carbon numbers on an industrial scale from petroleum derived feedstocks has been desired.

Microbial production of monocarboxylic acids and dicarboxylic acids is well known. In these reported reactions, normal paraffins contained in petroleum distillate are used as substrate for corresponding mono- and dicarboxylic acids.

A. Taoka and S. Uchida; U.S. Patent 4,275,158; June 23, 1981; assigned to Bio Research Center Co., Ltd., Japan have found that a yeast strain which belongs to Debaryomyces genus, *D. vanriji,* can produce dicarboxylic acids or dicarboxylic acids along with monocarboxylic acids by oxidizing hydrocarbons or mixtures of hydrocarbons and monocarboxylic acids having straight and long carbon chains.

Using this microorganism, this process produces, (1) dicarboxylic acids or a mixture of dicarboxylic acids and monocarboxylic acids corresponding to the hydrocarbons having straight and long carbon chains which have been used as the substrate, (2) dicarboxylic acids corresponding to natural as well as synthetic monocarboxylic acids used as raw material and (3) dicarboxylic acids corresponding to a mixture of hydrocarbons having long, straight carbon chains and monocarboxylic acids of similar chain length when used as substrate.

The microorganism, *Debaryomyces vanriji* (BR-308), employed in this process was collected from the soil near a petroleum refinery in Akita Prefecture and was isolated for use. The microorganism was deposited with the American Type Culture Collection with the accession number ATCC 20588.

Hydrocarbons having 10 to 18 carbon atoms are appropriate raw materials (substrate) for the production of the acid mixtures, hydrocarbons of 11 to 16 carbon chain length being particularly desirable. For the selective production of dicarboxylic acids, monocarboxylic acids and hydrocarbons having a skeletal length of 10 to 18 carbon atoms each are suitable precursors, hydrocarbon skeletal lengths of 11 to 16 carbons being preferred.

The oxidation reaction of this process is a typical resting cell phenomenon and can be carried out in an aqueous buffer solution, as for example a phosphate buffer solution of pH 7.

The reaction can also be carried out in a growth medium which contains nutrient for the yeast. The reaction then becomes a cultivation-oxidation reaction. In this mode of operation, the media contains the usual nutrient material including an assimilable carbon source, nitrogen source, and appropriate vitamins and minerals, all well-known to those skilled in the art.

When the oxidation is carried out as a resting cell reaction, it becomes necessary to grow up a healthy cell mass prior to oxidation. This cell mass is best prepared by growing the yeast culture in the above growth medium prior to oxidation. The whole cell culture including cells and nutrient material can be used in the oxidation reaction, or the cell mass can be removed from the spent nutrient by centrifugation or filtration prior to adding the cell mass to the substrate.

Thus, the strain of *D. vanriji* ATCC 20588, or a culture thereof or cells of the strain cultured previously are added to the medium containing the substrate to carry out the reaction and agitated, aerated through a nozzle, or shaken so that the microorganism can contact the components of the medium thoroughly.

The reaction temperature is kept at 25° to 35°C and pH is controlled at 3 to 9, preferably 4 to 8. The period of time of reaction depends on the substrate to be used, but usually a reaction takes 24 to 120 hours to finish completely.

It is advantageous to use a mixed substrate containing at least some monocarboxylic acid since the monocarboxylic acid tends to increase the solubility of the substrate in the aqueous reaction mixture and to suppress accumulation of additional monocarboxylic acid during oxidation. A readily available source of the monocarboxylic acid for this mixed substrate is from the accumulation of the acid from previous reactions.

When cultivation is carried out as described above, a substantial amount of dicarboxylic acids or a mixture of dicarboxylic acids containing monocarboxylic acids is produced and accumulated. These carboxylic acids are separated and purified by a conventional method such as extraction, solid-liquid separation, neutralization-extraction and fractional distillation, and then harvested as monocarboxylic acids and dicarboxylic acids, or a mixture of both.

Recovery of Acetic or Butyric Acid from Fermentation Broth

There are at present two basic methods for industrial production of acetic acid, both of which are based on natural gas, the cost of which is increasing rapidly. Organisms have been discovered, such as *Clostridium thermoaceticum* that can convert carbohydrate to acetic acid in yields of 80% or greater (100% in theory). Many organisms, which are attractive from a fermentation yield basis, cannot tolerate highly acid conditions. Hence, the organic acids must be partly neutralized as they are formed. This creates the problem of isolating a pure acid from an aqueous solution where it occurs primarily as the acid salt. In other cases as well, such as chemical process streams or waste streams, it is necessary to recover pure acids from aqueous solutions where the acid occurs primarily as the salt.

The approach used in the past has been to add a strong mineral acid, such as sulfuric acid, to the organic salt solution, and extract the resulting organic acid into an organic solvent for eventual recovery by distillation. This "consumables" approach necessitates adding an alkali to the fermentation liquor (or other solution) to maintain neutrality during culture (or other processing), adding sulfuric acid to the "waste liquor" to permit organic acid extraction, and then disposing of the resulting waste salt.

The process developed by *R.A. Yates; U.S. Patent 4,282,323; August 4, 1981; assigned to E.I. Du Pont de Nemours and Company* permits recovery of acids, such as acetic acid, from fermentation liquors without the use of any consumable salts or acids or the disposal problem for the resulting by-product salt. In the process as applied to fermentation, all materials with the exception of the fermentation substrates are recoverable and recyclable.

In this process, a fermentation substrate of carbohydrate, which preferably is a sugar such as glucose or xylose, is fermented in the presence of a bacterial microorganism such as *Clostridium thermoaceticum* and in the presence of an alkali metal bicarbonate, such as sodium bicarbonate, to form a salt of the acid (e.g., sodium acetate) plus carbon dioxide. The presence of the bicarbonate is advantageous since it can be used to buffer the fermenter wherever an organism cannot tolerate low pH. Alkaline earth salts can be used as well as alkali metal salts, provided that the organisms can tolerate the salt used. Generally this process applies when it is desirable to maintain the fermenter at pH 5 to 8.

The fermentation liquor may be extracted directly by solvent plus carbon dioxide under pressure or be passed through a solids extractor countercurrent to an ion exchange resin charged with bicarbonate to selectively extract acetate out of the liquor and return alkali metal bicarbonate to the fermenter. The preferred solvents form azeotropes with water. When an ion exchange resin is used, the organic acid form of the ion exchange resin can be extracted directly, countercurrently, by closing and opening appropriate valves, or be passed through a countercurrent solids extractor charged with carbon dioxide under pressure in the presence of a water-containing organic solvent, such as, but not limited to tert-butanol, 2-butanone, dimethyl ether or diethyl ether, to convert bound salt to free acid. Advantage can be taken of changes in relative affinity of resin for ions with change of temperature.

The bicarbonate resulting from the CO_2 generated carbonic acid is bound to the resin, and the resin thus reactivated in bicarbonate form is reused with the fermentation liquor. The free organic acid in the solvent is purified from the solvent by conventional means such as distillation or recrystallization. Solvent and water are returned to the system. Carbon dioxide is also recycled through the system, but a slight excess may be generated by fermentation connected with cell growth.

Figures showing flow sheets of the processes using (1) direct extraction from a fermenter with solvent and carbon dioxide and (2) countercurrent ion exchange extraction of acid from a fermenter are given in the patent, as are detailed descriptions of these methods.

Carbamyl Derivatives of Alpha-Hydroxy Acids

R. Olivieri, G.E. Bianchi, E. Fascetti, F. Centini, L. Degen and W. Marconi; U.S. Patent 4,332,905; June 1, 1982; assigned to E.N.I. Ente Nazionale Idrocarburi, Italy have developed a procedure for preparing α-hydroxy acids by the hydrolysis of 5-substituted-2,4-oxazolidinediones [formula(1)] to open the ring, producing compounds of formula (2) below, after which further hydrolysis yields the α-hydroxy acid of formula (3).

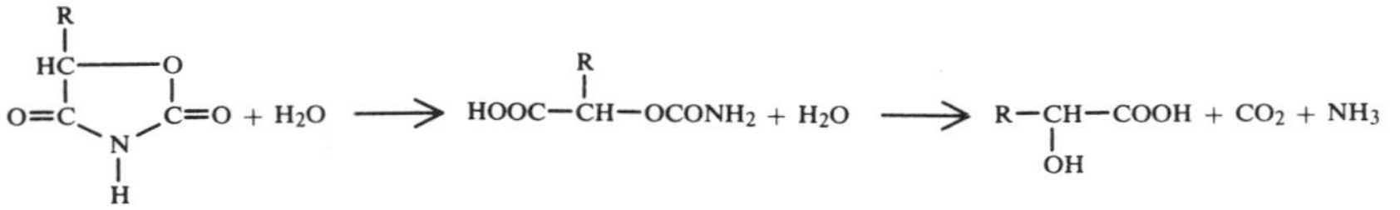

(1) (2) (3)

A further important subject of the process is the direct production of D-carbamyl-α-hydroxy acids by the stereoselective enzymatic hydrolysis of racemic mixtures of compounds of general formula (1). In this respect, it has been found possible to enzymatically hydrolyze 5-substituted DL-2,4-oxazolidinediones in such a manner as to give only the D-carbamyl-α-hydroxy acid, i.e. only one of the two possible optical isomers. The free α-hydroxy acid of D configuration can be obtained from the optically active carbamyl derivative by simple hydrolysis. Compounds such as those of formula (1) can be easily prepared from the corresponding α-hydroxy acids by reacting them with urea.

Resolution of the racemic mixture is effected by stereoselective hydrolysis of the oxazolidine ring carried out by enzymes easily obtainable from cultures of various microorganisms or from extracts of animal organs such as veal liver. One particular case which is extremely interesting is the preparation of D(–) mandelic acid, C_6H_5–CHOH–COOH, which is an important intermediate in the preparation of semisynthetic antibiotics.

This intermediate is generally prepared by resolution of the racemic mixture by means of the formation of diastereo-isomer salts with optically active natural bases such as brucine. These processes in any case give a maximum theoretical yield of 50% in that the L-enantiomer must be racemized before being recycled.

According to this process, the carbamyl derivative of D(–) mandelic acid can be advantageously prepared by the stereoselective enzymatic hydrolysis of the corresponding 5-phenyl-2,4-oxazolidinedione. A further considerable advantage of the process is that the enzymatic reaction substrate racemizes spontaneously under the hydrolysis conditions, so that at the end of the reaction the carbamate of the D-mandelic acid is obtained in a stoichiometric quantity with respect to the starting substrate.

The enzymatic activity required for preparing the D(–) mandelic acid carbamate has been found both in homogenized veal liver and in various microorganisms. Microorganisms of the following kinds have proved particularly suitable: *Agrobacterium radiobacter* NRRL B 11291, *Bacillus brevis* NRRL B 11080, *Bacillus stearothermophilus* NRRL B 11079, and various strains of Pseudomonas, such as sp. ATCC 11299, which are cultivated in a conventional manner.

Poly(Beta-Hydroxybutyric Acid)

K.A. Powell and B.A. Collinson; U.S. Patent 4,336,334; June 22, 1982; assigned to Imperial Chemical Industries Limited, England describe a microbiological process for the production of poly(β-hydroxybutyric acid), hereinafter referred to as PHB, and to microorganisms for use in such a process. They have found that certain strains of *Methylobacterium organophilum* will metabolize methanol to give relatively high proportions of high molecular weight PHB.

Accordingly they provide a method of producing PHB comprising aerobically culturing a purified *Methylobacterium organophilum* microorganism strain selected from strains designated by NCIB 11482 to 11488 inclusive and mutants and variants of such strains and CBS numbers 137.80 to 143.80 inclusive.

The microorganisms were obtained from the drains of a methanol-producing industrial plant and selected by growing on an aqueous methanol medium followed by culturing in the absence of any carbon source for several days. Surviving strains were isolated and purified by conventional techniques.

Mutants of the strains may be produced by conventional techniques, for example by exposure of the strains to UV radiation or by exposure to mutant inducing chemicals such as ethane/methane sulfonate or N-methyl-N-nitro-N-nitrosoguanidine. Mutants so produced can be separated from the wild-type strains by known techniques, e.g., penicillin and cycloserine enrichment techniques.

As indicated above, the microorganisms metabolize methanol. However they are also capable of metabolizing other carbon sources containing carbon-hydrogen bonds. Hydrocarbons can be metabolized, even methane in spite of its inertness. Oxygenated hydrocarbons, such as alcohols, ethers, etc. may be metabolized. Amines may be metabolized but will not give appreciable PHB concentrations when used as the sole carbon source, unless the accumulation of PHB is conducted under phosphorus starvation conditions, because of the need to maintain nitrogen and/or phosphorus starvation conditions for PHB accumulation. Carbohydrates may also be metabolized. For the production of PHB therefore, the assimilable carbon source should preferably be free of nitrogen.

In addition to the assimilable carbon source, sources of assimilable nitrogen and phosphorus are required for growth of the microorganism.

It has been found that PHB accumulation is favored by culturing the microorganism under nitrogen and/or phosphorus starvation conditions. In a preferred process, the microorganism is first grown in an aqueous medium containing sources of assimilable nitrogen and phosphorus and then culturing is continued under nitrogen and/or phosphorus starvation conditions. During the first stage the microorganism grows, i.e. multiplies, to give an economic concentration of the microorganism, preferably a concentration of 20 to 25 g/ℓ. Little or no PHB is accumulated during this stage. The culture produced may then be fed to a second vessel, where the second stage is conducted under conditions of starvation of one of the nutrients nitrogen or phosphorus. The second stage is conducted until the biomass is increased, mainly by PHB synthesis by from 30 to 50% by weight (33 to 55 g/ℓ).

Long-Chain Dicarboxylic Acids Using *Candida tropicalis*

In previously suggested methods for producing long-chain dicarboxylic acids by fermentation, yields have been poor, due to contamination of the medium with foreign microorganisms. Problems have also arisen in the steps of collection and separation of the acids from the fermentation broth.

The objectives of *K. Kato and N. Uemura; U.S. Patent 4,339,536; July 13, 1982; assigned to Nippon Mining Co., Ltd., Japan*, therefore, are to provide a process for production of long-chain dicarboxylic acids by fermentation techniques which produce good yields and advantageous means of separation of the acid products from the fermentation broth.

The salient features of their process are (1) use of a fungus of the *Candida tropicalis* strain for the fermentation and (2) adjusting the pH of the medium during the initial stage to 3.0 to 5.0 and in the remaining period, adjusting the pH to 6.5 to 7.5.

Another feature of the process is the addition of an alkaline material (such as caustic soda or caustic potash) to the culture after it is complete. This dissolves the dicarboxylic acids which have been formed. Bleaching powder and/or a hypochlorite are then added to check afterfermentation in amounts so that their concentration in the fermentation broth will be 20-200 ppm. Diatomaceous earth is then added and the culture broth is filtered under pressure, after which a mineral acid is added to the filtrate at a temperature of above 50°C to precipitate the acid, and the acid is then collected by filtration.

An example of *Candida tropicalis* mentioned in the patent as a suitable working strain for the fermentation has the number FERM-P 3291.

Production of Muconic Acid

Adipic acid is an important commodity in the chemical industry, particularly for consumption as a comonomer in the synthesis of polymers. Adipic acid can be obtained by oxidation of cyclohexane or cyclohexanol. Another prospective method is by the hydrogenation of muconic acid, which is a diolefinically unsaturated adipic acid derivative:

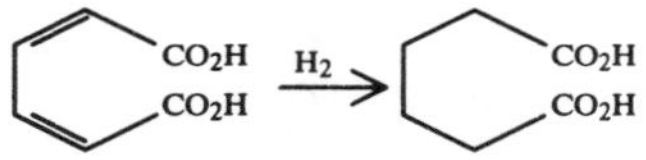

A potentially convenient source of muconic acid is by the microbiological oxidation of various hydrocarbon substrates.

No known naturally occurring microorganisms are known that metabolize an aromatic hydrocarbon substrate such as toluene by the ortho pathway via muconic acid and β-ketoadipate. Wild strains metabolize aromatic hydrocarbon substrates by the meta pathway via 2-hydroxymuconic semialdehyde instead of a muconic acid intermediate. Catechol 2,3-oxygenase is functional rather than catechol 1,2-oxygenase.

Thus, the potential of microbiological oxidation of toluene as a convenient source of muconic acid requires the construction of mutant strains of microorganisms which (1) metabolize toluene by means of the ortho pathway, and (2) allow the accumulation of muconic acid without further assimilation.

P.C. Maxwell; U.S. Patent 4,355,107; October 19, 1982; assigned to Celanese Corporation has accomplished this by a process which comprises (1) culturing a microorganism species selectively to provide strain A1 which metabolizes toluene by the ortho pathway via catechol to muconic acid, and which subsequently metabolizes the resultant muconic acid via β-ketoadipate to biomass and carbon dioxide; (2) continuously and selectively culturing strain A1 for rapid growth on toluene as the sole source of carbon to provide strain A2; (3) culturing strain A2 in selective enrichment cycles in a medium containing benzoate as the sole source of carbon and containing an antibiotic which kills only growing cells; (4) harvesting the strain A2 cells and diluting and culturing the cells in media containing a nonselective carbon source; (5) plating the strain A2 cells on a nutrient medium containing a limiting amount of a nonselective carbon source and excess benzoate; (6) isolating cells from single small colonies, and culturing the cell isolates and selecting a strain A3, wherein strain A3 converts toluene to muconic acid and lacks active muconate lactonizing enzyme.

The starting microorganism can be any organism capable of growth on toluene and possessing a catechol 1,2-oxygenase, e.g., a Pseudomonad.

Illustrative of the desired microorganisms are constructed strains of fluorescent Pseudomonads each of which has the following characteristics:

(a) possesses catechol 1,2-oxygenase enzyme with activity that is not inhibited in the presence of a low level of muconic acid in a growth medium;

(b) lacks substantially catechol 1,2-oxygenase enzyme;

(c) lacks functional muconate lactonizing enzyme;

(d) cells are rod shaped, vigorously motile and polarly flagellated; and

(e) cells grow well on p-hydroxybenzoate.

A new strain of *Pseudomonas putida* Biotype A, constructed as described above and having these characteristics, has been deposited with the American Type Culture Collection and has been designated as ATCC 31,916.

In a further embodiment, there is provided a process for the production of muconic acid which comprises feeding toluene to a buffered aqueous medium containing this ATCC 31,916.

The rate of toluene conversion typically is about 30 mg of muconic acid produced per dry weight gram of cells per hour. The conversion of toluene proceeds readily at a dry weight cell concentration of 50 g/ℓ, with a resultant muconic acid production rate of 1.5 g/ℓ/hr.

Under optimal conditions, the muconic acid accumulation limit can approach up to about 50 g of muconic acid per liter of growth medium. The microbiological oxidation process normally is conducted at ambient temperatures up to about 31°C.

USE OF METHYLOTROPHIC MICROORGANISMS

Methane is one of the most expensive carbon sources for microbial growth. It is known that there are many microorganisms capable of growing on a culture medium in the presence of methane as the principle carbon source. However, not all of these microorganisms share good growth characteristics. It is also known that methane-grown microorganisms can be used to convert methane to methanol under aerobic conditions.

These methane-utilizing microorganisms are generally known as "methylotrophs." The classification system for methylotrophs proposed by R. Whittenbury et al [*J. of Gen. Microbiology,* 61, 205-218 (1970)] is the most widely recognized. In their system, the morphological characteristics of methane-oxidizing bacteria are divided into five groups: Methylosinus, Methylocystis, Methylomonas, Methylobacter and Methylococcus.

Recently, Patt, Cole and Hanson [*International J. Systematic Bacteriology,* 26 (2) 226-229 (1976)] disclosed that methylotrophic bacteria are those bacteria that can grow nonautotrophically using carbon compounds containing one or more carbon atoms but containing no carbon-carbon bonds. Patt et al have proposed that methylotrophs should be considered "obligate" if they are capable of utilizing only carbon compounds containing no carbon-carbon bonds (e.g., methane, methanol, dimethyl ether, methylamines, etc.) as the sole sources of carbon and energy whereas "facultative" methylotrophs are those organisms that can use both compounds containing no carbon-carbon bonds as well as compounds having carbon-carbon bonds as the sources of carbon and energy.

In their paper, Patt et al disclosed a methane-oxidizing bacterium, which they identified as *Methylobacterium organophilum* sp nov (ATCC 27,886). This bacterium presumably differs from all previously described genera and species of methane-oxidizing bacteria because of its ability to utilize a variety of organic substrates with carbon-carbon bonds as sources of carbon and energy.

It is well recognized that there are two types of methylotrophic microorganisms based on their ability to grow on carbon-containing substrates. One type has been referred to as "methane-utilizers" and the other has been referred to as "methanol-utilizers." The methanol-utilizers are unable to grow in the presence of methane as the sole carbon and energy source, but will grow in the presence of methanol, methylamine, etc. The methane-utilizers are capable of growing on a plurality of C_1-type compounds, including methane, methanol, dimethyl ether, etc. Within the group of methane-utilizing methylotrophs and methanol-utilizing methylotrophs, there are obligate and facultative types of methylotrophic microorganisms.

The obligate methane-utilizer type methylotrophic microorganisms will only grow on C_1-type compounds, e.g., methane, methanol, dimethyl ether, methyl formate, methyl carbonate, etc. The facultative methane-utilizer type methylotrophic microorganisms will not only grow on the abovementioned C_1-type compounds, but will also grow on other organic compounds such as glucose. The obligate methanol-utilizer type methylotrophic microorganisms will grow on C_1 compounds, e.g., methanol, methylamine, but not on methane or on organic compounds such as glucose. The facultative methanol-utilizer type methylotrophic microorganisms will grow on the C_1-type compounds mentioned above (but not methane) and various other organic compounds such as glucose.

Microbiological Oxidation of C_{3-6} Alkanes and Secondary Alcohols to Ketones

C.-T. Hou, R.N. Patel and A.I. Laskin; U.S. Patents 4,250,259; February 10, 1981; and 4,268,630; May 19, 1981; both assigned to Exxon Research & Engineering Co., have devised processes for formation of the methyl ketones, i.e., acetone or 2-butanone from C_{3-6} alkanes or secondary alcohols. Their process comprises contacting a C_{3-6} alkane or a C_{3-6} secondary alcohol under aerobic conditions in a nonnutrient medium containing resting microbial cells derived from obligate or facultative methylotrophic microorganisms or enzyme preparations derived from those cells, in which the microorganisms have been previously cultivated under aerobic conditions in a mineral nutrient medium containing an oxygenase and/or dehydrogenase enzyme inducer as the growth and energy source. Examples of such inducers include methane (in the case of methane-utilizing methylotrophic microorganisms), methanol, dimethyl ether, methylamine, methyl formate, methyl carbonate, ethanol, propanol, butanol, etc.

The microbial cells or the enzyme preparations derived from the cells to be used in converting the C_{3-6} alkanes to the corresponding ketones may be derived from obligate or facultative methane-utilizing type methylotrophic microorganisms but not the methanol-utilizing type methylotrophic microorganisms. The microbial cells or the enzyme preparations derived from the cells to be used in converting C_{3-6} secondary alcohols to the corresponding ketones may be derived from either the obligate or facultative methane- or methanol-utilizing type methylotrophic microorganisms.

They have also found that methylotrophic yeast strains may be aerobically grown on a plurality of methyl radical donating carbon-containing compounds, such as methanol, methylamine, methyl formate, methyl carbonate, dimethyl ether, etc., to produce microbial cells or enzyme preparations derived therefrom and are capable of aerobically converting C_{3-6} linear secondary alcohols to the corresponding methyl ketones.

As an additional discovery a nicotinamide adenine dinucleotide (NAD+)-dependent secondary alcohol dehydrogenase (SADH) has been identified in cell-free extracts of various hydrocarbon-utilizing microbes, including bacteria and yeast. This enzyme is also found in cells grown on methanol. It specifically and stoichiometrically oxidizes C_{3-6} secondary alcohols to their corresponding methyl ketones. This enzyme has been purified 2600 fold and shows a single protein band on acrylamide gel electrophoresis. It has a molecular weight of 95,000±3000 daltons. The bacterial SADH consists of two subunits of 48,000 daltons and two atoms of zinc per molecule of enzyme protein. It oxidizes secondary alcohols, notably 2-propanol and 2-butanol. Primary alcohols are not oxidized by SADH.

As discussed previously, the obligate methane-utilizing methylotrophic microorganisms are capable of growing on methane, methanol and a plurality of methyl radical donating compounds, e.g., dimethyl ether, methyl formate, methyl carbonate, etc. The facultative methane-utilizing methylotrophic microorganisms not only grow on methane, methanol and various methyl radical donating compounds as mentioned above, but they are also capable of growing on various other organic compounds such as glucose.

The obligate methanol utilizing methylotrophic microorganisms are not capable of growing on methane, but are capable of growing on methanol and other methyl radical donating type compounds such as methylamine, methyl formate, methyl carbonate, etc. The facultative methanol-utilizing methylotrophic microorganisms, like the obligate methanol-utilizing methylotrophic microorganisms are not capable of growing on methane, but are capable of growing on methanol, the abovementioned methyl-containing and other organic compounds such as C_{2-6} alcohols and glucose. For the purpose of this process, however, the obligate or facultative methanol-utilizing microorganisms are to be grown on the alcohol dehydrogenase-inducing substrate, i.e., methanol, ethanol, propanol, butanol, methylamine and methyl formate, etc.

The microorganisms useful in this process are the strains having the following designations: *Methylosinus trichosporium* OB3b (NRRL B-11,196); *Methylosinus sporium* 5 (NRRL B-11,197); *Methylocystis parvus* OBBP (NRRL B-11,198); *Methylomonas methanica* S_1 (NRRL B-11,199); *Methylomonas albus* BG8 (NRRL B-11,200); *Methylobacter capsulatus Y* (NRRL B-11,201); *Methylococcus capsulatus* (Texas) ATCC 19069; *Methylobacterium organophilum* sp nov (ATCC 27,886); *Methylomonas sp* AJ-3670 (FERM P-2400); *Methylococcus* 999 (NCIB Accession No. 11,083); and *Methylomonas* SM3 (NCIB Accession No. 11,084).

The enzyme preparations mentioned are cell-free particulate fractions of the resting microbial cells.

Microbiological Oxidation of C_{1-6} Alkanes and Cycloalkanes

In a third patent by *R.N. Patel, C.-T. Hou and A.I. Laskin; U.S. Patent 4,269,940; May 26, 1981; also assigned to Exxon Research & Engineering Co.,* other methylotrophic microorganism strains are identified. These microbial cells and enzymes prepared from them are used to convert oxidizable substrates to oxidized products, e.g., C_{1-6} alkanes to alcohols, C_{3-6} alkanes to the corresponding C_{3-6} sec. alcohols and methyl ketones, C_{3-6} sec. alcohols to the corresponding methyl ketones, cyclic hydrocarbons to cyclic hydrocarbyl alcohols (e.g., cyclohexane to cyclohexanol), C_{2-4} alkenes selected from the group consisting of ethylene, propylene, butene-1 and butadiene to the corresponding 1,2-epoxides, styrene to styrene oxide, etc.

Several of these methylotrophic new strains grow well on a culture medium in the presence of oxygen and methane and methyl-radical donating compounds such as methanol, methylamine, methyl formate, methyl carbonate, dimethyl ether, etc. These strains of methylotrophic microorganisms are capable of producing microbial cells useful as feedstuffs when cultured under aerobic conditions in a liquid growth medium comprising assimilable sources of nitrogen and essential mineral salts in the presence of methane gas or the abovementioned methyl-radical donating carbon-containing compounds as the major carbon and energy source.

As another embodiment there are provided biologically pure isolates and mutants thereof of a plurality of newly discovered and isolated methane-utilizing microorganism strains. These biologically pure isolates are capable of producing microbial cells when cultivated in an aerobic nutrient medium containing methane or the abovementioned methyl-radical donating carbon-containing compounds as the major carbon and energy source.

A particularly preferred embodiment includes a process for producing propylene oxide from propylene by contacting propylene under aerobic conditions with microbial cells or enzyme preparation thereof wherein the microbial cells are derived from the newly discovered and isolated methane-utilizing strains of this process as described below and which have been previously grown under aerobic conditions in the presence of methane.

Another particular preferred embodiment includes a process for converting C_{3-6} linear secondary alcohols to the corresponding methyl ketones by contacting a C_{3-6} linear secondary alcohol under aerobic conditions with such microbial cells or enzyme preparations thereof (including cell extracts or purified SADH or NAD+).

The microbial strains used in this process are various methylotrophic microorganisms with NRRL numbers from NRRL B-11,205 to NRRL B-11,222 inclusive plus several strains of yeast: NRRL Y-11,328; NRRL Y-11,419 and NRRL Y-11,420.

Microbiological Oxidation of C_{2-4} n-Alkanes, Dienes and Vinyl Aromatic Compounds

In another patent by *C.-T. Hou, R.N. Patel and A.I. Laskin; U.S. Patent 4,347,319; August 31, 1982; assigned to Exxon Research & Engineering Co.,* the alkane oxidation process described is one for the microbiological epoxidation of C_{2-4} n-alkenes, dienes and vinyl aromatic compounds by microorganisms which have been aerobically cultivated in a nutrient medium containing methane. The microorganisms used are newly isolated methane-utilizing obligate and facultative methylotrophs.

These newly discovered and isolated methane and methyl-radical-utilizing (methylotrophic) microorganism strains are as follows:

Methylotrophic Microorganism Strain Name	ER & E Designation	U.S.D.A. Agriculture Research Center Designation
1. *Methylosinus trichosporium*	(CRL 15 PM1)	NRRL B-11,202
2. *Methylosinus sporium*	(CRL 16 PM2)	NRRL B-11,203
3. *Methylocystis parvus*	(CRL 18 PM4)	NRRL B-11,204
4. *Methylomonas methanica*	(CRL M4P)	NRRL B-11,205
5. *Methylomonas methanica*	(CRL 21 PM7)	NRRL B-11,206
6. *Methylomonas albus*	(CRL M8Y)	NRRL B-11,207
7. *Methylomonas streptobacterium*	(CRL 17 PM3)	NRRL B-11,208
8. *Methylomonas agile*	(CRL 22 PM9)	NRRL B-11,209
9. *Methylomonas rubrum*	(CRL M6P)	NRRL B-11,210
10. *Methylomonas rubrum*	(CRL 20 PM6)	NRRL B-11,211
11. *Methylomonas rosaceus*	(CRL M10P)	NRRL B-11,212
12. *Methylomonas rosaceus*	(CRL M7P)	NRRL B-11,213
13. *Methylobacter chroococcum*	(CRL M6)	NRRL B-11,214
14. *Methylobacter chroococcum*	(CRL 23 PM8)	NRRL B-11,215
15. *Methylobacter bovis*	(CRL M1Y)	NRRL B-11,216
16. *Methylobacter bovis*	(CRL 19 PM5)	NRRL B-11,217
17. *Methylobacter vinelandii*	(CRL M5Y)	NRRL B-11,218
18. *Methylococcus capsulatus*	(CRL M1)	NRRL B-11,219
19. *Methylococcus minimus*	(CRL 24 PM12)	NRRL B-11,220
20. *Methylococcus capsulatus*	(CRL 25 PM13)	NRRL B-11,221
21. *Methylobacterium organophilum*	(CRL 26 R6)	NRRL B-11,222
22. Pichia sp.	(CRL-72)	NRRL Y-11,328
23. Torulopsis sp.	(A_1)	NRRL Y-11,419
24. Kloeckera sp.	(A_2)	NRRL Y-11,420
and mutants thereof.		

The ER and E Designation in the table is that given by Exxon Research and Engineering Co. and the U.S.D.A. Designation is the accession number under which the cultures were deposited in the U.S.D.A. Northern Regional Research Laboratories.

The patent lists the morphological and taxonomical characteristics of each of the methylotrophic microorganisms given in the table.

Partial Degradation of Complex Cyclicorganics

I.J. Higgins; U.S. Patent 4,323,649; April 6, 1982; assigned to Imperial Chemical Industries Limited, England provides a process for the partial degradation of a complex organic compound (as hereinafter defined) into one or more other organic compounds when it is contacted with a methane-utilizing bacterium or with an extract thereof containing a methane monooxygenase and/or a dehalogenase enzyme.

The term complex organic compound as used here means a saturated or unsaturated (including aromatic) cyclic (including heterocyclic and multiple ring compounds) compound having at least one substituent which is a halogen atom, a group containing a halogen atom and/or a substituted or unsubstituted hydrocarbon group containing at least three carbon atoms, attached to a saturated or unsaturated ring.

Suitably the complex organic compound is either (A) a mono-, di- or trialkyl or alkenyl benzene or (B) a di- or trisubstituted benzene in which at least one substituent is a halogen atom or a group containing a halogen atom. Preferably type (A) compounds are monoalkyl benzenes having from 5 to 12 carbon atoms in the alkyl group, for example 1-phenylheptane which is degraded to yield benzoic and cinnamic acids and p-hydroxy-1-phenylheptane. Type (B) compounds contain e.g. alkyl or hydroxyl groups and halogen (particularly chlorine) atoms, for example m-chlorotoluene which, using washed cell suspensions, is dechlorinated and oxidized to benzyl alcohol and a hydroxy-benzyl alcohol.

The process may be carried out using whole cells (either viable or nonviable) of the bacteria or using appropriate enzyme extracts.

Any suitable methane-utilizing bacteria may be used in the process including both Type I and Type II bacteria. However Type II bacteria are preferred. Such bacteria fall into the family group Methylomonadaceae as described in Bergey's manual (1974) although the classification of such bacteria is not finally decided.

Particularly useful strains include strains of *Methylosinus trichosporium* (e.g. the OB 3b strain), *Methylocystis parvus, Methylomonas albus* (e.g. the BG 8 strain), *Methylomonas methanica* (e.g. the PM strain) and also strains of *Methylococcus capsulatus* including the Texas strain (ATCC 19069) and the Bath strain (NCIB 11132). Other strains which can be used include strains of *Methylobacterium organophilum.*

Cultures of *Methylosinus trichosporium* strain OB 3b have been deposited at The National Collection of Industrial Bacteria in Aberdeen, Scotland, (as NCIB 11131) and at The Fermentation Research Institute, Japan (as FERM P-4981).

For a given substrate the products obtained may vary depending upon the system used such as enzyme extracts or whole cells. Thus when a halogen-substituted compound is treated with whole cells or with an extract containing a monooxygenase and dehalogenase both oxidation and dehalogenation occur while with an extract containing only a dehalogenase or a monooxygenase then only dehalogenation or oxidation respectively occur. Other conditions affecting the products obtained include process conditions and the strain of microorganism. Optimum conditions can be readily determined by a worker experienced in this field. When, with whole cells or extracts, it is desired to achieve both oxidation and dehalogenation over lengthy periods it is necessary to provide a cofactor acting as a source of reducing power. Suitable cofactors include methane and particularly methanol. The dehalogenase can function without a cofactor in these circumstances and hence if dehalogenation only is required no cofactor is necessary.

Before use in the process cultures of *Methylosinus trichosporium* OB 3b can be grown under batch or continuous conditions with methane as a carbon source. The cells from the cultures can be harvested during the late logarithmic phase (batch culture) or during the steady state (continuous culture) by centrifuging at 3,000 *g* for 45 minutes. The cells are then washed twice with 20 mM sodium phosphate buffer (e.g. at pH 7.0) and, after resuspending in the same buffer, they may be stored at a low temperature, i.e. 0°C or less, until required for use.

A typical transformation using cells of *Methylosinus trichosporium* strain OB 3b prepared as described above may be performed in the following manner. A washed suspension, containing 70-80 mg dry weight of cells in 20 ml of 20 mM sodium phosphate buffer (pH 7.0) is shaken in a 250 ml conical flask for 12 hr at 30°C. The flask is sealed when containing an atmosphere of air or 50% v/v air/methane. The liquid substrates in 3 ml volumes are contained in center wells from which they can diffuse to contact the bacterium. Products obtained may be identified by combined gas chromatography and mass spectrometry.

In a useful application of the process, aqueous wastes comprising complex organic compounds (as hereinbefore defined) are treated with methane-utilizing bacteria or with extracts thereof containing a methane monooxygenase and/or a dehalogenase enzyme to cause partial degradation of the compounds and thereby to facilitate disposal of the wastes. This application is particularly useful where the compound is a chlorine-containing compound.

MISCELLANEOUS PRODUCTS

Acrylamide or Methacrylamide Using Corynebacterium or Nocardia

To produce acrylamide or methacrylamide, there has previously been used a process of reacting acrylonitrile (AN) or methacrylonitrile (MAN) with water using reduced copper as a catalyst. However, it has been desired to develop an industrially more advantageous process since the catalytic process involves a difficult catalyst preparation and regeneration, and the isolation and purification of the amide produced is onerous.

A catalytic process for producing acrylamide or methacrylamide utilizing microorganisms has been developed by *I. Watanabe, Y. Satoh, and T. Takano; U.S. Patent 4,248,968; February 3, 1981; assigned to Nitto Chemical Industry Co.,*

Ltd., Japan and bacteria having an extremely high activity for hydrolyzing acrylonitrile and methacrylonitrile to produce acrylamide or methacrylamide have been found. The strains N-771 and N-774 belonging to the genus Corynebacterium, and the strain N-775 belonging to the genus Nocardia have been located in the soils around a factory producing acrylonitrile and in the wastewater discharged from the factory. (Hereafter the aforementioned bacteria will be referred to as N-771, N-774 and N-775, respectively.)

The enzymatic nitrilase activity of these microorganisms is surprisingly high at low temperatures. As a result, a process for the hydrolysis of acrylonitrile and methacrylonitrile has been developed wherein the enzymatic activity of the bacterial cells is stably maintained at a high level for a long time with the accumulation of produced acrylamide or methacrylamide reaching concentrations as high as 10 wt % or more, which process does not require a difficult purifying step.

The process for continuously producing a highly concentrated acrylamide or methacrylamide aqueous solution by passing an aqueous solution of acrylonitrile or methacrylonitrile through a column or columns filled with immobilized bacterial cells having nitrilase activity, at a temperature ranging from the freezing point of the solution to 30°C at a pH of about 6 to 10, comprises:

(1) using a column having one or more feed inlets provided between the column inlet and the column outlet, continuously feeding an aqueous solution of acrylonitrile or methacrylonitrile via the column inlet and, at the same time, continuously feeding acrylonitrile or methacrylonitrile via the feeding inlet(s) in an amount soluble in the reaction medium; or

(2) using two or a plurality of columns connected to each other in series, and continuously feeding an aqueous solution of acrylonitrile or methacrylonitrile via the first column inlet and, at the same time, continuously feeding acrylonitrile or methacrylonitrile via the column inlet(s) of the successive columns in an amount soluble in the reaction mixture.

As the microorganisms to be used in the process, any one that has the ability to hydrolyze acrylonitrile or methacrylonitrile to produce acrylamide or methacrylamide may be used, as well as the aforesaid strains N-771, N-774 and N-775. For example bacteria from the genus Bacillus, the genus Bacteridium, the genus Micrococcus and the genus Brevibacterium as disclosed in U.S. Patent 4,001,081 may also be used. In addition, it is also possible to use the cellular extract prepared by destroying such bacterial cells, crude enzyme preparations, etc.

To culture the microorganism used in this process, ordinary culture mediums are used. The culture is aerobically conducted while maintaining the pH of the culture medium at about 6 to 9 at a temperature preferably about 25° to 30°C, for about 1 to 5 days.

The strains N-771, N-774 and N-775 are deposited at Fermentation Research Institute, as FERM P-4445, P-4446 and 4447, respectively.

In using a fermentation technique to produce acrylamide or methacrylamide from the suitable acrylonitrile, such as is described in the previous patent, the aqueous solution produced is unstable and susceptible to polymerization. When sufficient polymerization inhibitors are added to prevent this, the microbial reaction is adversely affected and the quality of the solution is poor.

As a result of studies to remove these defects, *I. Watanabe and Y. Satoh; U.S. Patent 4,343,899; August 10, 1982; assigned to Nitto Chemical Industry Co., Ltd., Japan* have found that a stable (meth)acrylamide aqueous solution can be obtained in high purity without polymerization upon production or concentration of the (meth)acrylamide solution, by treating the microorganisms with a water-soluble dialdehyde before feeding them to the reaction system of (meth)-acrylonitrile and water. Thus, many of the problems accompanying the use of microbial cells are overcome.

The microorganisms useful in the process are those described in the previous patent, particularly FERM P-4445, 4446 and 4447. If these microorganisms are to be immobilized any conventional process can be used, with an entrapping process using an acrylamide series polymer being particularly preferred. Immobilization can be conducted by suspending the aforesaid microbial cells in an aqueous medium containing a monomer or monomers like acrylamide and a cross-linking agent like N,N-methylenebisacrylamide, and conducting polymerization at a pH of about 5 to 10, preferably 6 to 8, and at a temperature of about 0° to 30°C, preferably 0° to 15°C, using a polymerization initiator, thus causing gelation. The content of microorganisms in the polymerization reaction solution varies depending upon the kind and the state of microorganisms used, but is typically about 0.1 to 50 wt %, preferably 1 to 20 wt %. The content of monomers in the polymerization reaction solution is about 2 to 30 wt %, preferably about 5 to 20 wt %.

The water-soluble dialdehyde treatment is carried out either on intact cells or immobilized cells. To be specific, these microbial cells (in the case of immobilized cells, after pulverizing to a suitable size) are suspended in a buffer solution such as 0.05 to 0.5 M phosphate solution, and a water-soluble dialdehyde is added thereto in an amount of about 0.1 to 10.0 wt %, preferably 0.5 to 5.0 wt %, based on the weight of dry cells. The reaction is conducted at a pH of about 5 to 10, preferably 6 to 8, at a temperature of about 0° to 30°C, preferably about 0° to 15°C, for 0.5 to 3 hours under stirring. In addition, immobilization of water-soluble dialdehyde treated cells can also be conducted after treating the cell suspension with the dialdehyde. Among the dialdehydes for use in the process, glyoxal and glutaraldehyde are commercially available and preferred.

In producing an aqueous solution of (meth)acrylamide, the dialdehyde-treated microbial cells (in the case of using immobilized cells, particles of suitable size) are filled in a reactor or a column, and are brought into contact with a (meth)-acrylonitrile aqueous solution under the aforesaid conditions. The reaction temperature is preferably about 0° to 15°C depending on retention of enzymatic activity. Additionally, the conversion of the reaction can be controlled by selecting the amount of cells, reaction time, flow rate of the substrate, and the like. Therefore, selection of proper conditions enables one to conduct the reaction conversion almost completely.

When such dialdehyde-treated microorganisms are used, the continuous column production of (meth)acrylamide goes smoothly, with no polymerization, and there is obtained an aqueous solution of the product which is stable to polymerization.

In the process of producing acrylamide from acrylonitrile by microorganic reactions in a batchwise or continuous method using granulated fixed cells such as has been described in the two previous patents, it has been noted that the fixed cells swell and lose their enzymatic activity. In further work, *I. Watanabe; U.S. Patent 4,343,900; August 10, 1982; assigned to Nitto Chemical Industry Co., Ltd., Japan* has found that this problem can be eliminated. They accomplish this by a process which is characterized in that at least one compound selected from the group consisting of alkali metal carbonates and alkali metal bicarbonates is added to the aqueous medium, singly or in combination with an organic carboxylic acid.

Alkali metal carbonates and alkali metal bicarbonates which are added to the aqueous medium according to this process include sodium carbonate, potassium carbonate, sodium bicarbonate, potassium bicarbonate, etc. The amount of the alkali metal carbonate or bicarbonate being added is from 0.05 wt % to about 0.5 wt %, based upon the weight of the aqueous medium containing the acrylonitrile.

When the organic carboxylic acid is added in combination with the alkali metal carbonate or bicarbonate, the amount of the alkali metal carbonate or bicarbonate to be added should be sufficient to neutralize the organic carboxylic acid and to keep the pH within the range of 7.5 to 8.5. For example, when sodium carbonate and acrylic acid are used, the weight ratio ranges of the sodium carbonate to the acrylic acid is from about 1.3:1 to about 1.7:1.

As such organic carboxylic acids, any can be used so long as they are water-soluble. Practically preferred examples of organic carboxylic acids which can be used in the process include formic acid, acetic acid, propionic acid, acrylic acid, etc. The amount of the organic carboxylic acid added is 0.005% or more, preferably 0.01% or more, based upon the weight of the aqueous medium containing the acrylonitrile.

In this process then, polyacrylamide gel-entrapped cells such as those described in the previous patent are pulverized to particles of a suitable size and, after being washed, are charged to a column reactor. By passing a substrate solution prepared by mixing the aqueous medium containing at least one compound selected from the alkali metal carbonates and bicarbonates, singly or in combination with the organic carboxylic acid, and acrylonitrile through the column, the aqueous acrylamide solution medium can be obtained as a column effluent.

Additionally, the reaction can be controlled by selecting the amount of the cells, the concentration of substrate acrylonitrile, the flow rate, etc., to obtain a conversion of nearly 100%. In this case, to maintain the nitrilasic activity of the fixed cells for a long period of time and to inhibit the formation of by-products such as acrylic acid, etc. it is preferred that the concentration of acrylonitrile be 5% by weight or less, that the reaction temperature be as low as possible within a range such that the substrate aqueous solution does not freeze, i.e., from just above the freezing point to 10°C, and that the pH be from 7.5 to 8.5.

Thus, a colorless, transparent acrylamide aqueous solution can be obtained as a reaction effluent. Since this acrylamide aqueous solution contains almost no impurities that exert adverse influences on the polymerization of acrylamide, it can be used, as is or after being concentrated, as a starting material for the production of acrylamide polymers for use in flocculants, paper strengthening agent, etc.

Production of D-Arabitol

In previously known fermentation processes for preparing D-arabitol, yeasts of the genera Candida, Saccharomyces, Hansenula, Debaryomyces, Torulopsis, Pichia or Endomycopsis were used and nutrient media containing carbohydrates such as glucose, sucrose or glycerol were the principal carbon sources employed. These known processes have various disadvantages, however. The substances such as glucose, glycerol and sucrose employed as nutrient media are expensive. Furthermore, it is very difficult to separate the D-arabitol formed in these processes from the sugars used as the carbon source.

A. Fujiwara and S. Masuda; U.S. Patent 4,271,268; June 2, 1981; assigned to Hoffmann-La Roche Inc. have found that a specific species of Pichia, namely *Pichia haplophila,* and mutants derived therefrom fermentatively produce D-arabitol in high yields from inexpensive hydrocarbons or ethyl alcohol. The arabitol thus obtained can advantageously be separated in a simple manner from the culture liquor.

The nutrient medium may contain additives other than a hydrocarbon or ethyl alcohol and the cultivation is suitably carried out at temperatures in the range of about 20° to about 40°C. The resulting D-arabitol is useful as a sweetening agent and as an intermediate.

Lipids Having a High Linoleic Acid Content

The process described by *O. Suzuki, Y. Jigami, S. Nakasato and T. Hashimoto; U.S. Patent 4,281,064; July 28, 1981; assigned to The Agency of Industrial Science and Technology, Japan* is one for producing lipids having a high linoleic acid content. In this process fungi of the Pellicularia genus are cultivated in a medium of a carbohydrate or vegetable fiber as a carbon source, and lipids (neutral lipids such as oil and fat and polar lipids such as phospholipid and glycolipid) having a high linoleic acid content are produced from the culture obtained by the above cultivation.

As for the fungi of Pellicularia genus employed in the molds such as *Pellicularia filamentosa* IFO 6476, 6523, 6675, 8985, 5879, 6262 and 6295, and *Pellicularia praticola* IFO 6253, etc. may be used, but any fungi of the Pellicularia genus can be employed. The fungi exemplified above are known molds which have been preserved at the Institute for Fermentation in Osaka.

As for the carbohydrates as a carbon source for the medium employed for cultivating the abovementioned molds, glucose, sucrose, starch, molasses, etc. are included. Vegetable fibers such as linter pulp, wood pulp, bagasse, etc. may be used. The carbon source is preferably used in an amount of 20-80 g/ℓ medium.

The cultivation of the abovementioned molds is carried out under aerobic conditions and usually by stationary cultivation, shaking cultivation, aeration agitation cultivation, etc., in a liquid medium. The pH of the medium is preferably in the range of 4.0 to 6.0, and cultivation is carried out usually at 15° to about 38°C for about 7 to 30 days.

The conventional inorganic or organic nitrogen sources are used in the medium together with the normal salts, trace elements, etc. The collection of lipids can be carried out according to a conventional manner, such as by solvent extraction.

Increasing Diacetyl Production of Suitable Bacteria

An important component of butter aroma is the compound diacetyl (2,3-butanedione). This compound is a yellow liquid having an extremely potent butter aroma. Diacetyl can be formed via two different methods. The first method involves chemical synthesis. For example, diacetyl can be synthesized from methyl ethyl ketone by conversion to an isonitroso compound which is then decomposed by hydrolysis with hydrochloric acid to diacetyl. A second method for producing diacetyl is by bacterial fermentation. For example, glucose can be fermented to methylacetylcarbinol which is then oxidized to form diacetyl.

J.A. Troller; U.S. Patent 4,304,862; December 8, 1981; assigned to The Procter & Gamble Company has devised a process for substantially increasing the diacetyl-producing capacity of diacetyl-producing bacteria per volume of nutrient medium by additions to the nutrient medium in which the bacteria are grown.

The diacetyl-producing composition includes an aqueous nutrient medium having a pH of from about 4.5 to 7.0 and containing a metabolizable amount of a diacetyl precursor. The composition further includes a bacterium selected from *Streptococcus diacetylactis, Streptococcus cremoris, Streptococcus lactis* or mixtures of these bacteria. Most importantly, the composition includes a humectant selected from glycerol, sucrose and mixtures thereof in an amount sufficient to lower the water activity (a_W) value of the nutrient medium to from about 0.95 to 0.99 and to increase the production of diacetyl by the bacteria per volume of nutrient medium. The diacetyl-producing composition is incubated at a temperature of from about 28° to 37°C to produce the diacetyl.

The precursor is preferably selected from citric acid and its bacteriologically acceptable salts such as sodium citrate. It is used in amounts of 0.1 to 1.0% by weight of the nutrient medium. The preferred nutrient medium is milk containing up to 2% butterfat.

Beta-Carotene by Use of *Phycomyces blakesleeanus*

The production of β-carotene by the fungus *Phycomyces blakesleeanus* depends on the culture media and conditions but is generally small. In the dark, this fungus produces about 50 micrograms of the pigment per gram of dry material (μg/g dry weight), an amount inappropriate for the desired objective.

Various ways of increasing this production are known. One of them consists in the addition of vitamin A to the normal culture medium of this organism. This addition causes the accumulation of up to 2,000 μg/g by weight of β-carotene, but the required concentrations of the vitamin are prohibitively high. Amounts up to 2,000 μg/g dry weight have been observed under the best conditions in the presence of β-ionone.

Another way of increasing the production of β-carotene is the formation of the strains containing nuclei of the two known wild sexual types of Phycomyces, which are represented by the symbols (+) and (-). The method for obtaining this type of strain, denominated intersexual heterokaryon, has been described.

These strains accumulate up to 500 μg/g dry weight of β-carotene, they have a peculiar morphology, with formation of small hyphae areas, or pseudophores, and they are unstable: they tend to segregate the components homokaryotically.

The object of the process by *F.J.M. Araujo, I.L. Calderon, I.L. Díaz, and E.C. Olmedo; U.S. Patent 4,318,987; March 9, 1982* is to obtain strains of the fungus *Phycomyces blakesleeanus* accumulated by the β-carotene pigment in an amount which permits the use of these strains in the production, on an industrial scale of the pigment.

The technique which is described presents the combination in a single strain of the three types of stimulations previously mentioned, furthermore making unnecessary the addition of vitamin A to the culture medium.

The process, therefore, comprises production of a biologically pure strain of *Phycomyces blakesleeanus* capable of producing β-carotene by cultivating the fungus *Phycomyces blakesleeanus* in a suitable medium, isolating a strain denominated C115 therefrom, treating the spores of C115 with N-methyl-N'-nitro-N-nitrosoguanidine (NTG) in a suitable cultivating medium; isolating a strain denominated S106 and a strain M1 therefrom; and constructing an intersexual heterokaryon from the strains of S106 and M1 to obtain the strain S106*M1; and then:

(a) introducing recessive lethal mutations in each one of the types of nuclei of the strain S106*M1 by treating the spores of mycelia S106*M1 with N-methyl-N'-nitro-N-nitrosoguanidine (NTG) to a survival level of approximately 0.1% and isolating the heterokaryons S218*S219 and S242*S243 therefrom; and

(b) treating the heterokaryons S218*S219 with N-methyl-N'-nitro-N-nitrosoguanidine (NTG) and isolating the stable strains S244*S245 and S246*S247 therefrom.

All these strains have been dispersed in the Public Depository and Strain Collection of the University of Sevilla, Spain.

ENZYMATIC PRODUCTION OF DRUGS AND ORGANIC CHEMICALS

ANTIBIOTICS

Antibiotic Desacetyl 890A$_{10}$

J.S. Kahan and F.M. Kahan; U.S. Patent 4,264,734; April 28, 1981; assigned to Merck & Co., Inc. have found that an effective antibiotic can be made by the enzymatic deacetylation of the compound 890A$_{10}$. The compound is produced by hydrolyzing the N-acetyl group of 890A$_{10}$ using an amidohydrolase capable of hydrolyzing the N-acetyl group. A convenient source of an amidohydrolase with this capability is amidohydrolase-producing strains of the microorganism *Protaminobacter ruber.* The particular enzyme produced by *Protaminobacter ruber* is N-acetyl-890A$_{10}$ amidohydrolase, a member of the enzyme subgroup designated E.C. 3.5.1 according to recommended nomenclature of the International Union of Pure and Applied Chemistry and the International Union of Biochemistry.

The microorganism capable of carrying out the deacetylation process was isolated from a soil sample and, based upon taxonomic studies, was identified as belonging to the species *Protaminobacter ruber* and has been designated MB-3528 in the culture collection of Merck & Co., Inc. A culture thereof has been placed on permanent deposit with the Northern Regional Research Laboratories, and has been assigned accession No. NRRL B-8143.

The compound desacetyl 890A$_{10}$ has the structure:

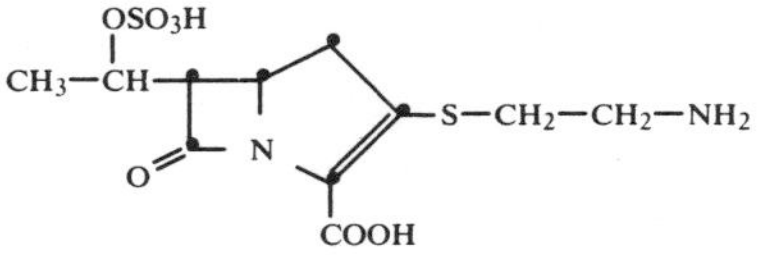

Desacetyl 890A$_{10}$ is a valuable antibiotic active against various Gram-positive and Gram-negative bacteria and, accordingly, finds utility in human and veterinary medicine. The compound can be used, for example, against susceptible strains of *Staphylococcus aureus, Proteus mirabilis, Escherichia coli, Klebsiella pneumoniae,*

Enterobacter cloacae and *Pseudomonas aeruginosa.* The antibacterial material may further be utilized as an additive to animal feeding stuffs, for preserving foodstuffs and as a disinfectant.

The parent compound $890A_{10}$ is produced during the aerobic fermentation, under controlled conditions, of suitable aqueous nutrient media inoculated with a strain of the organism, *Streptomyces flavogriseus.* Aqueous media, such as those employed for the production of other antibiotics, are suitable for producing $890A_{10}$.

Antibiotic Desacetyl $890A_1$ and $890A_3$

J.S. Kahan and F.M. Kahan; U.S. Patent 4,282,322; August 4, 1981; assigned to Merck & Co., Inc. provide a process for preparing two new antibiotics by enzymatic deacetylating of the compounds $890A_1$ and $890A_3$. Thus produced are desacetyl $890A_1$ (Formula 1) and desacetyl $890A_3$ (Formula 2) as shown below.

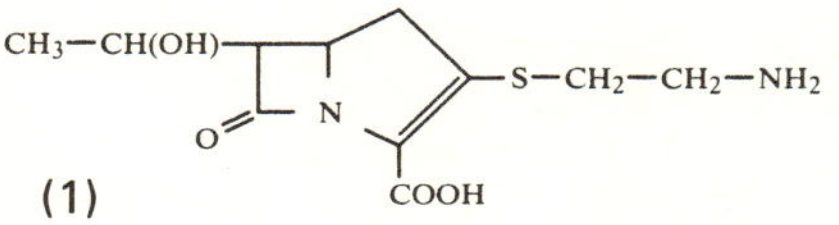

(1)

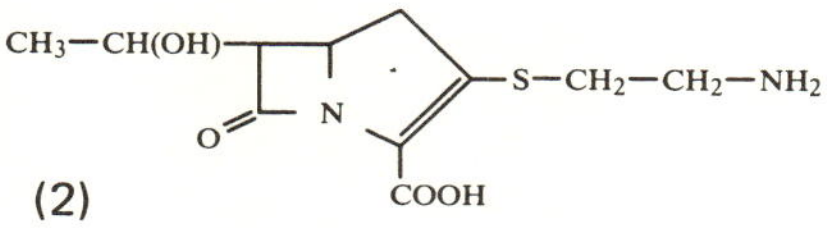

(2)

These antibiotic substances are produced by hydrolyzing the N-acetyl group of $890A_1$ and $890A_3$ using an amidohydrolase produced from strains of the microorganism *Protaminobacter ruber,* the microorganism designated MB-3528, which is the same as that used in the previous patent.

Cephalosporin Analogs

The following five patents, assigned to Kyowa Hakko Kogyo Co., Ltd. of Japan, all relate to the production of various optically active cephalosporin analogs. These compounds are all made in a similar manner—by treatment of a starting compound with an enzyme obtained from certain microorganisms. The preparation of the analogs will be described at the end of the section, to prevent repetition.

T. Hirata, Y. Hashimoto, T. Ogasa, S. Kobayashi, A. Sato, K. Sato and S. Takasawa; U.S. Patent 4,302,540; November 24, 1981; assigned to Kyowa Hakko Kogyo Co., Ltd., Japan have produced cephalosporin analogs of the general Formula (1) below by optically selective deacylation of compounds of the general Formula (2) below.

(1)

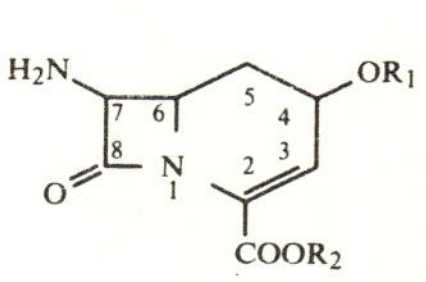

(2)

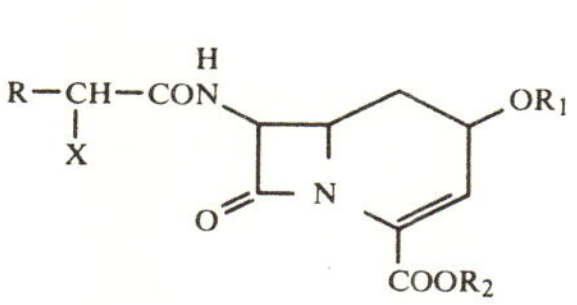

In Formulas (1) and (2), R represents a substituted or unsubstituted unsaturated 6-membered cyclic or 5- or 6-membered heterocyclic group, the substituent being a hydroxy, halo, nitro or methanesulfonamide group; X is hydrogen or an amino, hydroxy or lower alkyl group; R_1 represents a hydrogen, a lower (straight or branched chain, 1 to 5 carbon) alkyl or acyl group; R_2 is a hydrogen or a protective group of carboxylic acid; and the hydrogens in the 6- and 7-positions have cis configurations. The compounds of general Formula (1) may also be the salts of the analogs described.

In the second patent, *T. Hirata, Y. Hashimoto, I. Matsukuma, S. Yoshiie and S. Takasawa; U.S. Patent 4,302,541; November 24, 1981; assigned to Kyowa Hakko Kogyo Co., Ltd., Japan* describe optically active cephalosporins of Formula (1) prepared by an optically selective deacylation reaction using an enzyme and an optically inactive dl compound having an acyl group as the starting compound (general Formula 2).

(1)

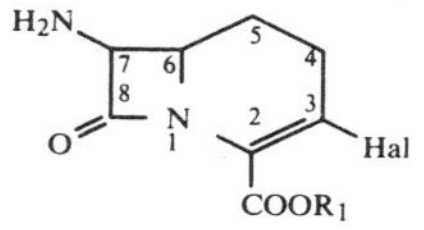

(2)

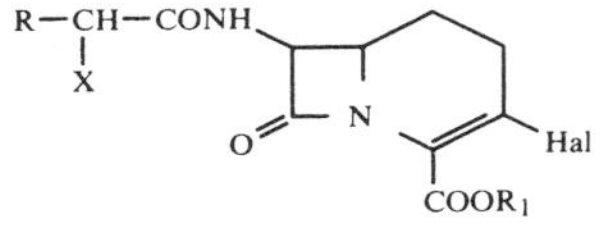

R above represents a substituted or unsubstituted, saturated or unsaturated six-membered carbocyclic or five-membered heterocyclic group, X represents a hydrogen, an amino group, a hydroxy group or a lower alkyl group, R_1 represents a hydrogen or a protective group of carboxylic acid, Hal represents a halogen atom, and the hydrogens at the 6- and 7-positions have cis configuration.

The optically active cephalosporins of this and the previous patent have stronger antimicrobial activity than their corresponding optically inactive dl-compounds. They are particularly useful as intermediates in the preparation of optically active acylated compounds which are strong antibacterial agents.

The optically active cephalosporin analogs of *T. Hirata, Y. Hashimoto, T. Ogasa, S. Kobayashi, I. Matsukuma and K. Kimura; U.S. Patent 4,316,958; February 23, 1982; assigned to Kyowa Hakko Kogyo Co., Ltd., Japan* have a general formula the same as that of Formula (1) of U.S. Patent 4,302,540, except that the $-OR_1$ group substituted in position 4 is here an $-R_1$ group, R_1 representing hydrogen or a lower alkyl group and R_2 being, again, a hydrogen or a protective group of carboxylic acid. These compounds are prepared by the process to be described at the end of this section, by optically selective deacylation of a compound represented by the general Formula (2):

(2)

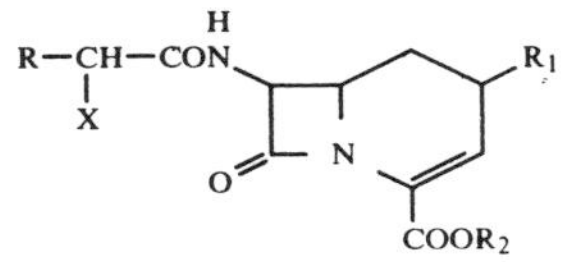

where R represents a substituted or unsubstituted unsaturated six-membered carbocyclic or five- or six-membered heterocyclic group, the substituent representing an hydroxy group, halogens, nitro group or methanesulfonamide group; X represents hydrogen, an amino group, a hydroxy group or a lower alkyl group, and R_1 and R_2 are as defined above, with the hydrogens at the 6- and 7 positions having cis configuration.

In the following two patents, the optically active cephalosporin analog intermediates previously described are further reacted with an α,α-disubstituted carboxylic acid in the presence of a microorganism such as will be detailed below to produce optically active acylated cephalosporin analogs.

Thus, *Y. Hashimoto, K. Kimura, T. Hirata, T. Ogasa, S. Kobayashi and I. Matsukuma; U.S. Patent 4,332,896; June 1, 1982; assigned to Kyowa Hakko Kogyo Co., Ltd., Japan* describe the treatment of compounds of the general Formula (1) of the previous patent by carboxylic acids of Formula (2) below to yield compounds of the Formula (3):

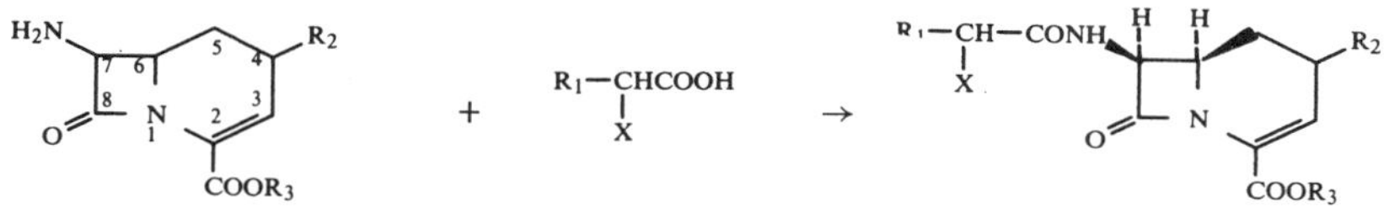

(1) (2) (3)

where R_1 represents a substituted or unsubstituted saturated or unsaturated six-membered carbocyclic or five-membered heterocyclic group; R_2 represents a hydrogen or a lower alkyl group; R_3 represents a hydrogen or a protective group of carboxylic acid; X represents a hydrogen, a lower alkyl group, a hydroxy group, a carboxy group or an amino group; and the hydrogens at the 6- and 7-positions have cis configuration.

These acyl derivatives which have an optically active carbacephem ring have strong antimicrobial activity as compared to the corresponding optically inactive acyl derivatives.

In the last patent of this group, *Y. Hashimoto, S. Takasawa, T. Hirata, I. Matsukuma and S. Yoshiie; U.S. Patent 4,335,211; June 15, 1982; assigned to Kyowa Hakko Kogyo Co., Ltd., Japan* describe the process of producing optically active acyl compounds of the cephalosporin analogs represented by general Formula (1) below by the selective acylation of compounds of the general Formula (1) of U.S. Patent 4,302,541 with an α,α-disubstituted carboxylic acid such as that described in the previous patent, in the presence of a microorganism or its reactive derivative such as will be detailed.

(1)

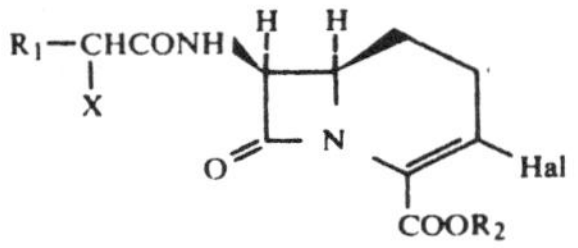

In the above, R_1, R_2, X and Hal have the same significance as previously defined.

Some suitable microorganisms for carrying out the five syntheses described in this section are as follows (with ATCC numbers, when available): *Aeromonas hydrophila; Achromobacter aceris; Arthrobacter simplex* ATCC 15799; Acetobacter sp. ATCC 21760; *Escherichia coli* ATCC 11105 and ATCC 13281; *Xanthomonas citri* and *X. physalidicola; Kluyvera citrophila* ATCC 21285; *Gluconobacter liquefaciens* ATCC 14835; *Clostridium acetobutylicum* ATCC 824; *Sarcina lutea* ATCC 9341; *Pseudomonas melanogenum* ATCC 17808; Flavobacterium sp. ATCC 21429; *Proteus rettgeri* ATCC 9250; and other strains of Staphylococcus,

Spirillum, Bacillus, Brevibacterium, Protaminobacter, Beneckea, Micrococcus, Mycoplana and Rhodopseudomonas.

For carrying out the optically selective reaction, the enzyme may be provided, more specifically, in any of the following forms: (1) As the culture liquor of the microorganism or treated matter thereof; (2) As cell bodies recovered from the culture broth by configuration which may be washed with saline water (usually about 1%), buffer solution and the like, or as a cell suspension; (3) As a disrupted cell suspension, i.e., a suspension of the cell bodies disrupted mechanically or chemically; (4) As a cell-free extract, i.e., a liquid obtained by removing the disrupted cell bodies from the disrupted cell suspension; or (5) As a purified enzyme solution which is obtained by recovering the enzyme protein with ammonium sulfate from the cell-free extract and subjecting the enzyme protein to gel filtration, ion-exchange cellulose column chromatography, ion-exchange Sephadex column chromatography, and the like. Cells or the purified enzyme immobilized by a conventional method may be used.

The reactions are carried out at a preferable temperature of 15° to 35°C and at a pH of 5 to 8 in an inactive solvent which does not affect the reaction.

As the solvent, water is most preferably used. Organic solvents such as acetone, methanol, ethanol, N,N-dimethylformamide, dimethylsulfoxide, and the like may be used alone or in combination with water. It is effective to add phosphate buffer, Veronal buffer or citric acid buffer to control the pH in the reaction. Reaction time, which is varied according to the kind and concentration of enzymes, the kind and concentration of substrates, reaction temperature or reaction pH, is generally 30 minutes to 24 hours. It is most preferable to terminate the reaction when the reaction ratio reaches maximum.

The concentration of cells is preferably 1 to 50 mg by dry weight per 1 ml of the reaction solution. When a purified enzyme is used, it is appropriate to use the amount of the enzyme having the same activity as that of the dry cell. The substrate compounds are used in an amount of 0.5 to 50 mg per 1 ml of the reaction solution.

After the completion of the reaction, isolation of the desired compound is carried out by a conventional method employed in the isolation and purification of organic compounds from culture liquors such as absorption using various carriers, ion-exchange chromatography, gel filtration, liquid-liquid extraction, and the like.

Enzyme Produced by *S. capillispira* Which Deesterifies Cephalosporin Methyl Esters

B.J. Abbott and D.R. Berry; U.S. Patent 4,316,955; February 23, 1982; assigned to Eli Lilly and Company provide a method of deesterifying cephalosporin methyl esters by contact with an enzyme derived from fermentation.

This enzyme is produced by *Streptomyces capillispira* when cultured under enzyme-producing conditions. It deesterifies cephalosporin methyl esters of the following formula where R is hydrogen or $R^3CH(NH_2)CO–$; R^1 is methyl, chloro, acetoxymethyl, methoxy, methoxymethyl, aminocarbonyloxymethyl, methylthiadiazolylthiomethyl or methyltetrazolylthiomethyl; R^2 is hydrogen or methoxy; R^3 is phenyl, cyclohexadienyl, cyclohexadienyl monosubstituted with hydroxy, or phenyl mono- or disubstituted with halo, hydroxy, C_1 to C_3 alkyl or C_1 to C_3 alkoxy; to form the corresponding carboxylic acid.

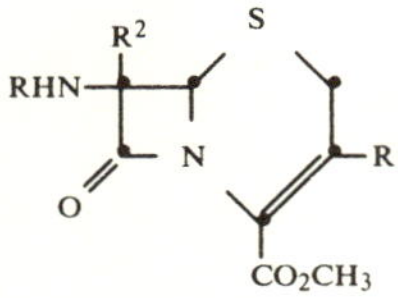

The microorganism which produces the enzyme of this process is a new species, *S. capillispira*, which has been designated A49492. It was isolated as a substantially biologically pure culture from a sample of soil collected in Sweden, and has been deposited at the Northern Regional Research Laboratory under the number NRRL 12279.

A49492 produces the enzyme of this process when it is cultured under proper enzyme-producing conditions. The organism can utilize many carbohydrate sources, including invert sugar, corn syrup, glucose, fructose, maltose, starch, and others.

Useable nitrogen sources include peptones, soybean meal, amino acid mixtures and the like. The customary soluble inorganic salts may be used as sources of trace elements in the culture; such salts include those capable of yielding iron, sodium, potassium, ammonium, calcium, phosphate, chloride, carbonate and like ions.

It has been found, however, that the organism produces the enzyme best when it is grown in defined media. In particular, the identity and amount of the nitrogen source must be chosen with some care for best enzyme production. It has been found that amino acids, such as proline, serine, glycine and the like, may be used, but that ammonium ion is a somewhat more satisfactory nitrogen source.

The pH of the culture should be held in the range from about 6 to 8, most preferably from about 6.9 to 7.1.

A49492 requires an adequate supply of air for growth. It is most convenient to aerate small-scale cultures in shaking flasks, and, in large scale, by submerged aerobic culture.

A49492 will grow well over a temperature range between about 25° and 40°C. It is preferable, however, to grow the organism at a temperature from about 30° to 37°C.

When the enzyme has been produced by a culture of A49492, it may be used by simply adding the substrate compound to the fermentation mixture, or it may be harvested and purified for use in the process.

To make multiple reuse of the enzyme and to increase its stability, it is preferable to immobilize the enzyme by a suitable method. The immobilization of enzymes for such processes is known in the art, and the usual expedients have been found to be effective with the enzyme produced by A49492. The preferred immobilization method is ionic binding to diethylaminoethylcellulose (DEAE Sephacel, Pharmacia Fine Chemicals Corp.). Another preferred immobilization method is ionic binding to DEAE cellulose in the form of beads prepared by the process of U.S. Patent 4,063,017.

The immobilized enzyme is efficiently contacted with the substrate compound by either column or stirred-tank procedures. If a column is to be used, it is packed with the immobilized enzyme, and an aqueous solution or suspension of the substrate compound is passed through the column. When a stirred tank is used, it may be operated either continuously or batchwise according to the usual manner in the art.

After the substrate compound has been in contact with the enzyme, whether immobilized or not, for an adequate period of time to produce the desired yield of cephalosporin acid, the acid is isolated by the common organic chemical procedures. For example, it may be converted to an appropriate salt, especially a sodium or potassium salt for therapeutic use, and isolated by evaporating the water from the product solution under vacuum at moderate temperatures.

It has been found that the deesterification process is preferably carried out at a pH of from about 7.5 to about 8.5. The optimum temperature for the process is from about 25° to about 30°C, although temperatures from about 15° to about 45°C may be used. Since the enzyme is not particularly stable at elevated temperatures, it is advisable to use the optimum temperature. It has also been found that the enzyme, when in a partially purified state, can be stored for extended periods of time by freezing or lyophilization. The fermentation broth in which the enzyme is produced, and cell mass isolated from the fermentation broth, can be stored by freezing.

Penicillins and Cephalosporins by Action of an Acylase

Processes are well known for preparing penicillins or cephalosporins by reacting a methyl or ethyl ester of an acid having an acyl moiety with 6-aminopenicillanic acid or 7-aminocephalosporanic acid or reactive derivatives thereof. The acyl source of these processes is invariably a lower alkyl ester, especially methyl ester. Many lower alkyl esters are sparingly soluble in water and attain only low concentration insufficient for effective acylation, however, and as a result, synthesis with immobilized enzymes often is difficult due to two-phase formation or plug formation in the column.

To solve these problems, *E. Kondo, T. Mitsugi, T. Fujiwara and R. Muneyuki; U.S. Patent 4,340,672; July 20, 1982; assigned to Shionogi & Co., Ltd., Japan* have developed a process for preparing β-lactam antibacterials by reacting an ester of Formula (1) with aminoazetidinone carboxylic acid of Formula (2) in the presence of acylase in an aqueous medium to produce a β-lactam antibacterial of Formula (3):

$$\mathrm{RCOO(CH_2\underset{\underset{X}{|}}{C}HO)_nY} \qquad (1)\ [\text{acyl source}]$$

$$\mathrm{H_2N{-}Q} \qquad (2)\ [\text{amino source}]$$

$$\mathrm{RCONH{-}Q + HO(CH_2\underset{\underset{X}{|}}{C}HO)_nY} \qquad (3)\ [\beta\text{-lactam antibacterial}]$$

where RCO– is an acyl group; X– is a hydrogen atom, lower alkyl group or hydroxy-lower alkyl group; Y– is a hydrogen atom or a lower alkyl group; n is a positive integer; and Q– is a group of the formula:

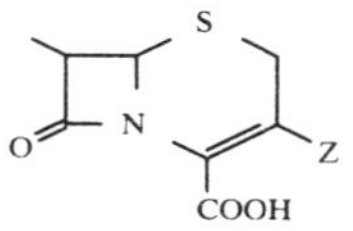

or

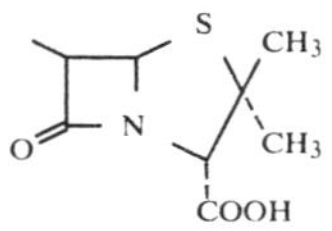

in which Z is a hydrogen or halogen atom, a nucleophilic group, a methyl, halomethyl or methyl substituted by a nucleophilic group.

Enzymes of bacterial or fungal origin are especially important as the acylase for this process from the viewpoint of production, efficiency, cost and stability. Especially suitable bacteria include acylase-producing bacteria belonging to genus Micrococcus, Arthrobacter or Bacillus, including specific strains of *Micrococcus roseus* M-1054-1 (FERM-P 3744; ATCC 31251); *Arthrobacter globiformis* (FERM-P 3743; ATCC 31250); and *Bacillus circulans* (FERM-P 5153). These fungi or bacteria can be used in the form of cells, crushed cells, crude enzyme, pure enzyme, immobilized enzyme or like preparations.

The cells are produced usually under aerobic condition, e.g., by liquid propagation with aeration. The propagation medium is an aqueous solution at pH 6 to 8 containing conventional nitrogen and carbon sources and inorganic salts. Propagation is carried out at 20° to 40°C for 10 to 60 hours. The acylase is collected in a conventional manner and is preferably immobilized by adsorption or chemical bonding onto a carrier.

By using this method, the following compounds can be successfully prepared: cefaclor, cefacetrile, cefazolin, cefatrizin, cefadroxyl, cefapyrin, cefamandole, cefalexin, cefaloglycin, cefalotin, cefaloridine, cefaclomezin, cefsulodin, ceftezol, cefradin, CGP-9000, phenylacetamidocephalosporanic acid, phenoxyacetamidocephalosporanic acid, amoxicillin, ampicillin, carbenicillin, phenoxymethylpenicillin, phenoxypropylpenicillin and benzylpenicillin.

Penicillin Derivatives

The process developed by *M. Cole and R.A. Edmondson; U.S. Patent 4,347,314; August 31, 1982; assigned to Beecham Group, Limited, England* is one for the preparation of penicillin derivatives and in particular for the preparation of α-carboxy-6α-methoxypenicillin derivatives by the enzymatic hydrolysis of an esterified derivative.

In this process, compounds of Formula (1), where R represents a phenyl group or a 2- or 3-thienyl group, are formed by subjecting compounds of Formula (2), where R^1 is an aryl radical to the action of an enzyme selected from bromelin, papain, gelatase, trypsin, pancreatin or an esterase-producing strain of *Escherichia coli, Pseudomonas aeruginosa, Aspergillus niger,* or Saccharomyces sp.

(1) 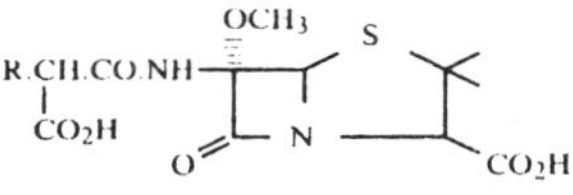

(2) 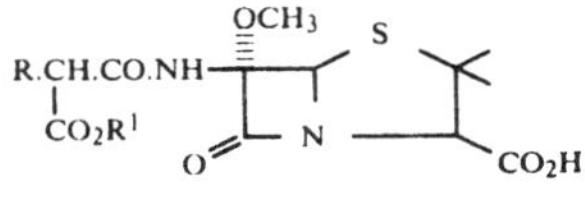

Preferred aryl groups R^1 include phenyl, and mono-, di- and tri-(C_{1-6})-alkyl substituted phenyl such as o-m-, or p-methylphenyl, ethylphenyl, n- or isopropylphenyl, or n-, sec-, iso- or tert-butylphenyl.

Suitable esterase forming strains of the abovementioned microorganisms and molds include *E. coli* K12 (NCIB 10112) and BRL 1873 (ATCC 9723), *Ps. aeruginosa* A (NCIB 10110) and R59 (NCIB 10111), *Aspergillus niger* BRL 822 (IMI 130783), *Saccharomyces cerevisiae* BRL 611 and *Saccharomyces carlsbergensis*, BRL 622. A preferred esterase is that produced by *Aspergillus niger*.

The esterase enzyme can be prepared by culturing the microorganism or mold in a conventional manner, especially under aerobic conditions in a suitable liquid or semisolid media. Preferred conditions are 20° to 30°C at a pH of 5 to 9, suitably about pH 7, for 1 to 10 days. The cultured microorganism containing the esterase is employed for the process in the form of the cultured broth, separated cells or isolated enzyme.

M. Cole and R.A. Edmondson; U.S. Patent 4,346,168; August 24, 1982; assigned to Beecham Group Limited, England have also prepared the α-carboxy-6α-methoxy penicillin derivatives of the general Formula (1) of the previous patent by the action of the enzyme α-chymotrypsin or an esterase-producing strain of Streptomyces on compounds of general Formula (2) of the previous patent.

Suitable esterase-producing strains of Streptomyces include *S. olivaceus* ATCC 3335 and *S. clavuligerus* ATCC 27064. These microorganisms are cultivated in the conventional manner.

OTHER PHARMACEUTICALS

Hexuronyl Hexosaminoglycane Sulfate for Thrombosis Prevention

F. Fussi and G. Fedeli; U.S. Patent 4,264,733; April 28, 1981; assigned to Hepar Chimie SA, Switzerland have found that a heteropolysaccharide, more particularly a hexuronyl hexosaminoglycane sulfate, as obtained by extraction from animal organs, is not only endowed with anticoagulant, antithrombotic and clearing activity, but can be administered both orally and parenterally, can be absorbed through the intestinal barrier and topically, and shows a ratio between antithrombotic activity and anticoagulant activity which is favorable in comparison with heparin.

The extraction process for obtaining the above-identified product is characterized by the following steps:

(a) Hydrolysis of an animal organ, ground and suspended in water, with a proteolytic enzyme;

(b) Precipitation of the limpid liquid by a water-miscible solvent;

(c) Solubilization of the precipitate in a solution of a salt of a strong mineral acid and of a mono- or divalent cation (e.g., sodium, ammonium, calcium), the saline solution having a ion concentration corresponding to that of 0.8 M NaCl;

(d) Addition at a temperature not higher than 80°C of an excess of quaternary ammonium halide, selected from those having in their molecule at least one aliphatic group with more than 12 atoms, whereby a complex is formed which remains in solution, whereas the simultaneously formed precipitate is separated;

(e) Precipitation of the complex by dilution with water until an ion concentration not higher than that of 0.4 M NaCl is obtained;

(f) Isolation of the precipitated complex and solubilization thereof in a salt solution having the same characteristics of that of the step (c);

(g) Precipitation of the desired product by a water-miscible solvent and desiccation.

The animal organ which is subjected to the hydrolysis, is preferably fresh or deep-frozen. Among the useful animal organs, lungs and duodenum are preferred. The proteolytic enzyme is preferably selected from vegetal (papain, ficin, bromelin) or bacterial endopeptidases, and the conditions in which the hydrolysis is carried out depends on the type of enzyme. It is particularly preferred to operate the hydrolysis under mild heat, for periods not less than 3 hours, until the value of the α-amino nitrogen (as determined by means of the Soerensen Method) no longer varies.

For the precipitation of the limpid liquid after hydrolysis the solvent to be used is selected from acetone, dioxane, methanol, ethanol and the like.

The quaternary ammonium halide of step (d) is preferably selected from chlorides and bromides of cetylpyridinium and cetyltrimethylammonium. The related excess is of at least 0.5 g of quaternary ammonium salt per kg of starting animal organ. Lastly for the precipitation of the final product (step g) the use of acetone is preferred and the precipitate is vacuum-dried or lyophilized.

Amino Acid Solution Which Produces Cysteine in Living Cells

A. Meister and J.M. Williamson; U.S. Patent 4,335,210; June 15, 1982; assigned to Cornell Research Foundation have devised a method of restoring the glutathione level of numerous tissues where 5-oxoprolinase is present, particularly the human liver. It also provides a method of combating poisoning associated with the lessening or depletion of the glutathione content of cells. It is particulaly applicable to the treatment of patients suffering from overdoses of acetoaminophenols.

It is known that 5-oxo-L-prolinase (L-pyroglutamate hydrolase) catalyzes the adenosine-triphosphate-dependent cleavage of 5-oxo-L-proline to glutamate. It has been found that a sulfur analog of 5-oxo-L-proline, viz., L-2-oxo-thiazolidone-4-carboxylate is also acted upon by this enzyme but the products include cysteine rather than glutamate. In the latter reaction the carboxylate is cleaved to yield an adenosine diphosphate/cysteine ratio of 1:1. The enzyme which exhibits an affinity for the analog similar to that for the natural substrate is inhibited by the analog, both in vitro and vivo. Thus, the carboxylate serves as a potent inhibitor of the gamma-glutamyl cycle at the step of 5-oxoprolinase. Administration of L-2-oxo-thiazolidine-4-carboxylate to mice that has been depleted of hepatic glutathione led to the restoration of normal hepatic glutathione levels. Since L-2-oxo-thiazolidine-4-carboxylate is an excellent substrate of the enzyme, it has been found to be useful as a component of base amino acid nutritional solutions and as an intracellular delivery system for cysteine. It is a good therapeutic agent for correcting conditions causing glutathione lessening or depletion, e.g., that resulting from poisoning induced by excessive amounts of acetaminophenol in the system.

Approximately 50% of hospitalized patients require some degree of nutritional support to counteract malnutrition and to assist in disease recovery. Many of such patients cannot take food directly because of G.I. tract dysfunction or poor oral intake. One of the primary means for establishing total parenteral nutrition is the infusion of basic amino acid solutions via central vein or peripheral vein administration. Typical base amino acid solutions contain eight essential amino acids and seven nonessential amino acids and are normally prepared and sold commercially in 10% aqueous solutions.

Of the amino acids not contained in such solutions the most important is cysteine. Cysteine cannot be administered intravenously due to its toxic effects on the system. Cysteine is quite important to human metabolism since, of all the amino acids contained in the basic solutions, the only one containing sulfur is L-methionine. Methionine is metabolized only in the liver so that if a patient has a malfunctioning or partially functioning liver, his system becomes devoid of sulfur bearing protein. Meister and Williamson provide a method for introducing cysteine which can be metabolized as well in organs other than the liver.

In accordance with one aspect of their process the basic amino acid solutions are modified by the incorporation therein of 0.25 to 2.5 g/dl of L-2-oxothiazolidine-4-carboxylate.

A standard base amino acid formulation for peripheral vein injection is a 3.5% by weight solution of the essential and nonessential amino acids and electrolytes, such as sodium, potassium, magnesium, chloride and acetate.

To the above nutrition support solutions there is added according to this method, L-2-oxothiazolidine-4-carboxylate, preferably in the form of its neutral salt, e.g., sodium, potassium, magnesium, etc., in the amount specified above.

4-Substituted-1-B-D-Ribosyl Imidazo-(4,5-c) Pyridines

4-Substituted-1-B-D-Ribosyl Imidazo-(4,5-c) pyridines are compounds of pharmacologic interest—for use by themselves or as intermediates. For example, 3-deaza-adenosine has been found to be an antifungal agent and an immunosuppressant.

These compounds, however, have not been easily prepared. *T.A. Krenitsky and J.L. Rideout; U.S. Patent 4,347,315; August 31, 1982; assigned to Burroughs Wellcome Co.* have found that 4-substituted 1H-imidazo-(4,5-c) pyridines may be readily ribosylated by an enzymatic method, with the advantages over the prior art chemical methods that it is stereospecific, adaptable to large-scale production and offers improved yield in the ribosylation reaction. The method comprises the reaction of a 4-substituted-1H-imidazo-(4,5-c) pyridine with a ribose donor system comprising ribose-1-phosphate and a phosphorylase-type enzyme.

The 4-substituent may be any substituent required in the final product. Of particular interest are halogens, amino, thiol, alkylthio, substituted amino (including lower alkyl amino and protected amino such as benzyl amino and benzhydryl amino) substituents.

Of these compounds, 4-amino-1-B-D-ribofuranosyl-1H-imidazo-(4,5-c) pyridine is of particular interest and may be prepared directly by ribosylation of 4-amino-1H-imidazo-(4,5-c) pyridine, a method not practicable by the prior art chemical means.

Although the ribose-1-phosphate required may be provided by synthetic processes, it can be convenient or even advantageous if it is generated enzymatically in situ from a ribosyl donor and an inorganic phosphate to obtain the required ribose-1-phosphate. Although the reactions may be carried out separately, i.e., by isolating the ribose-1-phosphate from an enzymatic reaction or chemically synthesizing ribose-1-phosphate and using it as a starting material, it has been found advantageous to carry out both reactions in a "one pot" process by forming the ribose-1-phosphate intermediate in situ. The net effect of the coupled reactions therefore is the transfer of the ribosyl moiety of the donor ribonucleoside to the free 3-deazapurine base, thereby producing the desired ribonucleoside.

The ribosyl moiety donor may be a purine ribonucleoside, for example, adenosine, a pyrimidine ribonucleoside, for example, uracil ribonucleoside, or a mixture of various ribonucleoside and nonnucleoside material. However, for this purpose it is preferable that the ribosyl moiety donor is substantially free from nonnucleoside material and also that it is a pyrimidine ribonucleoside.

It has been found that both reactions described hereinabove are catalyzed by various enzymes which are present in many different microorganisms and mammalian tissues. The phosphorolysis of the donor ribonucleoside is catalyzed, for instance, by purine nucleoside phosphorylase if the donor is a purine ribonucleoside, or by pyrimidine nucleoside phosphorylase, thymidine phosphorylase or uridine phosphorylase if the donor is a pyrimidine ribonucleoside. The second reaction, by which the desired 3-deazapurine ribonucleoside is synthesized from the 3-deazapurine and ribose-1-phosphate, is catalyzed by purine nucleoside phosphorylase.

The required ribosyl transferring enzyme system, therefore, may consist of the latter phosphorylase alone, or in combination with any one of the former type if the ribosyl donor is a ribonucleoside of a pyrimidine or pyrimidine analogue.

Although the enzymes required for the catalysis of these reactions occur in many different microorganisms, aerobic bacteria such as *B. stearothermophilus* and especially *E. coli* ATCC 11303 were found to be excellent sources of such enzymes. The bacteria which provide the enzymes may be cultured under a variety of conditions. However, media which contained large quantities of glucose were found to be undesirable since the levels of the nucleoside phosphorylase enzymes in the bacterial cells were depressed in the presence of glucose.

It has been found that crude enzyme preparations are less suitable than purified preparations. In most cases, therefore, it is desirable to purify the crude enzyme preparations before addition to the reaction mixture. This may be achieved in a number of ways known in the art.

The enzymes, provided in a sufficiently effective state and concentration, may then be used to catalyze the abovementioned reactions. A typical reaction mixture contains a ribosyl donor, a 3-deazapurine base, inorganic phosphate, for example, dipotassium hydrogen phosphate (K_2HPO_4), and the appropriate enzyme or enzymes in an aqueous medium or in a medium containing up to 50% of an organic solvent such as methanol, ethanol, propanol, butanol, acetone, methyl ethyl ketone, ethyl acetate, toluene, tetrahydrofurn, dioxane, dimethyl sulfoxide, trichloromethane, or Cellosolve. The preferred concentration is from 1 to 200 mM. The reaction is performed at near neutral pH, i.e., in the pH range of about 5 to 9, preferably 6.0 to 8.5 and at a temperature of 3° to 70°C.

The desired purine ribonucleosides may be recovered or isolated by any of the known means for separating mixtures of chemical compounds into individual compounds.

Cultivation of Chondrocytes to Produce Antiinvasion Factor

Certain tissues have a high resistance to invasion by foreign cells, and cartilage tissues have been found to have especially high resistance to invasion by foreign cells such as cancer cells and blood capillary cells.

Described in U.S. Patent 4,042,457 (Kuettner) is a protein fraction obtained from cartilage tissues which inhibits cell proliferation and tissue invasion. This protein fraction will hereinafter be referred to as antiinvasion factor, or AIF.

AIF has been shown to express antitumor properties both in vivo and in vitro. Tumors grown in vivo shrink in the presence of AIF. Tumorous mice injected with sufficient AIF demonstrate regression of the tumors and a lack of new daughter tumors. Accordingly, AIF holds promise as a cancer-treating agent. However, AIF must be made more readily obtainable for its full drug potential to be realized.

Current methods of obtaining AIF involve obtaining waste animal parts from slaughterhouses, separating the cartilage tissue and extracting the AIF, a long and tedious procedure. As a first step toward increasing the available amount of AIF, it would be desirable to obtain by some other method, such as by culturing chondrocytes (cartilage cells) or chondroblasts (cartilage-forming cells).

Culturing of cartilage cells containing significant amounts of AIF is complicated by the inherent nature of cartilage tissue. Cartilage cells in vivo produce an extracellular matrix which separates one cell from another. One of the main functions of the cells comprising cartilage would appear to be the production of the extracellular matrix. While an animal is growing, the cartilage is growing and many cartilage cells retain their ability to reproduce, but in an adult animal, the majority of cartilage cells are embedded within the extracellular matrix and are nonreproductive.

Accordingly, it is an objective of *K.E. Kuettner; U.S. Patent 4,356,261; Oct. 26, 1982; assigned to Rush-Presbyterian-St. Luke's Medical Center* to culture mammalian chondrocytes to produce cartilage tissue with high amounts of AIF. This is achieved by plating, in high density, chondrocytes obtained from mammals which have a high percentage of viability so that, generally, each active chondrocyte is in surface contact with another active chondrocyte. While it has been recognized that high density plating tends to produce cartilage tissue, previous attempts have been less than successful due to the failure to concentrate reproductively viable cells in sufficient density.

Kuettner has found that the chondrocytes of weight-bearing articular cartilage, as are found in the ankles and toes of adolescent mammals, have a very high viability and high degree of biological activity. In particular, calf fetlock cartilage contains a high percentage of reproductively and metabolically active chondrocytes. A growing calf gains weight very rapidly, and to maintain the calf's weight, the weight bearing limbs, such as the hooves and fetlocks, enlarge proportionally, and the cartilage tissue proportionally grows and expands.

Calf fetlocks are generally a waste product of slaughterhouses and may therefore be obtained with relative ease. After the animal is slaughtered, the fetlocks are transported to the laboratory. The skin is removed, the first phalangal joint is carefully opened under aseptic conditions and the synovial fluid is removed. The cartilage of the exposed joint is carefully shaved off so as to obtain as much cartilage tissue as possible and yet avoid contamination by other tissues such as bone. The shavings are collected in Ham's F-12 medium or any other tissue culture medium suitable for tissue survival but enriched in antibiotics and antifungal agents, e.g., 50 μg gentamycin and 5 μg of beta-amphotericin per ml.

The cartilage shavings contain both the chondrocytes and the extracellular matrix. In order that the cells may be cultured, the extracellular matrix must be removed. This is done enzymatically by various enzymes which digest the proteins of the extracellular matrix. The shavings are first digested with pronase in an amount between about 0.5 and 2.0% and most preferably about 1% w/v in Ham's F-12 medium or other suitable tissue culture medium containing 5% bovine serum for 90 minutes. The pronase will digest most of the proteins of the extracellular matrix including proteoglycans and some collagen. The shavings after washing and centrifugation are further digested in medium containing 5% bovine serum and between about 0.1 and 1% w/v collagenase for a period of about 1 to 2 hours to digest the remaining collagen from around the cells.

The cells, after washing and centrifugation, are further digested with between about 0.1 and 3% w/v trypsin or testicular hyaluronidase and most preferably about 0.25% trypsin or hyaluronidase for between about 5 and 10 minutes in the presence of 5% bovine serum. The trypsin digests any remaining extracellular proteins and testicular hyaluronidase digests the sugar moieties attached to any remaining extracellular protein, thereby denuding the chondrocytes.

Mammalian serum, preferably fetal serum, is then added to the digestive mixture so that the denuded cells may be coated and protected with serum immediately after being denuded.

The isolated cells are collected by centrigugation at about 900 rpm for 10 minutes. The cell pellet is washed in phosphate buffered physiological saline pH 7.4 as described in *J. Cell Biol.* 49: 451(1971), and the cells are suspended in Ham's F-12 medium without serum. The cells are then passed through a 90 μm Nitex screen to separate the cells which are resuspended in Ham's F-12 medium containing about 5 to 10% serum. Sufficient cells are added to the medium to achieve a concentration of at least 10^6 cells per ml.

In order to produce cultured material in large quantities, it is desirable to use roller bottles. The application of culturing techniques to roller bottle technology is necessary if a culture is to be mass produced. Roller bottles have large surface areas and permit easy exchange of the medium and gases. Because of the sensitivity and difficulty of culturing chondrocytes, roller bottle technology was previously considered unsuitable for the culturing of cartilage tissue. However, it has been found that good growth of cartilage tissue may be achieved in roller bottles from cells denuded according to the above methods.

The cultures are maintained in Ham's F-12 medium supplemented with the antibiotic and antifungal agents described above as well as with 50 μg of ascorbic acid per ml and 15% bovine serum. The cells are cultured at 37°C in a humidified atmosphere containing at least 5% carbon dioxide. The medium is changed every other day. After about 10 days, nodules appear which contain extensive extracellular matrix chemically identified as being cartilaginous in nature.

AIF may be extracted from the chondrocyte cultures as well as the collected growth media according to the methods described in U.S. Patent 4,042,457.

AMINO ACIDS

Enzymatic Hydrolysis of Racemic Hydantoins to Give Optically Active Amino Acids

Racemic forms of compounds having the general formula (1) below may be subjected to enzymatic hydrolysis by hydropyrimidine hydrolase extracted from calf liver, or hydropyrimidine hydrolase produced by microorganisms of the Pseudomonas genus to produce compounds of general formula (2):

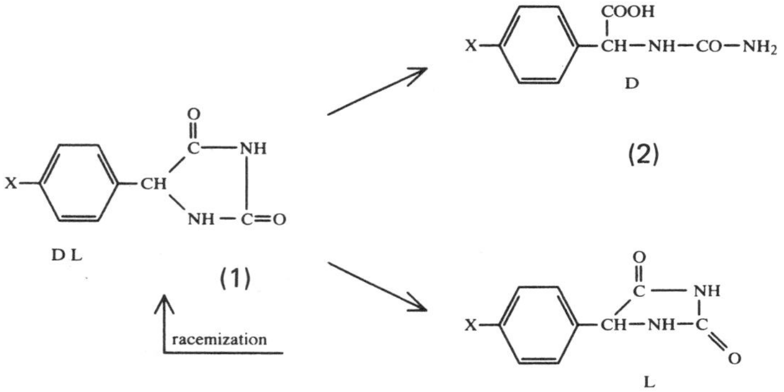

where X may be –H, –OH or $-OCH_3$.

A. Viglia, E. Fascetti, E. Perricone and L. Degen; U.S. Patent 4,248,967; Feb. 3, 1981; assigned to Snamprogetti SpA, Italy have found that this reaction may also be accomplished by hydrolysis using thermophilic microorganisms of the Bacillaceae family. These hydrolysis reactions can be conducted at such comparatively high temperatures as those from 40° to 60°C.

The strains found useful for the hydrolysis of the hydantoins are *Bacillus brevis* and *B. stearothermophilus.* Two suitable strains have been designated NRRL B-11079 and B-11080. The hydrolysis of the hydantoins does not take place exclusively in the presence of microorganisms in their phases of growth or in the presence of intact cells thereof or the relative spores, but also in the presence of extracts of the microorganisms mentioned above. The microorganisms, for example, can be cultured in a liquid nutrient medium to obtain an accumulation of hydrolase in the cells and the D,L-hydantoins can be added at a subsequent time to the broth-cultures.

The enzymic hydrolysis can also be conducted with the resulting cells. In the latter case, the bacterial cells recovered from the broth-culture and thoroughly washed are slurried in a system which is appropriate and has been properly buffered and the racemic hydantoin is added to it.

A further technical and economical improvement can be achieved by immobilizing the enzyme through combinations thereof with macromolecular compounds via the formation of chemical bonds with the matrix, or via bonds of the ionic type, or by physical immobilization.

Continuous Enzymatic Conversion of α-Ketocarboxylic Acids into Corresponding Amino Acids

A process for the enzymatic conversion of water-soluble α-ketocarboxylic acids into their corresponding amino acids in a membrane reactor has been refined by *C. Wandrey, R. Wichmann, W. Leuchtenberger, M.-R. Kula and A. Bückmann; U.S. Patent 4,304,858; December 8, 1981; assigned to Degussa AG, Germany.*

In their process, the membrane has a mean pore diameter of 1 to 3 nm and there is present a solution of the formate dehydrogenase, the substrate specific dehydrogenase and from 0.1 to 10 mmol/ℓ of NAD^+/NADH present bound to a polyoxyethylene having an average molecular weight between 500 and 50,000 and preferably between 1,200 and 10,000. An aqueous solution is continuously supplied which has 50 to 100% of the maximum amount soluble, but not over 2,000 mmol/ℓ, of the α-ketocarboxylic acid to be converted in the form of a water-soluble salt as substrate; an amount of ammonium ion about equimolar to the substrate amount; and from 100 to 6,000 mmol/ℓ of a formate. A differential pressure over the membrane of 0.1 to 15 bars is maintained and behind the membrane a filtrate stream containing the amino acid formed is continuously drawn off. The pressure is preferably 0.2 to 3 bars.

The process permits water-soluble α-ketocarboxylic acids to be converted continuously and with high space-time yields into the corresponding aminocarboxylic acids and is therefore useful for a cost-favorable production of these amino acids.

The membrane reactor contains a solution of a formate dehydrogenase, a substrate specific dehydrogenase and NAD^+/NADH greatly enlarged in molecular weight. The formate dehydrogenase is suitably employed in such an amount that its activity is at least 12,000 μmol/ℓ-minute. Upwardly the amount of its addition suitably should be so limited that the protein concentration is maximally about 20 g/ℓ. The substrate specific dehydrogenase is suitably added in such an amount that the ratio of the activities of formate dehydrogenase and substrate specific dehydrogenase is between 1:1 and 1:5.

The required NAD^+/NADH as coenzyme in the process must be enlarged to such an extent in molecular weight through bonding to a polyoxyethylene that it is still water-soluble, in order to permit a homogeneous catalysis, but is still held back by the membrane.

The coenzyme having an enlarged molecular weight is added in such an amount that the concentration of NAD^+/NADH is preferably 1 to 7 mmol/ℓ.

Suitable formate dehydrogenases for carrying out the process can be isolated, for example, from *Candida boidinii* or from *Pseudomonas oxalaticus*. An example of a substrate specific dehydrogenase usable in the process is the frequently employed L-alanine dehydrogenase which can be obtained from *Bacillus subtilis*. Using this enzyme, for example, pyruvic acid can be converted into L-alanine, 2-oxo-4-methylvaleric acid into L-leucine, 2-oxo-3-methylvaleric acid into L-isoleucine, 2-oxo-3-methylbutyric acid into L-valine or 2-oxo-valeric acid into L-norvaline. Suitably the reacting α-ketocarboxylic acids are employed in the form of their sodium or potassium salts, e.g., sodium pyruvate or potassium 2-oxovalerate.

Since in the conversion of α-ketocarboxylic acids, e.g., α-ketoalkanoic acids into the corresponding aminoacids, e.g., aminoalkanoic acids an optically active center is newly formed, the product concentration in the filtrate stream can be measured continuously with the help of a polarimeter. The aminoacid formed can be obtained from the filtrate in known manner, e.g., with the help of an acidic ion-exchanger.

In further work, *C. Wandrey, R. Wichmann, W. Leuchtenberger, M.-R. Kula and A. Bückmann; U.S. Patent 4,326,031; April 20, 1982; assigned to Degussa AG, Germany* have found that the use of the same type of membrane reactor and the same process, with one exception, can be used to convert ketocarboxylic acids to their corresponding α-hydroxycarboxylic acids. The exception is that a different substrate-specific dehydrogenase is used, in this case D- or L-lactate dehydrogenase. The preferred polymer to be united to the nicotinamide-adenine-dinucleotide is polyethylene glycol. Other polymers useful for increasing the molecular weight of the NAD^+/NADH are dextran, polyether polyol, polyethyleneimine, polyacrylamide or a methyl vinyl ether-maleic anhydride mixed polymer.

By this method, for example, pyruvic acid can be converted into lactic acid, phenyl pyruvic acid into L- or D-phenyl lactic acid, 2-oxo-4-methylvaleric acid into L- or D-2-hydroxy-4-methylvaleric acid, 2-oxo-3-methylvaleric acid into L- or D-2-hydroxy-3-methylvaleric acid, or 2-oxo-3-methylbutyric acid into L- or D-2-hydroxy-3-methylbutyric acid or 2-oxo-valeric acid into L- or D-2-hydroxyvaleric acid.

Production of D-Amino Acids from Hydantoins and/or Racemic Carbamoyl Derivatives

R. Olivieri, A. Viglia, L. Degen, L. Angelini and E. Fascetti; U.S. Patent 4,312,948; January 26, 1982; assigned to Snamprogetti SpA, Italy describe a process for producing D-aminoacids starting from racemic mixtures of their N-carbamoyl derivatives or from the corresponding hydantoins by employing stereospecific enzymic complexes obtained from microorganisms of the Agrobacterium genus.

A few D-aminoacids, especially phenylglycine and p-hydroxyphenylglycine are important intermediates for the preparation of compounds which are widely used in the pharmaceutical industry.

The chemical methods previously used for the separation of the optical antipodes which are based on the use of camphosulfonic acid are very expensive and give low yields. Another method consists in hydrolyzing the D-acylamino acid with the D-acylase enzyme, but this method produces a product having a poor optical purity.

This new procedure is characterized by the fact that D-aminoacids can be obtained starting from racemic compounds having the following general formula:

$$\begin{array}{l} R-CH-NH-CO-NH_2 \\ \quad\;\; | \\ \quad\;\; COOH \end{array}$$

in which R is a member selected from the group consisting of substituted and unsubstituted aliphatic and aromatic radicals, the enzyme carbamoylase being stereoselective towards the D-forms. The enzymic preparations used are completely devoid of any activity on the L-enantiomeride. This circumstance permits that D-aminoacids having an absolute optical purity may be obtained.

The enzymic complexes used in this process are produced by microorganisms of the Agrobacterium genus, isolated from agricultural land and identified as 1302, 1303 and 1304. Strain 1302 has been deposited with the Northern Regional Research Center where it has been given the accession number NRRL B 11291.

Microorganisms of the genus Agrobacterium are cultured under aerobic conditions in a conventional culturing medium which contains sources of nitrogen, carbon, phosphorus, and mineral salts at a temperature preferably between 25° and 35°C for a period of preferably between 20 and 30 hours, at a pH preferably of 7 to 7.5.

D(−)aminoacids are directly produced in the fermentation media which contain DL-hydantoins or DL-N-carbamoyl derivatives of aminoacids, both as single sources of nitrogen and integrated with the usual nitrogen sources. D(−)aminoacids can also be produced by directly using the microbial slurry as resting cells or using extracts of it.

The extraction, from the bacterial slurry, of the enzyme complexes takes place with the usual methods adopted in enzymology. For this purpose, the cells are disintegrated with appropriate apparatus, e.g., French Pressure Cell Press, Manton Gaulin Homogenizer, rotary disintegrators and also with ultrasonic vibrators.

The hydrolysis of the hydantoins or of the N-carbamoyl derivatives of the aminoacids can be carried out by adding to the reaction mixture the enzyme in the following forms: fresh cells, freeze-dried cells, toluenized cells, acetonic powder or raw or purified extracts.

An additional technical and economical improvement can be achieved by immobilizing the enzymes by combination with macromolecular compounds by formation of chemical bonds with the matrix or bonds of ionic type or by physical immobilization.

Thus, D-aminoacids can be produced from racemic mixtures of their N-carbamoyl derivatives or from their corresponding hydantoins. Starting materials may thus be, e.g., DL-5-phenylhydantoin, which yielded D(−)phenylglycine; and 5-p-hydroxy-DL-phenylhydantoin and N-carbamoyl-DL-p-hydroxyphenylglycine, both of which yielded D-p-hydroxyphenylglycine upon treatment with enzymes prepared from Agrobacterium sp. NRRL-B 11291.

Use of Tryptophan Synthetase or Tryptophanase

There are presently several good processes for preparing tryptophan from indole and serine by enzymatic synthesis. The indole used can be produced industrially at relatively low cost. The serine for the process is most cheaply produced by an organic synthesis which, however, produces only DL-serine. Since only L-serine is useful in the enzymatic synthesis of tryptophan, the D-serine must undergo treatment under high pressure and temperatures to affect racemization, and the racemization product is used for the enzymatic reaction.

By repeating the enzymatic reaction and racemization alternately, DL-series can finally be converted to L-tryptophan substantially completely. This process, however, has various serious defects.

It is, therefore, a primary objective of *Y. Asai, M. Shimada and K. Soda; U.S. Patent 4,335,209; June 15, 1982; assigned to Mitsui Toatsu Chemicals, Inc., Japan* to provide a process for preparing L-tryptophan by reacting indole with L-serine in the presence of tryptophan synthetase or tryptophanase, in which DL-serine or D-serine can be effectively utilized for the preparation of L-tryptophan.

Their described process for the preparation of L-tryptophan comprises reacting indole with L-serine in the presence of tryptophan synthetase or tryptophanase, wherein indole and DL-serine or D-serine are used and a serine-racemizing enzyme is included in the reaction system to convert at least a part of D-serine to L-serine.

The serine racemizing enzyme that is used in the process will be referred to as serine racemase hereinafter. As the serine racemase-producing microorganism, the following can be mentioned, for example, *Pseudomonas putida* IFO 12996, *Brevibacterium leucinophagum* MT-10072, *Pseudomonas desmolytica* MT-10170, *Pseudomonas fragi* MT-10173 and *Pseudomonas taetorolens* MT-10186.

Extraction of the low substrate specificity amino acid racemase from *Pseudomonas putida* IFO 12996, one of serine racemase producing strains, is accomplished by the process comprising the following steps: (1) crude extraction; (2) ammonium sulfate precipitation; (3) first chromatography using diethylaminoethyl cellulose; (4) second chromatography using diethylaminoethyl cellulose; (5) chromatography using Sephadex G-200 (Pharmacia Fine Chemicals Co.) and (6) crystallization.

As the strain producing tryptophan synthetase that is used in this process, there can be mentioned, for example, various strains of *Escherichia coli* and *Neurospora crassa* ATCC 14692.

As the strain capable of producing tryptophanase that is used in the process, there can be mentioned, for example, *Proteus vulgaris* IFO 3167, *Aerobacter aerogenes* IFO 12019, *Klebsiella pneumoniae* ATCC 8724 and *Bacillus alvei* ATCC 6348.

Any of the normal synthetic and natural culture media may be used for culturing strains producing tryptophan synthetase, tryptophanase or serine racemase, so far as it contains a carbon source, a nitrogen source, an inorganic substance and, if necessary, a small amount of a nutrient. In culturing the tryptophan synthetase producing strain, it is ordinarily necessary to add a minute amount of tryptophan, anthranilic acid or indole to a culture medium. In culturing the tryptophanase producing strain, since tryptophanase is an inducing enzyme, if tryptophan is added in an amount of about 0.1 to 0.5% by weight to a culture medium, cultured cells having a high tryptophanase activity can be obtained.

Best results are achieved if all the enzymes in this process are used in the form of immobilized enzymes.

MISCELLANEOUS ORGANIC CHEMICALS

Epoxides and Glycols from Alkenes

S.L. Neidleman, W.F. Amon, Jr. and J. Geigert; U.S. Patent 4,247,641; Jan. 27, 1981; assigned to Cetus Corporation describe a process for the production of epoxides and glycols from alkenes wherein an enzyme is used to produce an intermediate halohydrin and the halohydrin is converted to an epoxide or glycol by an enzymatic or chemical process.

This enzymatic halogenating process has several advantages including the use of the inexpensive, less dangerous, inorganic halide, rather than the elemental halogen, i.e., bromide ion rather than bromine; use of ambient temperature; and use of standard or close to standard atmospheric pressure. In additon, this enzymatic process involves the use of dilute H_2O_2, not necessarily purified. The H_2O_2 may be added directly or generated in situ by an enzymatic or chemical reaction. This reduces the cost of the H_2O_2 as compared to the cost of concentrated, purified material; increases the safe usage of the substance; and extends the life of the halogenating enzyme.

The olefins useful in the process can be broadly defined as any hydrocarbon containing a carbon to carbon double bond.

The process makes use of certain peroxidases which have catalytic activity with respect to breaking the double bond of olefin compounds and promoting hydroxylation on one of the carbons while promoting halogenation on the adjacent carbon under particular reaction conditions and are referred to herein as "halogenating enzymes."

A preferred halogenating enzyme is derived from the microorganism *Caldariomyces fumago*. Other sources of halogenating enzyme include seaweed, milk (lactoperoxidase), thyroid (thyroid peroxidase), leukocytes (myeloperoxidase) and horseradish (horseradish peroxidase). Certain of these peroxidases are commercially available.

The microorganism *Caldariomyces fumago* may be grown as a static or agitated, submerged culture in Czapek-Dox medium at room temperature for 3 to 10 days by conventional methods. The halogenating enzyme, chloroperoxidase, is prepared from an aqueous homogenate of the mycelial pads of the microorganism grown under static conditions or from the filtrate of the microorganism grown under static or agitated submerged culture conditions. It may also be used in an immobilized form.

In addition to the halogenating enzyme, a source of inorganic halide and an oxidizing agent are required in the reaction mixture. A preferred oxidizing agent is hydrogen peroxide, which may be added directly to the mixture in a single batch addition, or in a continuous slow feed, or may be generated as a slow feed in situ by the use of a hydrogen peroxide-producing enzyme system. Such enzyme systems are well known in the art, and include glucose oxidase in the presence of glucose, D- and L-amino acid oxidases in the presence of D- and L-methionine, methanol oxidase in the presence of methanol, etc.

With in situ generation of hydrogen peroxide using glucose oxidase or methanol oxidase, coproducts include gluconic acid (in the case of glucose oxidase) and formaldehyde (in the case of methanol oxidase). Although each of these coproducts is commercially useful, it is conceivable that if this process were adopted on a large scale for the production of epoxides and glycols, the amount of coproduct produced by in situ hydrogen peroxide generation could exceed market demand by a substantial amount. Under such circumstances, the process might be developed or modified in such a way as to produce a coproduct which is capable of being absorbed by relatively higher market demand.

For example, the enzyme used for the production of hydrogen peroxide in situ might be glucose-2-oxidase, which could be used to catalyze the reaction of D-glucose to D-arabino-2-hexosulose which could be readily converted to D-fructose by simple chemical hydrogenation—the process being that described in U.S. Patent 4,246,347.

S.L. Neidleman, W.F. Amon, Jr. and J. Geigert; U.S. Patent 4,284,723; Aug. 18, 1981; assigned to Cetus Corporation also detail the procedure for producing epoxides or glycols from gaseous alkenes. Here again an enzyme is used to produce an intermediate halohydrin.

Example: The process for producing propylene oxide from propylene in an integrated immobilized enzyme/cell system in a continuous-flow column configuration is demonstrated in this example.

One column configuration includes a first region which contains the initial mixture (50 ml) of reaction components: propylene gas continuously bubbled in, 0.42 M KBr, 0.01 M glucose, and phosphate buffer (pH 4.4, 0.3 M). The mixture is slowly dripped onto a column of immobilized chloroperoxidase (CP) and glucose oxidase (GO) prepared by mixing 5 ml CP-beads and 1 ml GO-beads. This is the first stage of the system to produce propylene bromohydrin.

The details of immobilizing glucose oxidase and chloroperoxidase follow: The glucose oxidase (1,460 units/ml) is purchased from Sigma Chemical Company. The insoluble beads of AH-Sepharose 4B are obtained from Pharmacia Fine Chemicals Company. Both enzyme and beads are adjusted to pH 5.0. To immobilize the enzyme onto the beads, 10 ml glucose oxidase and 10 ml beads are mixed. The coupling reaction is initiated by the addition of 2 ml N-cyclohexyl-N'-[2-(4-methylmorpholino)-ethyl] carbodiimide solution (100 mg/2 ml). The reaction mixture is incubated at 4°C overnight. The beads are then washed with 0.03 M phosphate buffer (pH 4.4). The glucose oxidase-AH-Sepharose 4B beads are stored at 4°C for use.

The chloroperoxidase is immobilized by mixing equal volumes dialyzed chloroperoxidase with hydrated DEAE-Sephadex A50 beads. The enzyme beads complex are washed three times with 0.03 M phosphate buffer (pH 4.4). The activity of the beads is estimated to be 2.17 units/ml with monochlorodimedon assay. The immobilized chloroperoxidase is stored at 4°C for future use.

The resulting eluate, containing excess reagents as well as propylene bromohydrins and gluconic acid, is adjusted to pH 6 with phosphate buffer (pH 6, 1.0 M) to allow for substantial activity with halohydrin epoxidase contained in the next column region. The immobilization of halohydrin epoxidase is performed by immobilizing the intact cells of *Flavobacterium* sp. on an acrylamide support.

The polymerized gel is then blended into beads with a Waring blender (low speed, 20 seconds). This is the second stage of the system, producing propylene oxide from the propylene bromohydrins of the first stage.

The results obtained are 75 μg/ml propylene bromohydrin and 5 μg/ml propylene epoxide at a flow rate through the column of 1 ml/hr.

It may be seen, therefore, that this process is capable of producing epoxides and glycols from gaseous alkenes with minimum energy requirements, relatively high safety, and relatively low cost.

Coproduction of Hydrogen Peroxide and 2-Keto-D-Gluconic Acid

In another process for production of hydrogen peroxide which yields a useful coproduct, *S.L. Neidleman, W.F. Amon, Jr. and J. Geigert; U.S. Patent 4,351,902; assigned to Cetus Corporation* have developed a like process in which the coproduct is 2-keto-D-gluconic acid.

2-Keto-D-gluconic acid has a variety of commercial uses. As the calcium salt, it is used in photography, principally in developer formulations. Also, it can be readily converted to other commercially useful products such as furfural, D-isoascorbic acid, D-arabinose and D-ribulose. It is also useful as a food preservative.

In the preferred form, the process utilizes D-glucose as a starting material. D-glucose may be obtained from any of a variety of relatively low cost sources. In the first step of the process, the D-glucose is reacted with air or oxygen at the second carbon to effect an enzymatically catalyzed oxidation. This first step produces the intermediate product D-glucosone and, in addition, yields hydrogen peroxide. This reaction is preferably conducted in an aqueous solution at about neutral pH, but can be conducted within the pH range of from about 3 to 8 with the use of appropriate buffers or other pH control. Preferably, the conversion is effected at ambient temperature, but can be conducted within a temperature range of from about 15°C to 65°C. Pressure conditions are preferably atmospheric pressure.

Known enzymes which have the capability of catalyzing this reaction are glucose-2-oxidase or pyranose-2-oxidase. Pyranose-2-oxidase is produced by the microorganism *Polyporus obtusus* and glucose-2-oxidase is produced by the Basidiomycete *Oudemansiella mucida*. The microorganisms may be grown in agitated submerged culture at room temperature by conventional methods. Further details of the sources and preparation of these enzymes are described in the previous patent.

Preferably, the enzymes are used in an immobilized form, although free enzymes can also be used.

Reactions employing immobilized enzymes may be run in columns or reaction tanks or other suitable reactors.

After producing the intermediate product D-glucosone, a further reaction takes place wherein the D-glucosone is oxidized at the first carbon in an enzymatically catalyzed oxidation to form hydrogen peroxide and 2-keto-D-gluconic acid. The enzyme glucose-1-oxidase, produced, for example, from the microorganism *Aspergillus niger*, reacts with D-glucosone to yield 2-keto-D-gluconic acid and hydrogen peroxide. The pressure, temperature and pH ranges for the second step of the reaction are essentially the same as those described in connection with the first.

The result of the two-step conversion process is that two mols of hydrogen peroxide and one mol of 2-keto-D-gluconic acid are produced for every mol of D-glucose converted.

A second embodiment of the process is also described. As was the case in the first embodiment, the coprocess is a two-step process. In the first step, the first carbon of D-glucose is oxidized by utilizing the enzyme glucose-1-oxidase under conditions essentially the same as those described in connection with the first embodiment. The result is the production of the intermediate product D-glucono-δ-lactone.

The second step of this embodiment operates to oxidize the second carbon of the D-glucono-δ-lactone in an enzymatic oxidation in which the enzyme pyranose-2-oxidase is employed. The reaction, because D-glucono-δ-lactone is an acidic substance, requires pH control and, therefore, buffering in order to keep the reaction mixture closer to a neutral pH for the reaction with the pyranose-2-oxidase. The resulting product is a mixture of 2-keto-D-gluconic acid and D-isoascorbic acid. In both enzymatic oxidation steps of this embodiment hydrogen peroxide is also produced. Therefore, for every mol of D-glucose utilized, two mols of hydrogen peroxide are produced as well as one mol equivalent of a mixture of 2-keto-D-gluconic acid and D-isoascorbic acid. In both embodiments, air or oxygen is added for both the first and second step reactions.

D-Isoascorbic acid is a commercial product useful as an antioxidant for food products, or as an intermediate for producing other useful end products such as furfural. In fact, the mixture of the 2-keto-D-gluconic acid and D-isoascorbic acid can be converted to furfural without separation of the two products.

Ascorbic Acid from Lactose

J.P. Danehy; U.S. Patent 4,259,443; March 31, 1981; assigned to Bernard Wolnak and Associates, Inc. has developed a procedure for synthesizing ascorbic acid from lactose, rather than from glucose, the usual commercial starting material.

Utilizing lactose as the starting material in the synthesis of ascorbic acid provides an inexpensive initial reagent for the production of that vitamin. Deproteinizing whey readily provides a source of lactose. In particular, whey represents an underutilized by-product in the production of cheese and thus has minimal cost as an initial reagent in the ascorbic acid synthesis.

The synthesis begins with the hydrolysis of lactose to produce a first mixture of D-galactose and D-glucose. Generally, this reaction proceeds efficiently with the assistance of the enzyme lactase.

The mixture of D-galactose and D-glucose then undergoes oxidation to produce D-galacturonic acid and D-glucuronic acid in a further mixture. Where the oxidation process may attack the reactants' terminal hydroxyl groups, they may receive the protection of an alkyl or aryl group to form their glycosides. After the oxidation, these additional groups must undergo removal through hydrolysis to leave the desired acids. An oxidizing agent, such as permanganate, can effect the actual oxidation.

The process further proceeds with the reduction of the D-galacturonic and D-glucuronic acids, while mixed together, to provide L-galactonic and L-gulonic acids respectively. Removing the water from a solution containing the latter then forces their conversion to a mixture of their corresponding γ-lactones.

The reverse reaction, which would reconvert the γ-lactones to the acids, proceeds at a relatively slow rate. Accordingly, the γ-lactones may undergo dissolution in water in preparation for the last step. Finally, then, the γ-lactones of L-galactonic and L-gulonic acids, while mixed together, are oxidized to ascorbic acid. Specifically, the oxidation of these acids produces 2-keto-L-galactonic acid and 2-keto-L-gulonic acid. Both then rearrange to ascorbic acid.

The actual oxidation process may conveniently take place in the presence of an enzyme obtainable from animal or vegetable tissue. The enzyme should have the capability of transferring electrons from the γ-lactones to molecular oxygen. In particular, pea seeds can provide the enzyme for this second oxidation step. To obtain it, 30 g of pea seeds are soaked in water at 25° to 30°C for about 12 to 48 hours. They then undergo crushing in a cooled aqueous solution, preferably at 0° to 4°C in 40 ml of an aqueous solution having 0.4 M sucrose, 0.1 M sodium phosphate, a pH of 7.5, and 0.1% by weight of magnesium sulfate. Centrifugation then removes starch and cellular debris from the mixture. The turbid supernatant solution contains very fine particles in which the enzymatic activity resides. Its pH value is adjusted to about 7.4 to 7.5.

The equimolar mixture of the γ-lactones of L-galactonic acid and L-gulonic acid as previously described above, receives the supernate containing the enzyme. This solution then remains at about 36° to 38°C for about 4 to 6 hours. Adding additional amounts of lactones will permit the production of further product.

The oxidation with the enzyme actually produces, initially, 2-keto-L-galactonic acid and 2-keto-L-gulonic acid. However, both of these rearrange rapidly to give ascorbic acid. The reverse rearrangements occur only to a minimal, if appreciable, extent. Thus, ascorbic acid represents the discernible product from the oxidation reactions.

The solution containing the ascorbic acid undergoes filtration to clarify it. It should then be evaporated in a rotary evaporator heated in a water bath at 60°C and at a negative partial pressure of around 30 mm Hg until the volume reduces to about one-fifth of its original. The resulting syrup then cools to room temperature and is thoroughly mixed with an equal volume of acetone. The crystals of ascorbic acid, which have appeared by this point, are recovered by filtration. Cooling the filtrate to 0°C permits the recovery of the additional product.

Stereoselective Resolution of DL-Phenylglycine Derivatives by Enzyme Resins

H. Schutt; U.S. Patent 4,260,684; April 7, 1981; assigned to Bayer AG, Germany has devised a new enzymatic process for the stereoselective resolution of a DL-phenylglycine derivative. He has found that such enzymatic resolution of phenylglycine derivatives can be carried out at a substantially higher concentration when N-acyl-DL-phenylglycine esters or amides, as the starting material, are subjected to enzymatic hydrolysis, in solution, on enzymes bonded to carriers.

The compounds to be resolved correspond to the general formula (1) below:

(1)

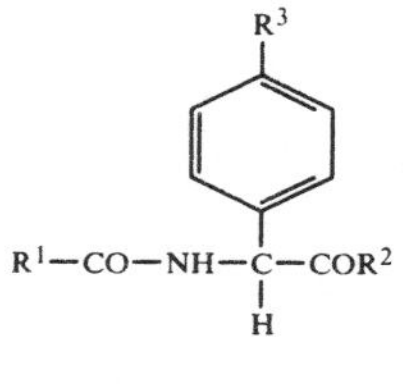

DL-Form

Preferred compounds of the formula (1) are, on the one hand, those in which R^1 denotes hydrogen, mono-, di- or trichloromethyl, trifluoromethyl or ω-carboxy-C_2-C_6-alkyl, R^2 denotes methoxy or ethoxy and R^3 denotes hydrogen; and, on the other hand, those in which R^1 denotes hydrogen, methyl, mono-, di- or trichloromethyl, trifluoromethyl or ω-carboxy-C_2-C_6-alkyl, R^2 denotes methoxy or ethoxy and R^3 denotes hydroxyl, methoxy, ethoxy or acetoxy.

Possible solvents for these esters and amides are anhydrous or water-containing organic solvents, in particular water-miscible solvents such as dioxane, ethanol and acetonitrile.

Possible enzymes are, in particular, proteolytic enzymes, preferably serine proteases and sulfhydrylproteases, preferably subtilin, α-chymotrypsin, papain, ficin or bromelin. Subtilin is preferred. About 100 mg to 1 g of enzyme bonded to a carrier are preferably employed per mol of compound to be resolved, and this enzyme can be reused at least twenty times.

Proteolytic enzymes which are isolated from *Bacillus subtilis* and *Bacillus licheniformis* and which are added to washing agents to remove protein residues are particularly suitable. These industrial enzymes are known chiefly as Maxatase (Gist-Broacdes NV), Optimase (Miles-Kali-Chemie) and Alcalase (Novo Industrie AS).

Particularly preferred carriers are a copolymer of a methacrylate, methacrylic acid and maleic anhydride, optionally crosslinked by glutardialdehyde, and especially in the case of subtilin, α-chymotrypsin, papain, ficin and bromelin, a copolymer of tetraethylene glycol dimethacrylate, methacrylic acid and maleic anhydride, optionally crosslinked by glutardialdehyde.

Industrial processes using enzymes bonded to carriers are particularly economical if the processes can be carried out at high concentrations of substrate. In a high concentration in aqueous dioxane, ethanol or acetonitrile, the N-acyl esters and amides of phenylglycines are in the form of clear solutions. It is possible to obtain a 10% strength solution of N-acetyl-DL-phenylglycine methyl ester in a dioxane/water mixture (1:3, by volume).

Enzymatic resolution of the N-acyl-DL-phenylglycine esters is preferably carried out at a temperature of 20° to 40°C in a pH range of 6 to 8, the pH value preferably being kept constant at 7.8 by adding a strong base. The substrate is stable under the reaction conditions. The course and the end point of the enzymatic reaction can be determined by neutralization of the H^+ ions formed. Neutralization can be effected by inorganic bases as well as by organic bases.

After the enzymatic reaction has ended, the products, i.e., the N-acyl-D-ester or the N-acyl-D-amide and the N-acyl-L-acid of the phenylglycines, are extracted from the aqueous reaction medium by shaking or stirring with organic solvents. The D-esters or D-amides are separated from the L-acids in the customary manner by rendering the reaction solution alkaline and extracting the esters or amides with an organic solvent, such as chloroform, ethyl acetate, methylene chloride or butyl acetate. The aqueous phase which remains is then rendered acid, for example, with sulfuric acid, and again extracted with ethyl acetate.

D-Phenylglycine and D-4-hydroxyphenylglycine are used as starting substances for the preparation of semisynthetic antibiotics of the penicillin series. L-Phenylglycine is a starting substance for L-aspartyl-L-phenylglycine methyl ester, which is used as a sweetener.

Aldonic Acids by Enzymatic Oxidation from the Aldose

It is known to oxidize oses to the corresponding aldonic acids by an enzymatic method, by employing the enzymes which are specific for these reactions and are generally referred to as oxidases. Thus, glucose is oxidized to gluconic acid with the aid of glucose oxidase. Although very selective, this reaction nevertheless exhibits the disadvantage that it gives a low degree of conversion per unit of enzyme employed, and this results in a very low productivity and makes the process rather uneconomic.

The objective of *G. Coppens; U.S. Patent 4,345,031; August 17, 1982; assigned to Solvay & Cie., Belgium* was to provide a process for the manufacture of aldonic acids by oxidizing the corresponding oses, which process does not exhibit the disadvantages of the conventional processes and makes it possible, in particular, to improve the productivity substantially. For this purpose, the molar ratio of oxygen to ose in the reaction mixture is kept above 0.1. A molar ratio of oxygen to ose of more than 0.2 is preferably used. Good results have been obtained when this molar ratio is between 0.2 and 50. Molar ratios of more than 50 are also suitable but are less advantageous from an economic point of view.

The oxygen used to oxidize the ose to the corresponding aldonic acid can be employed in various forms. In general, oxygen gas is dissolved in the reaction mixture. The oxygen can be employed in the form of pure oxygen or in the form of mixtures of oxygen with inert gases. In general, a gaseous mixture containing essentially oxygen, preferably in a proportion of at least 90% by weight, is used.

The amount of oxygen dissolved in the medium is generally more than 20 mg/ℓ. Advantageously, the conditions used are such that the amount of oxygen dissolved in the reaction medium is between 40 and 10,000 mg/ℓ. The oxygen partial pressures are thus fairly high and are generally at least 0.5 bar. They are preferably between about 1 and 100 bars. Good results have been obtained by using pressures of 3 to 80 bars.

The process applies to the oxidation of various types of oses. Thus, it is possible to oxidize glucose, galactose and lactose. Excellent results have been obtained by applying the process to the oxidation of glucose to gluconic acid.

Oxidation of oses to the corresponding aldonic acids is carried out by an enzymatic method with the aid of the oxidases which are specific for the reactions in question. The enzymes can be in various forms, i.e., either in the form of free enzymes or in the form of immobilized enzymes or also contained in cells, the latter being the case of, for example, microorganisms. Good results have been obtained when employing immobilized enzymes.

The amount of enzyme employed can vary within wide limits. It depends on the nature of the enzyme, the technical grade used and also the form in which it is employed (namely whether or not it is immobilized). In the case of the oxidation of glucose, the glucose oxidase is generally used at a rate of between 1 and 1,000 I.U./g of glucose to be converted.

The reactions involving oxidation of oses to corresponding aldonic acids are generally carried out in a solvent medium. Various solvents can be used for this purpose. Water, by itself or mixed with other solvents which are inert with respect to the reaction, such as alcohols, is generally used as the solvent. Good results have been obtained by using water as the solvent.

In addition to the reactants and the other constituents which may be present in the starting ose, the reaction products, the solvents and the enzyme, the reaction mixture can also contain other constituents. Among the latter, there may be mentioned agents for adjusting the pH, agents for separating off or neutralizing the aldonic acids formed, bactericidal or fungicidal agents such as ethylene oxide (for controlling possible bacterial pollution) and additives which make it possible to improve the activity of the enzyme. If necessary, agents for removing the hydrogen peroxide formed can also be added. Catalase is very suitable for this purpose.

When employed in the immobilized form, the enzymes can be arranged in the reactors in the form of a fixed bed or a fluidized bed or also in the form of a moving bed or a turbulent bed. The immobilized enzymes can be completely immersed in the liquid reaction mixture or it is also possible to cause the latter to run over the immobilized enzymes.

When the enzymes are employed in the free form, mixer reactors are generally used. If they operate continuously, these reactors are generally provided with devices for preventing the enzyme from leaving the reactor during the removal of production. These devices can be filters used for ultrafiltration. After the reaction, the enzymes can be separated from the reaction mixture in accordance with various techniques which are in themselves known, such as decantation or filtration.

Glyceraldehyde from Glycerol with Methanol Dehydrogenase

Glyceraldehyde is an industrially important chemical compound used by manufacturers of cosmetic products. It is currently produced by an osmium tetroxide-catalyzed conversion of acrolein to glyceraldehyde; however, a problem of osmium recovery is encountered with this process.

Microorganisms capable of oxidizing a hydroxyl group of a polyalcohol molecule to form a keto compound are well documented, for example, glycerol to dihydroxyacetone. However, this oxidation reaction is limited to the secondary alcohol function of the substrate.

H.J. Wolf; U.S. Patent 4,353,987; October 12, 1982; assigned to The Upjohn Company has devised processes for preparing glyceraldehyde without the use of potentially carcinogenic reagents. This desirable feature is achieved by use of a natural enzyme, methanol dehydrogenase, acting upon glycerol. The enzyme can be used in the pure or crude forms. Further, a microbial process using a methanol dehydrogenase-producing microbe can be the method for contacting the enzyme with glycerol to produce glyceraldehyde.

Species of several genera are suitable to use as a source of methanol dehydrogenase in the microbiological conversion of glycerol to glyceraldehyde. These genera are included in the family Methylomonodaceae described below. Some members of the genus Pseudomonas (family Pseudomonoadaceae) that can utilize methanol as the sole carbon and energy for growth can also be used. The preferred microorganism is *Methylobacterium organophilum,* NRRL B-12486, ATCC 27886.

The microorganism is grown in a conventional aqueous mineral salts medium using methane or methanol as the carbon source. Additionally, it can be grown on mono- and disaccharides, tricarboxylic acid cycle intermediates, or broth rich in organic nutrients. Methanol is the preferred substrate.

The organisms require oxygen present in the atmosphere for growth. The temperature range for growth is 5° to 55°C with an optimum of 28° to 31°C for *M. organophilum.* The organisms will grow in media having a pH in the range of 5.5 to 8.5, preferably around neutral pH.

Methanol dehydrogenase is synthesized in high levels by cells grown on methane or methanol. To increase the low enzyme levels found in cells grown on more complex organic substrates, methanol is added to the culture medium at any stage of growth as an inducer of enzyme synthesis.

Several types of preparations can be used to supply methanol dehydrogenase activity for the bioconversion, including whole cells, cell-free extracts or partially purified enzyme. Whole cell preparations may be used directly from the growth medium, harvested, stored by freezing or drying or immobilized. The cells may be immobilized by standard techniques, such as entrapment within polyacrylamide or covalent coupling of the cells to a polyelectrolyte carrier.

The bioconversion of glycerol to glyceraldehyde is accomplished by exposing the enzyme methanol dehydrogenase to glycerol in an aqueous medium. The medium is a weakly buffered solution with a pH in the range of 5 to 11 that contains 0 to 20% glycerol. The bioconversion mixture is incubated for 0 to 7 days at a temperature in the range of 5° to 50°C. During incubation, the mixture has access to atmospheric oxygen and is preferably stirred. Peak production of glyceraldehyde occurs early during the bioconversion. Conversion of 35% of the glycerol present has been observed.

ENZYMATIC PRODUCTION OF NATURAL AND ARTIFICIAL SWEETENERS

PRODUCTION OF FRUCTOSE FROM GLUCOSE

Recovery of Crystalline Fructose

The production of fructose has until now not been commercially feasible because of the expense of separating the residual glucose from the fructose-rich syrup produced by the known methods of making the fructose from glucose or sucrose.

S.L. Neidleman, W.F. Amon, Jr., and J. Geigert; U.S. Patent 4,246,347; January 20, 1981; assigned to Cetus Corporation describe an economical process for producing crystalline fructose from glucose by way of a two-step process: enzymatic oxidation followed by chemical hydrogenation.

According to the process, generally stated, glucose in aqueous solution is enzymatically converted to D-glucosone with an appropriate enzyme such as carbohydrate oxidase or glucose-2-oxidase. This conversion is allowed to proceed spontaneously, rapidly and substantially completely. D-glucosone, without prior isolation, is then converted to fructose by suitable chemical hydrogenation. This conversion is also allowed to proceed rapidly and essentially completely. The resulting fructose, substantially free of glucose and all other saccharides, is recovered either as an aqueous solution or in solid form.

A preferred carbohydrate oxidase enzyme is derived from the microorganism *Polyporus obtusus.* Sources of glucose-2-oxidase include several other microorganisms such as the two molds *Aspergillus parasiticus* and *Aspergillus flavus-oryzae,* mollusca and red alga.

The enzyme is preferably used in an immobilized form, although free enzyme can also be used. The processes for enzyme immobilization are familiar to those skilled in the art, and consist of reacting a solution of the enzyme with one of a broad range of surface treated or untreated organic and inorganic supports. Included among these are polyacrylamide, ethylenemaleic acid copolymers, methacrylic-based polymers, polypeptides, styrene-based polymers, agarose, cellulose, dextran, silica, porous glass beads, charcoal or carbon black wood and sawdust, hydroxy apatite and aluminum or titanium hydroxide. Enzymes in this form have increased stability, extended life and usefulness and recoverability. Reactions employing immobilized enzymes may be run in columns or reaction tanks or other suitable reactors.

In addition to the carbohydrate oxidase or glucose-2-oxidase enzymes, a source of oxygen is needed. Also, a method of hydrogen peroxide removal or utilization is required in the reaction to convert glucose to D-glucosone most efficiently. This is because H_2O_2 oxidizes certain critical sites on the enzyme molecule, damaging its function. Procedures of hydrogen peroxide removal include (1) decomposition by the enzyme, catalase, (2) decomposition by known chemical means, and (3) decomposition by using decomposing matrices such as manganese oxide or carbon black as the immobilizing support for the oxidoreductase enzyme. In a preferred alternative method, the hydrogen peroxide produced, rather than being decomposed, may be consumed to produce a valuable coproduct. The enzymatic conversion of glucose (from any suitable source) is preferably conducted in water at neutral pH, and ambient temperature and pressure.

After conversion of substantially all of the glucose to D-glucosone, the D-glucosone is readily and essentially completely converted to fructose using molecular hydrogen and an appropriate catalyst, preferably palladium or nickel.

Glucose Isomerase Immobilized by Adsorption onto Colloidal Silica

Since glucose isomerase was discovered as a product of microorganisms, the industrial production of glucose isomerase has been increasing abruptly especially in recent years. Glucose isomerase is the general name for the enzyme class which converts glucose (dextrose) into fructose (levulose). Glucose isomerase is mainly used for the production of levulose-bearing syrup (LB syrup) from glucose on an industrial scale.

At present, industrial production of LB syrup is conducted by keeping glucose in contact with microbial cells containing glucose isomerase, at a temperature between about 60° and 70°C for about 2 to 3 days. When using a batch system however, this conventional method has the disadvantage that the efficiency of utilization of the enzyme is low. Another disadvantage is that product purification costs are expensive, because coloration occurs during the reaction conducted at high temperatures for a long time. Furthermore, this method requires expensive equipment. In order to solve these problems, good methods for the preparation of an immobilized glucose isomerase are being sought and various methods have been developed.

S. Enokizono and S. Ushiro; U.S. Patent 4,252,899; February 24, 1981; assigned to CPC International Inc. have developed (1) a process for immobilizing glucose isomerase which comprises adsorbing a glucose isomerase in a colloidal silica by contacting the enzyme with the colloidal silica, then solidifying the glucose isomerase adsorbed silica by freezing, as it is after gelatinization; (2) a process for immobilizing glucose which comprises contacting continuously a glucose-containing solution with the immobilized glucose isomerase as defined in (1) to isomerize the solution.

The colloidal silica to be used in the process covers all colloidal silica. For example, the Ludox products of Du Pont—HS-30, HS-40, AM and TM; and the Snowtex products of Nissan Chemical Co.—Snowtex 20, 30 and N, may be used.

Any available glucose isomerase enzyme may also be used such as those prepared from the microorganism *Lactobacillus brevis* and that originating from the cells of the ray fungi, *Streptomyces olivochromogenes.*

This enzyme can be used in three different forms; namely, (1) crude glucose isomerase extracted from the cells of the glucose isomerase producing microorganisms either by autolysis or by ultrasonic treatment, and separated from the cellular debris, (2) partially purified glucose isomerase treated with protamine to precipitate the nucleic acids, (3) crystalline glucose isomerase obtained by crystallization from partially purified glucose isomerase after fractionation with ammonium sulfate. One unit of enzyme activity is defined as the amount of enzyme which forms 1 mmol fructose in one minute when incubated with a 0.1 M glucose solution in the presence of 0.01 M $MgCl_2$ and 0.001 M $CoCl_2$ (CPC unit). The process is illustrated by the following examples.

Example 1: 100 ml of commercial colloidal silica, Ludox HS-30 (Du Pont) was introduced into a beaker (200 ml volume) and was diluted with 160 ml of deionized water. The pH of this solution was then adjusted to 7.0 with 1 N HCl. After that, 35 ml of enzyme solution containing 41,500 units of glucose isomerase was added to the solution. The enzyme solution had been prepared from the cells of *Streptomyces olivochromogenes,* liquid cultured for 50 hours, through digestion with lysozyme and partial purification using isopropyl alcohol.

Next, the mixture was gently agitated at room temperature for 15 minutes. Then 2 ml of 25% commercial glutaraldehyde solution was added to the resulting solution. The solution was then gently agitated at room temperature for 60 minutes. Then the mixture was placed in a freezer kept at a temperature of -20°C. After 12 hours, the frozen mixture was taken out of the freezer and was stirred at room temperature. The mixture was separated in the beaker into two portions, a solid phase and a liquid phase. The mixture was then placed again in the freezer for 12 hours. Then, the mixture was stirred again at room temperature, and was separated again into two portions, the solid phase and the liquid phase. The liquid phase was removed from the mixture by decantation. Forty grams of the solid material was obtained. The solid material which contains immobilized glucose isomerase, has 420 units of glucose isomerase activity per gram of solid material (as is).

Example 2: 40 g of the immobilized glucose isomerase (moist) (bed vol, 48 ml) prepared as in Example 1 was packed in a jacketed glass column (2.5 x 20 cm). With the column held at a temperature of 60°C, a 50% glucose solution containing 5 mM $MgCl_2$ and adjusted to pH 8.0 was passed through it at a constant flow rate of SV 2. The fructose content of the effluent was 51.2% of the solid substrate at first. This value was held for 2 days and then decreased to 25.6%, half of the initial value, after 30 days.

Treatment with Ion Exchange Material Before Isomerization

In the process for isomerizing glucose in a glucose-containing liquor to fructose, the use of microbial and fungal enzymes adsorbed onto or bonded to inert carriers to provide immobilized biological catalysts has largely superseded older methods whereby soluble enzymes or whole cells of microorganisms were utilized. In general, immobilized enzymes provide a number of significant advantages over soluble or cell-bound enzymes particularly in commercial systems for carrying out continuous conversion processes. Because of the economics of these systems, it is of the utmost importance that the enzymes not be substantially inactivated or denatured by the process used to affect immobilization. It is equally important that the conditions under which the immobilized enzymes are utilized are such that the stability of the enzymes is maintained over a period sufficient to permit conversion of large quantities of substrate. Thus, for example, the presence in the substrate of materials which in some manner interfere with or inactivate glucose isomerase may have a deleterious effect on the stability of the immobilized enzyme and shorten its effective life to a significant degree.

Generally, prior to isomerization, glucose-containing liquors are refined by conventional means, e.g., by treating the liquors with carbon and ion exchange materials in order to remove interfering metals and carbohydrate by-products which might inactivate or denature glucose isomerase in an uneconomically short period. It has been found, however, that although such treatments provide some prolongation of the effective life of immobilized glucose isomerase, the stability of the enzyme is not as great as is desirable.

S.P. Barrett and W.J. Nelson; U.S. Patent 4,288,548; September 8, 1981; assigned to Standard Brands Incorporated however, provide a process for treating a glucose-containing substrate with anion exchange resin in the bisulfite/sulfite form whereby the stability or effective life of immobilized glucose isomerase utilized to convert a portion of the glucose to fructose is increased.

A number of types of ion exchange materials may be utilized in the process, the only requirement being that they be capable of being converted to the bisulfite/sulfite form. Exemplary of such materials are anion exchange cellulose and Sephadex and anion exchange resins. Resins are preferred since they can simply be placed in a column or columns and the glucose solution passed therethrough in a continuous manner. Moreover, they can be relatively easily regenerated. The preferred resins are of the weak base and strong base types. Exemplary of suitable strong base resins are Dowex 1, 2 and 21K (Dow); Duolite A-101D and A-102 (Diamond Shamrock); Ionac A-535 and A-540 (Ionac) and Amberlite IRA-900 (Rohm & Haas). Suitable weak base resins are exemplified by Duolite A-6, A-7, A-30B and ES-561 (Diamond Shamrock) and Ionac A-300 (Ionac). The preferred resins are those in which the matrix is principally composed of polystyrene and which have a relatively low degree of crosslinking.

In order to convert the resin to the bisulfite/sulfite form, it is preferred that the resin first be in the OH form. This may be accomplished by treating the resin with a suitable hydroxide. After washing to remove excess salts, the resin can be contacted with a source of sulfite and bisulfite ions. Solutions of salts which will provide a source of these ions may be utilized. Exemplary of suitable salts are sodium and potassium sulfite and sodium and potassium bisulfite. In general, excess quantities of these salts will be used to convert the resin to the desired form to insure that the full exchange capacity of the resin is achieved. Thereafter, the resins are rinse or otherwise washed to remove the excess generant.

The conditions under which the glucose-containing liquor is contacted with the resin may vary widely but, typically, the pH of the solution will preferably be in the range of from about 3.5 to 6.5.

Temperatures in the range of from about 25° to about 70°C have provided satisfactory results. The preferred temperature range is from about 50° to about 65°C.

The presence of enzyme activators is generally desirable during enzymatic isomerization processes. In commercial practice, small amounts of salts such as $MgSO_4$ and $NaHSO_3$ are typically added to the substrate prior to isomerization for this purpose.

Although the flow rate at which the liquor is passed through a column or columns of the resin is not critical, satisfactory results have been obtained at a rate of from about 0.08 to about 0.21 gal/min/ft^3 of resin. The liquor may be passed through the column or columns in an up-flow or a down-flow direction.

Using a Strain of Streptomyces to Produce the Enzyme

L. Degen, P. Branduzzi, R. Olivieri and N. Cimini; U.S. Patent 4,291,123; September 22, 1981; assigned to Snamprogetti, SpA, Italy describe a method for the production of fructose and syrups containing fructose and glucose by using an isomerizing enzyme obtained from microorganism of the Streptomyces genus.

For the enzymic isomerization of glucose into fructose, glucose isomerase (D-xylose-ketol-isomerase, 5:3:1.5) is added to a solution of glucose, such as corn syrup, and the reaction conditions are controlled in such a way that a fraction of glucose is converted into fructose, the amount of glucose which is converted into fructose being a function of an equilibrium constant which, at 60°C, is 1.

A number of procedures for such an enzymatic process are known, using enzymes extracted from a number of species. The enzymes used in this process are of the genus Streptomyces, and were isolated from wood lots in the botanical garden of Pavia, Italy. They have the NRRL numbers 11120 and 11121.

They may be cultured by the customary methods using conventional nutrient solutions.

An outstanding advantage of these Streptomyces strains is their ability to hydrolyze xylans, which are widespread in naturally occurring materials such as wheat straw, bagasse, woods of various kinds and oily seed hulls.

The glucose can be isomerized by using a fresh culture of microorganisms, or freeze-dried mycelia; or preparations in which the isomerase has been obtained from the mycelia. Lastly, a further technical and economical improvement can be achieved by immobilizing the enzyme by combinations of macromolecular compounds by forming chemical bonds with the matrix, or bonds of an ionic type, or by physical immobilization.

Using a Strain of Ampullariella to Produce the Enzyme

A number of glucose isomerases have been isolated from different species of microorganisms. Disadvantages of using an enzyme conversion of glucose to fructose reside in the cost of producing the enzyme and the sensitivity of the enzyme to environmental parameters. Isomerase preparations are usually inactivated by higher temperatures, so relatively low temperatures must be used to carry out the isomerization. Thus, longer overall reaction times are necessary than would be required if the isomerization could be carried out at a higher temperature. In addition, immobilization techniques necessary to recover the isomerase during processing are limited to those employing only mild temperature conditions. Therefore, a glucose isomerase having good heat stability is especially preferred.

S.E. Foley, P.J. Oriel and C.C. Epstein; U.S. Patent 4,308,349; December 29, 1981; assigned to The Dow Chemical Company have found a method of producing glucose isomerase from a microorganism belonging to the genus Ampullariella which comprises growing the Ampullariella in a culture medium and recovering the glucose isomerase therefrom. The glucose isomerase produced using this method has demonstrated superior thermal stability. Samples of the isomerase derived from some strains have been shown to retain essentially all of their original activity after heating at 75°C for a period of 24 hours.

The process for converting D-glucose to D-fructose includes the step of treating an aqueous solution containing glucose with an isomerase produced by a microorganism belonging to the genus Ampullariella thereby converting a portion of the glucose to fructose. Due to the difficulty of isolating members of the genus only four species of Ampullariella has been previously described in the literature.

All the known species have been found to produce glucose isomerase. In addition, new organisms belonging to the genus have been isolated and found to produce glucose isomerase. These isolates although belonging to the genus Ampullariella cannot be placed into any of the known species within the genus, and therefore represent previously unknown species of Ampullariella. These isolates are referred to as Ampullariella species 3876 (ATCC 31351), 3877 (ATCC 31352), 3965 (ATCC 31353), and 3966 (ATCC 31354). Especially preferred for the production of glucose isomerase is Ampullariella species 3876, the enzyme of which has demonstrated superior heat stability.

Ampullariella are aerobic and grow well on various nutrient medium. Although for the production of glucose isomerase it is preferred that the culture medium contain xylose, the isomerase will be produced in lesser amounts in the absence of xylose. The amount of xylose required in the medium to give satisfactory enzyme production for commercially recoverable quantities are generally between about 0.1 and 10% by weight. Members of the genus generally grow best at temperatures between about 15° and 45°C. The glucose isomerase is produced intracellularly and is generally not excreted into the surrounding culture medium. To isolate the isomerase it is, therefore, necessary to disrupt the cell walls to release the enzyme. Disruption of the cell walls may be accomplished by techniques well known in the art, such as, for example, by ultrasonic rupture.

As recognized by one skilled in the art, the conversion of D-glucose to D-fructose using the glucose isomerase produced above is generally carried out in an aqueous medium. Aqueous solutions may preferably contain from about 30 to 60% glucose by weight. The final fructose concentration of the isomerized solution will depend on a number of factors well understood in the art, and the process is generally terminated when the percent weight of fructose is between about 40 to 50% of the total combined dry weight of glucose and fructose.

The syrup may be treated with glucose isomerase using either a batch or continuous process. In either case, it is generally preferred that the enzyme be immobilized by a water-insoluble support, usually a polymer, either by physical absorption, covalent bonding, or by entrapment.

Using Immobilized Glucose Isomerases of Bacillus Origin

R.S. Leiser; U.S. Patent 4,310,628; January 12, 1982; assigned to A.E. Staley Manufacturing Company has provided a process to improve productivity of the process of producing fructose syrup from dextrose (glucose) by action of glucose isomerase enzyme. The isomerization takes place within an immobilized glucose isomerase bed in which the isomerase is obtained from the Bacillus genera and the process exhibits an enhanced rate of isomerizing dextrose to fructose when the glucose isomerization reaction is conducted: (a) in the presence of Co^{2+} ions; and (b) at a temperature greater than about 60°C. The process comprises:

(a) providing a refined dextrose syrup which is essentially free from Co^{2+} ions and containing on a total dry solids weight basis at least 90% monosaccharide;

(b) isomerizing the dextrose syrup to a fructose syrup by passing the dextrose syrup through a bed of immobilized glucose isomerase maintained at a temperature of no more than about 60°C and a pH between 7.0 and 7.5;

(c) recovering the fructose syrup while replenishing the bed with fresh dextrose syrup; and

(d) continuing the isomerization of the dextrose syrup in the bed for more than 3,000 hours; and

(e) terminating the isomerization when the total isomerase bed activity has been reduced to a value of less than 20% of its original activity.

The process employs a refined dextrose syrup having a high monosaccharide solids content and essentially free from Co^{2+} ions. The reactor may produce the desired fructose syrup product in a single pass or the syrup may be recycled through a reactor at a higher flow rate until the desired interconversion is achieved or a plurality of reactors may be connected in series so that the fructose content is incrementally increased as the syrup flows through each reactor in the series.

Productivity of the isomerization process is generally enhanced by employing feed syrups of a high monosaccharide content. High dextrose conversion syrups containing more than 90% dextrose (dry solids weight basis) or 95% dextrose or higher (especially at about 97 to 99%) are particularly useful feed syrups.

The yield of fructose is improved when the feed syrup has a magnesium ion content of 0.003 to 0.01 M and a dry solids weight content from 45 to 55%.

Although the isomerization process generally applies to immobilized isomerases which are derived from the Bacillus genera, this process is particularly adapted for use with isomerases obtained from the *Bacillus coagulans* family (e.g., NRRL B-5305 and NRRL B-5351) and immobilized in accordance with Netherlands Patent 73/12525 filed on September 11, 1973. These isomerases are more stable against inactivation when used in an isomerization process with a stabilizing amount of Co^{2+} ion (e.g., between 0.0015 to 0.004 M), possess a pH optimum of about 8.5 and optimum isomerization temperature well above 60°C.

With Glucosone as an Intermediate Step

Commercial methods for the production of fructose, a commercially important sweetener, primarily involve a two-step process, the first, hydrolysis of a polysaccharide such as starch to produce glucose and the second, isomerization of the so-produced glucose to form fructose. The latter step, as is well-known, produces a mixture of glucose and fructose from which it is difficult to separate the desired product, fructose. The commercial separation method involves the use of crystallization and/or fractionation techniques which are costly and time-consuming.

Glucose can also be converted to fructose by the action of an enzyme, designated glucose-2-oxidase, to form glucosone (D-arabino-2-hexulose) which in turn can be reduced to fructose with zinc and acetic acid. However, fructose produced in this way can be contaminated with significant amounts of by-products from both the enzymatic conversion of glucose and the alkene conversion reaction. In particular, the latter reaction produces halohydrins and alkylene oxides, e.g., ethylene oxide, which are highly toxic materials even at levels in the region of parts per million. Thus, fructose produced by such a process will require careful and costly purification to attain food-grade purity.

J.A. Maselli and R.O. Horwath; U.S. Patent 4,321,323; March 23, 1982; assigned to Standard Brands Incorporated have devised a process which can produce food-grade fructose from glucose. In this method, glucose is enzymatically oxidized to glucosone in a reaction zone from which hydrogen peroxide is removed by use of a hydrogen peroxide-permeable membrane into a second reaction zone.

In accordance with one embodiment, the second reaction zone contains a reducing agent for the hydrogen peroxide which migrates through the semi-permeable membrane. The presence of the reducing agent in the second zone encourages a faster migration of the hydrogen peroxide out of the first zone. Reducing agents for this embodiment are well known to those skilled in the art and include a variety of systems such as organic reducing agents, anions, cations and enzymes. Organic reducing agents are exemplified by aldehydes which are readily oxidized to corresponding carboxylic acids. Reducing anions include, for example, oxalate, sulfite, phosphite and iodide ions. Reducing cations include a wide variety of cations which can exist in variable valence states such as the transition metals Fe, Co, Ni, Cr and the like. Reducing enzymes are readily available from a variety of natural products and include catalase and peroxidase. Catalase is found in yeast, eggs and blood while peroxidase is found in horseradish.

The membranes employed are for the purpose of establishing two separate zones and permitting migration of hydrogen peroxide from the first to the second zone. The membranes therefore should be of suitable pore size to selectively permit hydrogen peroxide migration, but to preclude passage of larger molecules in the first reaction zone. Such membranes are readily available commercially and can be defined in terms of the molecular weight of solute particles to pass through the membrane. In this process, membranes which permit substances of a molecular weight of less than about 100 are to be used, and preferably less than 50.

The glucose-2-oxidase enzyme can be provided in the form of the enzyme solution in water, immobilized enzyme or immobilized cells or mycelia or the free cells or mycelium. Most commonly since the enzyme is intracellular, the cells or mycelia of the selected microorganism are used by merely suspending them in the reaction solution. Promoters and protectors for the enzyme can also be present. For example, as described in *Folia Microbiol.* 23, 292–298 (1978), the presence of fluoride ion promotes the enzymatic oxidation of glucose with *Oudemansiella mucida.* Protectors for enzymes can also be used, e.g., Co, Mn and Mg salts.

A wide variety of microorganisms can be used to produce the glucose-2-oxidase employed in this process. Some examples described in the literature for this purpose are: *Aspergillus parasiticus* [*Biochem. J.* 31, 1033 (1937)]; *Iridophycus flaccidum* [*Science* 124, 171 (1956)]; and *Gluconobacter roseus* [*J. Gen. Appl. Microbiol.* 1, 1152 (1955)].

In further work, *J.A. Maselli and R.O. Horwath; U.S. Patent 4,321,324; March 23, 1982; assigned to Standard Brands Incorporated* have suggested that the hydrogen peroxide be reacted in the second reaction zone with an alkene, which is then (1) converted to a glycol by the hydrogen peroxide, or (2) converted to a halohydrin and then to an alkylene oxide or glycol corresponding to the original alkene reactant by reaction with hydrogen peroxide, a halogenating enzyme and a halide ion source, to form the halohydrin which is then converted to an epoxide or glycol, by methods described in European Patent 7176.

Example: Mycelia of *O. mucida* are grown in accordance with Example 1 of Czechoslovakian Patent 175897 and the equivalent of 15 g (dry weight) of the mycelia are suspended in 3 ℓ of 2.5% glucose solution 0.05 M NaF in one zone of a 10 ℓ reactor fitted with a hydrogen peroxide-permeable membrane to form two zones. In the second zone, ethylene gas is bubbled through an aqueous solution of chloroperoxidase and halide ion buffered with a phosphate buffer (0.1 M potassium phosphate) as described in European Patent 7176 (Examples 1 through 18).

The suspension in the first zone is mixed at 25°C and aerated with oxygen. After 24 hours the mycelia are then separated from the solution in the first zone and resulting clear solution is then hydrogenated over Raney Ni at 500 atmospheres hydrogen gas and 100°C. The aqueous mixture is filtered clear of the catalyst, decolorized with carbon, deionized with ion-exchange (anionic and cationic) and concentrated to a fructose syrup at reduced pressure. Alternatively, the aqueous mixture is concentrated and fructose allowed to crystallize.

The fructose obtained as either syrup or crystalline product is of food-grade quality. The halohydrins obtained by the reaction in the second zone are converted to the corresponding epoxides by treatment with sodium hydroxide.

Essentially the same results are obtained when *O. mucida* is replaced with the following organisms: *Polyporus obtusus, Radulum casearium, Lenzites trabea, Irpex flavus, Polyporus versicolor, Pellicularia filamentosa, Armillaria mellea, Schizophyleum commune and Corticium caeruleum.*

Glucose Isomerase Immobilized by Chemical Bonding

Glucose isomerase is an enzyme capable of converting glucose into fructose and vice versa and is used for producing fructose from glucose. Many proposals have been made to stabilize and/or to immobilize glucose isomerase for the multiple reuse.

The enzymatically active product which is useful in the process for isomerization of glucose into fructose comprises an organic polymeric material comprised of at least 50% by weight, based on the weight of the organic polymeric material, of a monovinyl aromatic compound in polymerized form. The polymerized monovinyl aromatic compound has a β-aminopropionamidomethyl group as a side chain represented by the formula shown below.

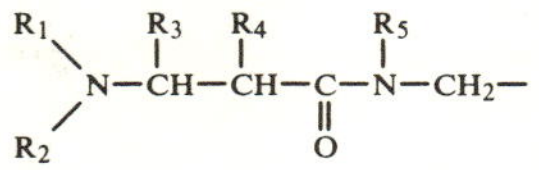

In the above formula R_1 is selected from hydrogen, an alkyl group having 1 to 6 carbon atoms and a hydroxyalkyl group having 2 to 6 carbon atoms; R_2 is selected from an alkyl group having 1 to 6 carbon atoms, a hydroxyalkyl group having 2 to 6 carbon atoms;

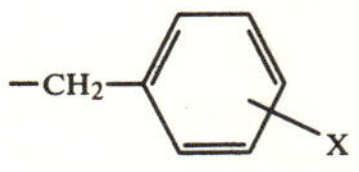

and

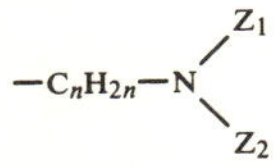

(where X, Z_1 and Z_2 are selected from hydrogen and an alkyl group having 1 to 6 carbon atoms, and n is an integer of from 2 to 6); or R_1 and R_2 are bonded, a heterocyclic structure represented by the formula:

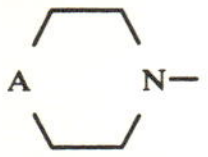

where A is $-CH_2-$, $-O-$ or $-NR_6-$ (R_6 is H or an alkyl group having 1 to 6 carbon atoms); and R_3, R_4 and R_5 may be the same as or different from each other and are selected from hydrogen and a methyl group. The organic polymeric material has glucose isomerase immobilized with the β-aminopropionamidomethyl group side chain or the organic polymeric material.

The monovinyl aromatic compound may be selected from styrene, α-methylstyrene, vinyltoluene, vinylxylene and chlorostyrene. The monovinyl aromatic compound polymer may be employed in combination with other polymeric materials such as polyolefins, polyamides, polyesters and their copolymers.

The shape of the article with which glucose isomerase is immobilized is not particularly limited, but one preferable form of the article is a fiber comprised of the monovinyl aromatic compound polymer and another fiber-forming organic polymeric material. The fiber may be either a blend fiber made from a uniform blend of the two polymeric materials, or a core-sheath type or islands-in-a-sea type composite fiber, the sheath or sea ingredient being predominantly comprised of the monovinyl aromatic compound polymer and the core or island ingredient being comprised of the fiber-forming organic polymeric material. Of these fibers, the blend fiber and the islands-in-a-sea type composite fiber are most preferable because of their good resistance to separation.

The enzymatically active product is prepared by the following process. In the first step, an article comprised of either the monovinyl aromatic compound polymer or a combination of the monovinyl aromatic compound polymer and the other polymeric material is treated with an acrylamidomethylating agent such as N-methylolacrylamide and its carboxylic acid esters and alkyl ether derivatives; N-methylolmethacrylamide and its carboxylic acid esters and alkyl ether derivatives, and N,N'-(oxydimethylene)bisacrylamide.

The acrylamidomethylated product is then treated with an amino compound whereby the acrylamidomethyl groups are converted to β-aminopropionamidomethyl groups.

Preferable amino compounds are organoamino compounds (including multiamino compounds) having at least one primary or secondary amino group and forming little or no crosslinking. Such amino compounds include, for example, diethylamine, dihexylamine, dipropanolamine, N-methylaminoethanol, ethylenediamine, morpholine, piperidine and piperazine.

The aminated product is then brought into contact with a glucose isomerase-containing solution or dispersion, whereby glucose isomerase is immobilized with the β-aminopropionamidomethyl group side chains of the aminated product.

Glucose isomerase can be obtained from various microorganisms which include, for example, actinomycetes such as *Streptomyces phaseochromogenus* and *Streptomyces albus,* and bacteria such as *Bacillus coagulans, Bacillus megatherium, Lactobacillus brevis,* the Pseudomonas genus and the Aerobacter genus.

In order to enhance the degree of activity retention of the resulting enzymatically active product, the isomerase-immobilized product may be treated with a solution containing a crosslinking agent, whereby the immobilized glucose isomerase is crosslinked with the crosslinking agent. Instead of treating the isomerase-immobilized product, the crosslinking agent may be incorporated in the glucose isomerase-containing solution used for immobilization. The crosslinking agent used is one which is capable of crosslinking a protein and is popularly called a multifunctional protein modifier. Such a crosslinking agent includes, for example, polyglutaraldehydes, such as glutaraldehyde dialdehyde starch, and, polyisocyanates, such as tolylene diisocyanate and hexamethylene diisocyanate.

STARCH SACCHARIFICATION

Using Beta-Glucanase Enzyme to Produce High Fructose Sweeteners

None of the most widely used enzyme processes to convert starch to sweeteners produce pure and highly concentrated syrup (that is, having an excess of about 95 DE) directly from the enzyme treatment of impure starch flour slurries. Most of them require a starch recovery step to isolate the purified starch prior to the enzyme treatment.

K.W.R. Schoenrock, T.H. Henscheid and H.G. Rounds; U.S. Patent 4,247,636; January 27, 1981; assigned to The Amalgamated Sugar Company, however, have developed a process which eliminates the need for this starch recovery step, therefore resulting in an economical conversion of low cost impure barley starch flours into valuable concentrated sweetener syrups. It also produces a high protein meal and cereal germ oils as by-products.

Typically, the impure starch source is first ground into flour and slurried with water. Usually, roughage, such as grain hulls, is mechanically removed prior to slurrying the flour. Beta-glucanase (preferably bacterial in origin) is desirably next added to the slurry and allowed to react. Alpha-amylase is added to the reacted slurry to liquefy the starch mixture. In practice, commercially available bacterial beta-glucanase contains significant quantities of alpha-amylase and proteinase enzymes. Alpha-amylase is not present in sufficient amount to liquefy the starch, however, when a practical amount of the beta-glucanase is used. The residual solids and cereal germ oils are then separated from the resulting aqueous solution and processed into high protein meal. The glucose solution is then saccharified by treatment with glucoamylase to a DE above 90. However, according to certain embodiments of the process, liquid/solids separation and the purification procedures may be carried out after partial saccharification with soluble glucoamylase, preferably after a DE of about 50 to 70 is reached. In that event, completion of the saccharification step with either soluble or immobilized glucoamylase to a DE of above about 90 follows the purification steps. The saccharified mixture is then filtered, purified and concentrated. Its pH is adjusted with active MgO to within the range of 8.0 to 8.5, and it may then be isomerized in conventional fashion by reaction with glucose isomerase to produce a high fructose sweetener.

Steeping of the whole barley in the presence of SO_2, e.g., with sulfurous acid (H_2SO_3), prior to milling further facilitates purification of the starch hydrolysates and improves the quality of the final product.

The process is particularly well adapted to impure starch sources having high concentrations of beta-glucans. These materials, in combination with the glutens of the grains, have been difficult to separate from the starch in the starch source by conventional techniques. The use of beta-glucanase enzyme early in the process, the specific sequencing of processing steps, and the use of active MgO for pH adjustment prior to isomerization represent important advances in the state of the art. The steeping procedures are also of particular advantage when barley grain is processed.

Production of a Highly Thermostable Glucoamylase

At present, batch processes are applied to the production of dextrose by saccharification of starch hydrolyzates with a glucoamylase. Many of the commercially available glucoamylase enzymes produced today are derived from microorganisms of the genera Rhizopus and Aspergillus. When these enzymes are used to produce dextrose, they are generally reacted at 55° to 60°C for 2 to 4 days. If glucoamylase can be immobilized and the saccharification can be conducted continuously through a column, the reaction time is reduced and no large reaction tank is necessary thereby saving labor and energy. Glucoamylase enzymes produced by Rhizopus and Aspergillus can be immobilized by ion exchange processes, physical adsorption, covalent bonding, gel entrapment, etc. Any glucoamylase enzymes immobilized by any of these processes are, however, inactivated when used at above 50°C.

A microbial strain has been discovered belonging to the genus Talaromyces which produces a glucoamylase having an optimum reaction temperature of 75°C and characterized as being capable of retaining at least about 90% of its initial glucoamylase activity when held at 70°C and pH 4.5 for 10 minutes.

M. Tamura, M. Shimizu and M. Tago; U.S. Patent 4,247,637; January 27, 1981; assigned to CPC International Inc. have developed a method for the production of this glucoamylase wherein a microorganism of the genus Talaromyces, which produces the glucoamylase, is cultured in a medium and the enzyme is recovered from the culture broth. The strain was identified as *Talaromyces duponti* strain G45-632 which is stored at the Fermentation Research Institute, Chiba City, Japan, as Deposit No. 4566.

The glucoamylase enzyme is prepared by culturing cells of the strain of *Talaro-*

myces duponti FERM 4566, in a nutrient medium and isolating the glucoamylase enzyme preparation from the culture medium.

The glucoamylase enzyme preparation has a maximum glucoamylase activity at about 75°C as measured by a 10-minute reaction on a 2% maltodextrin solution at pH 4.5 and retains at least about 90% of its initial glucoamylase activity when held at 70°C for 10 minutes at pH 4.5.

The use of this enzyme, particularly in an immobilized state, permits this saccharification of a liquefied starch solution to be carried out in a continuous process at temperatures above 60°C without any decline in the dextrose yield.

Production of Glucoamylase Active at Neutral pH

At present, when producing dextrose industrially from starch, the principal glucoamylases employed for the saccharification process are those produced by microorganisms belonging to the genera Rhizopus and Aspergillus. The glucoamylases are employed at a pH of 4.5 to 5.0.

It has been found that higher percentages of dextrose can be produced if the reaction can be conducted at a nearly neutral pH. *M. Tamura, M. Shimizu and M. Tago; U.S. Patent 4,254,225; March 3, 1981; assigned to CPC International Inc.* have, therefore, found a microbial strain which produces a glucoamylase having optimum activity at a pH of 6.0 to 6.5 and good thermostability. The glucoamylase is capable of converting a 30% by weight solution of a 10 DE liquefied starch to a product containing at least about 96% dextrose when reacted with the starch hydrolyzate at pH 6.0 to 6.5 at 55°C.

The microorganism which produces this glucoamylase has been identified as a member of the genus Stachybotrys, and has been designated *Stachybotrys subsimplex* FERM 4377. It is cultivated by normal techniques used in the culture of molds. The glucoamylase can be recovered from the culture material and purified by conventional methods.

This glucoamylase, when used for the saccharification of liquefied starch to produce dextrose, at a pH of 6.0 to 6.5, gives an increased yield of dextrose over methods using conventional glucoamylases and saccharification carried out under acidic conditions.

Use of Immobilized Cyclodextrin Glucanotransferase

S. Okada, S. Kitahata, S. Yoshikawa and K. Miyake; U.S. Patent 4,254,227; March 3, 1981; assigned to KK Hayashibara Seibutsu Kagaku Kenkyujo, Japan have developed processes for producing syrups or syrup solids containing fructose-terminated oligosaccharides (hereinafter abbreviated SFTO), characterized by subjecting a mixture containing liquefied starch and either fructose or sucrose to the action of immobilized cyclodextrin glucanotransferase E.C. 2.4.1.19 (hereinafter abbreviated as CGT).

As previously described CGT is produced by bacteria of this genus Bacillus such as *Bacillus macerans, Bacillus megaterium, Bacillus circulans, Bacillus polymyxa* and *Bacillus stearothermophilus,* and the genus Klebsiella such as *Klebsiella pneumoniae.*

It is also known that SFTO are produced by subjecting a mixture containing liquefied starch and either fructose or sucrose to the action of a CGT solution whereby the glucosidic residues of the dextrin molecules in the liquefied starch are transferred to the fructose or sucrose molecules.

It has been found that at least three reactions occur when a mixture containing liquefied starch and either fructose or sucrose is subjected to the action of an aqueous CGT solution to produce SFTO. They are: (1) the reaction that produces cyclodextrins from the liquefied starch; (2) the reaction that transfers the glucosidic residues in the cyclodextrins formed by (1) to the fructose or sucrose molecules; and (3) the reaction that transfers the glucosidic residues in the liquefied starch directly to the fructose or sucrose molecules. It has been found that SFTO can be produced more effectively and advantageously by relying mostly on (3), and concentrated efforts were made on improvement of this reaction.

It was found that the dextrinogenic activity per unit weight CGT protein decreases to about 10 to 30% when the enzyme is immobilized by any conventional method, whereas the alpha-cyclodextrin-decomposing activity hardly decreases or increases to about 80 to 130% as compared with that of the intact enzyme.

It was also found that when a given substrate solution containing liquefied starch and either fructose or sucrose in the same proportions and concentrations is allowed to react under the same conditions with either immobilized or intact enzyme in the same amount of alpha-cyclodextrin-decomposing activity per unit weight liquefied starch, the use of the immobilized enzyme results in a remarkably faster rate of transfer action per unit weight CGT protein than intact enzyme. In other words, it was found that a much higher rate of SFTO formation per unit weight enzyme protein is realizable with the employment of immobilized enzyme.

Using Heat- and Acid-Stable Alpha-Amylases

When alpha-amylase enzymes are used to treat starch containing materials, they are used as the initial step in the production of a number of starch hydrolysate materials, such as malto-dextrins, corn syrups, dextrose, levulose, maltose and others. The alpha-amylase enzyme hydrolyzes starch molecules to break them down into a variety of intermediate molecular weight fragments known as malto-dextrins. The malto-dextrins are subsequently treated with one or more additional enzyme preparations including glucoamylase, beta-amylase and glucose isomerase in order to produce the desired final product. Alternatively, a plurality of these enzyme preparations can be introduced into a slurry of the starch material simultaneously to directly produce the desired starch hydrolysate.

Alpha-amylase enzymes are available from a wide variety of sources. Most alpha-amylase enzymes are produced from bacterial sources such as *Bacillus subtilis, Bacillus licheniformis, Bacillus stearothermophilus* and others which are cultivated in an appropriate culture medium, the cells produced therefrom are then destroyed and the enzyme preparation is thereafter separated from the broth and purified.

Many of the commercially available alpha-amylase enzymes produced today are derived from *Bacillus subtilis* microorganisms. When these enzymes are used to convert starch to starch hydrolysates, they will generally have an optimal temperature ranging from about 80° to about 85°C, and an optimal pH of about 6.0. The conditions of temperature and pH necessary for efficient use of the enzyme have two disadvantages. Firstly, if starch is converted with the enzyme at a pH of about 6 and at a temperature of about 80° to about 85°C, a part of the reducing end-groups of the starch is isomerized, and in the subsequent conversion process, maltulose is produced which reduces the degree of recovery of the desired product, e.g., dextrose, levulose, or maltose. Secondly, the optimum pH of glucoamylase used in the conversion and saccharification process is generally about 4.5 in the case of *Aspergillus niger*-type enzymes and a pH of about 5.0 in the case of Rhizopus-type enzymes. Therefore, upon completion of the liquefaction step using the alpha-amylase enzyme, it has been necessary to adjust the pH from about 6 to 4.5 or 5.0. This pH adjustment increases the ion concentration and as a result, increases the load and consequent refining expense using the ion exchange resins used in the purification of the final product.

It is a principal object of *M. Tamuri, M. Kanno and Y. Ishii; U.S. Patent 4,284,722; April 18, 1981; assigned to CPC International Inc.* to produce alpha-amylase enzymes which have good heat-stability as well as good stability under acidic conditions, particularly at pH values to render their use under conditions compatible with other amylases such as glucoamylase in the process of converting starch to starch hydrolysates.

According to this process heat and acid-stable alpha-amylase enzymes which are characterized as: (1) capable of retaining at least about 70% of their initial alpha-amylase activity when held at 90°C and at a pH of 6.0 for 10 minutes in the absence of added calcium ion; (2) capable of retaining at least about 50% of their initial alpha-amylase activity when held at 90°C, at a pH of 6.0 for 60 minutes in the absence of added calcium ion; and (3) capable of retaining at least about 50% of their initial alpha-amylase activity at a temperature of 80°C and at a pH of 4.5 in the presence of 5 mmols of calcium ion for 10 minutes. The enzymes are also capable of retaining at least about 50% of their initial activity at a temperature of 85°C and at a pH of 4.55 for 30 minutes in the presence of 5 mmols of calcium ion and 22.5 weight percent starch. These enzymes are derived from a Bacillus microorganism and preferably *Bacillus stearothermophilus.*

These enzymes are produced by cultivating in a suitable medium a *Bacillus stearothermophilus* microorganism, preferably a strain selected from the group consisting of *Bacillus stearothermophilus* ATCC numbers 31,195; 31,196; 31,197; 31,198; or 31,199, and thereafter recovering the alpha-amylase enzyme produced.

These alpha-amylase enzymes can be used in the liquefaction and conversion of starch in the starch saccharification industry, in desizing processes in the textile industry and as additives in detergent formulations. They are particularly suited to the liquefaction and conversion of starch in the production of malto-dextrins, and the subsequent production of dextrose using glucoamylase since isomerization of the end group of the molecules can be avoided because the enzyme can be efficiently used at an acidic pH (i.e., a pH of 4.5 to 5.0), thereby increasing the dextrose yield. The use of these enzymes also reduces the ion exchange load in the refining process because no pH adjustment is required prior to saccharification and conversion with the glucoamylase enzyme.

Preparation of Low DE Starch Hydrolysates and Syrups

There is a large potential market for syrups and syrup solids with bland taste, low sweetness and low hygroscopicity at a low dextrose equivalent (DE) level. Such syrups, hydrolysates and syrup solids are useful as basis for the preparation of food items as well as for bodying agents and as additives having nonsweet, water holding, nonhygroscopic characteristics.

F.C. Armbruster; U.S. Patent Reissue 30,880; March 9, 1982; assigned to Grain Processing Corporation has provided such a process for preparing a low DE starch hydrolysate. This process comprises subjecting a mixture of starch and water, having a solids content of less than about 50%, to the hydrolytic action of acid to obtain a starch hydrolysate having a DE between about 5 and 15, subjecting the acid hydrolysate to the hydrolytic action of bacterial alpha-amylase to a DE between about 10 and 25, the increase in DE being at least about 5, to produce a starch hydrolysate having a dextrose content of less than 4%, and that is also characterized by having a sum of the percentages of saccharides therein, dry basis, having a degree of polymerization of 1 to 6, divided by the DE, provide a ratio greater than about 2.0. This ratio is referred to hereinafter as the characteristic or descriptive ratio.

This process also provides a method for preparing a low DE syrup by the concentration of the starch hydrolysate described above, to produce a syrup having a solids content greater than 50%. The syrup, i.e., concentrated hydrolysate, may or may not be refined by conventional means.

One preferred method of practicing the process involves the steps of slurrying cornstarch in water to a density between 5° and 30° Baumé, adjusting the pH of the slurry to between 1 and 3 and raising the temperature of the starch slurry to between 70° and 160°C to solubilize and hydrolyze the starch to a DE between 5 and 15. The pH is adjusted to between 6 and 8 and the hydrolysate is dosed with bacterial alpha-amylase. The mixture is then hydrolyzed under the proper conditions to a DE between 10 and 25, preferably between about 15 and 25 when it is desired that the final product be a haze-free syrup. The enzyme conversion step is carried out at a temperature in the range between about 50° and 95°C. The mixture is held at the conversion temperature for a period of time ranging from a few minutes to as long as 1 or 2 hours or perhaps more.

The resulting hydrolysate may be concentrated and/or refined by conventional procedures to yield a stable corn syrup, which is substantially haze-free and highly soluble in water. The syrup may be spray dried to yield corn syrup solids with low hygroscopicity and high water solubility.

Suitable starches include cereal starches such as corn, grain sorghum and wheat, waxy starches such as, waxy milo and waxy maize, and root starches, such as potato starch and tapioca starch.

The preferred enzyme used for the conversion of the acid hydrolysate to low DE hydrolysates is the type commonly referred to in the art as bacterial alpha-amylase. It is a starch liquefying, heat resistant, hydrolytic alpha-amylase. Suitable bacterial alpha-amylase may be produced by certain strains of *Bacillus subtilis, Bacillus mesentericus* and the like by conventional fermentation methods. HT-1000, a bacterial alpha-amylase (Miles Chemical Laboratories), is an example of an enzyme preparation that is suitable for use in this process. Other suitable bacterial alpha-amylases include Rhozyme H39 (Rohm & Haas) and CPR-8 (Wallerstein Division of Baxter Laboratories, Inc.).

Using a Mixture of Enzymes

The objective of *B.E. Norman; U.S. Patent 4,335,208; June 15, 1982; assigned to Novo Industri A/S, Denmark* is to provide a saccharification process for starch hydrolysates which produces syrups with a higher dextrose (D-glucose) content than normally obtained, e.g., more than 30% solids by weight.

According to this process maltodextrin solutions for example, starch hydrolysates of 8 to 12 DE, are saccharified by a combination of glucoamylase and isoamylase at a pH from 3 to 5, and at temperatures of 50° to 60°C.

The glucoamylase dosage herein contemplated is significantly lower than industry has customarily used to make glucose syrups, being in the range of 0.05 to 0.3 AG units/g of DS (dry substance), preferably 0.075 to 0.15. The isoamylase dosage range is 25 to 500/IA units/g DS, preferably 100 to 200/IA units/g DS. One AG unit of glucoamylase activity is the amount of enzyme which hydrolyzes one micromol of maltose per minute at 25°C and pH 4.3. A commercially available liquid form of glucoamylase from *Aspergillus niger* (Amyloglucosidase Novo 150, Novo Industri A/S, Denmark) contains 150 AG units per ml.

One 1A unit of isoamylase activity is the amount of enzyme which causes an increase in absorbency of 0.01 at 610 nm under the following standard conditions: A reaction mixture consisting of 1 ml of a suitably diluted enzyme solution, 5 ml of 1% amylopectin (waxy-maize starch, Snowflake 04201, CPC) solution, and 1 ml of 0.5 M acetate buffer solution pH 3.5 is incubated at 40°C for 30 minutes. 0.5 ml of the reaction mixture is withdrawn and mixed with 0.5 ml of 0.01 M iodine solution and 15 ml of 0.02 N H_2SO_4 solution. After 15 minutes the absorbency at a wavelength of 610 nm is determined. An experimental isoamylase preparation from *Pseudomonas amyloderamosa* (Hayashibara, Japan) contained 500,000 IA units per gram.

The reaction time is usually in the range of 24 to 96 hours, and the substrate concentrations contemplated are 20 to 50% DS, preferably more than 30% of dry substance by weight.

ENZYMATIC TRANSFRUCTOSYLATION OF SUCROSE

R.E. Heady; U.S. Patent 4,276,379; June 30, 1981; assigned to CPC International Inc. describes three processes having to do with the enzymatic transfructosylation of sucrose. The first process (1) is one for the production and isolation of a new fructosyl transferase enzyme from the fermentation broth of *Pullaria pullulans* ATCC 9348. The second and third processes relate to (2) the enzymatic transfructosylation of sucrose to produce a fructose-polymer-containing substrate, and (3) the production of fructose syrups containing greater than 55% fructose by using a glucose isomerase enzyme on this substrate.

The enzymes referred to above are preferably used as immobilized enzymes, being coupled to porous ceramic alkali-resistant carrier materials. Such immobilized isomerase and fructosyl transferase enzymes are described in U.S. Patents 3,850,751 and 3,868,304.

In processes (2) and (3) above, sucrose is subjected to the action of the cell-free fructosyl transferase enzyme, which converts the sucrose to a mixture comprising glucose, fructose, and polysaccharides containing at least 66% by weight of fructosyl moieties where the fructosyl moieties are linked by (2-1)-beta linkages. The mixture is then subjected to the action of an immobilized glucose isomerase enzyme, and the polysaccharides in the mixture are hydrolyzed in the absence of active isomerase enzyme.

The first substrate should be one which is an aqueous solution of sucrose at a concentration of at least 10 g of sucrose per 100 ml of water. The temperature may range from 25° to 65°C and the pH from 4.5 to about 6.5.

In further work, *R.E. Heady; U.S. Patent 4,317,880; March 2, 1982; assigned to CPC International Inc.* found that an improvement in the process described in the last patent could be made by contacting the sucrose-containing substrate with a preparation containing a mixture of the fructosyl transferase and glucose isomerase enzymes.

The reaction product is a syrup which contains fructose polymers plus glucose and fructose. This syrup is useful as a specialty carbohydrate for sweetener and other applications. It may also be hydrolyzed to yield a syrup whose principal sugar is fructose. The fructose content of the sugars in these syrups is generally higher than 60% by weight and ranges up to about 75% and even higher, depending upon the composition of the sucrose substrate and the reaction conditions employed.

This simple process is surprisingly effective in producing high fructose syrups. The concurrent action of fructosyl transferase and glucose isomerase enzymes on sucrose gives fructose polymers with a substantially higher average percentage (up to 15%) of fructose than those obtained by the sequential reaction of the same enzymes. The reaction is carried out at a temperature of from about 50° to 60°C and at a pH of from about 6.3 to 6.7.

In another variation of the process of contacting a sucrose-containing substrate with a fructosyl transferase enzyme, *R.E. Heady; U.S. Patent 4,335,207; June 15, 1982; assigned to CPC International Inc.* provides a 2-step process for the production of ethyl alcohol and fructose polymers. The process involves contacting a sucrose-containing substrate with a fructosyl transferase enzyme and fermenting the resulting product with a yeast preparation. Purification of the reaction product by removal of cellular debris (e.g., yeast cells) and ethanol yields a syrup containing fructose polymers. This syrup is useful as a specialty carbohydrate for sweetener and other applications. It may also be hydrolyzed to yield a syrup whose principal sugar is fructose. The fructose content of the sugars in these syrups generally is higher than 66% by weight and ranges up to about 75% and even higher, depending upon the composition of the sucrose substrate and the reaction conditions employed.

An alternative embodiment of the process involves contacting a sucrose-containing substrate with a mixture of fructosyl transferase and glucose isomerase enzymes. The resulting product is then fermented with a yeast preparation in a second step of the process. This alternative process produces ethyl alcohol and fructose polymers containing a very high percentage of fructose units. These fructose polymers may be hydrolyzed to yield a high fructose syrup containing over 80% fructose.

This process is a distinct improvement over prior processes for the preparation of fructose polymers. In this process, it is not necessary to separate the fructose polymer syrup from glucose which is formed concurrently. By fermenting the glucose in the secondary substrate with a yeast preparation that does not ferment the fructose polymers, separation of the glucose is obviated and a valuable, easily separated by-product, ethyl alcohol, is obtained.

MISCELLANEOUS PROCESSES

Preparation of Malt High in Alpha-1,6-Hydrolase

It is well known that cereal grains such as barley, rye, oats and wheat can be germinated, i.e., malted, to modify the kernel structure, composition and enzyme content. The resulting malts have many important uses in foods for animals and humans. Most important of all, however, is malted barley which is a basic material used in the brewing and distilling industries, and in the saccharification of starch.

G.W. Pratt, T.W. Chapple and M.J. Fahy; U.S. Patent 4,251,630; February 17, 1981; assigned to Kurth Malting Corporation have provided a malt which has a higher alpha-1,6-hydrolase content. Whether this malt is in the form of green malt, partially dried malt or fully dried or kilned malt it has at least 55, and generally at least 60, units of alpha-1,6-hydrolase per one gram of malt. One unit of alpha-1,6-hydrolase is defined as equal to 1 mg of maltose equivalent produced by 1 gram of malt from a 0.5% pullulan substrate in 60 minutes at 50°C and pH 5.

The green malt which is obtained from the germination step will generally contain a minimum of 75 units of alpha-1,6-hydrolase per one gram of malt. Conventional brewer's green malt usually has no more than 17, while distiller's green malt has about 44 units of alpha-1,6-hydrolase per one gram of malt. The green malt has too much water to be suitable for shipping. However, it can be used immediately, without drying, in the production of malt syrup for the production of maltose or in brewing and in distilleries.

When the green malt is dried, the level of the alpha-1,6-hydrolase has been found to rise for the first 10 to 24 hours, from 75 units to 90 units or higher when water content is 50 to 85%. This partially dried malt, after slow drying for 24 hours at temperatures of 75° to 95°F, can be used soon after production. The drying is then completed at higher temperatures until the fully dried malt has only 4 to 10% moisture, at which point the alpha-1,6-hydrolase content has decreased to 55 to 60 units.

This enzyme malt may be further characterized as generally having an acrospire length of from about 1¼ to 3 or more times the average kernel length. This ratio is generally applicable for green malt, partially dried malt or fully dried malt. The acrospire length is a visible indication as to the state of the modification of the kernel and of alpha-1,6-hydrolase development. The dry enzyme malt is very friable and easily ground to a flour. Enzyme malt, in addition to high levels of alpha-1,6-hydrolase, also possesses alpha-amylase and beta-amylase so that these useful enzymes are also present for hydrolytically converting starch and starch degradation products to sugars.

To produce enzyme malt it is important following steep-out to continue germination until the acrospire length, measured from the kernel base, is from at least

1¼ to 3 or more times the length of the kernel. This germination period may be for about 5 to 15 days, but generally will extend for about 8 to 11 days.

It is important during germination to maintain the barley well moisturized by watering it 2 to 3 times per day and by using well moistured air during aeration of the barley. In general, it is best to have the moisture content of the malt continually increasing during germination. The germinating barley should be watered sufficiently to prevent the acrospire and rootlets from becoming dry. Increasing the moisture content of the malt, from that present at steep-out, about ½ to 2½% per day is desirable. The temperature during germination should be controlled and, while it may be kept in the range of normal malting practice, it is desirable to let the temperature rise near the end of germination. The first part, or first one-half, of the germination period is desirably effected at 55° to 75°C, and the last 1 to 5 days of germination at 75° to 90°F.

In addition to conducting germinating as described, it is generally beneficial to effect steeping of the barley under carefully controlled pH conditions. Specifically, the pH of the steep water during most or all of the steeping should be kept in the range of about pH 3.0 to 7.5, and advisably in the range of pH 4.0 to 6.0. Any suitable organic acid or inorganic acid may be used to obtain an acid pH. Preferably, the steep water is acidified using about 100 to 2,000 ppm of sulfuric acid based on the barley weight.

About 0.05 to 5 ppm of gibberellic acid, previously used in malting, may be used in this process to stimulate enzyme production during malting and is most beneficially applied to the barley as it moves from the steep tanks to the germination compartments. About 1,000 to 4,000 parts of sodium or potassium metabisulfite or sodium or potassium bisulfite per million parts of barley on a dry weight basis may also be added to the steep water to suppress microorganism growth. After germination, the green malt is dried carefully by kilning at temperatures mentioned above.

Enzymatic Compositions for Isomerizing Glucose to Levulose

The use of glucose isomerase to isomerize glucose to levulose is well known. Various methods of immobilizing the enzyme have been suggested, but although these various processes make it possible to obtain compositions having a satisfactory initial enzymatic activity or good abrasion resistance, they do not improve the lifetime of the preparations in question, from the point of view of the enzymatic activity. This lifetime does not generally exceed about two weeks.

Furthermore, it is known that, in order to give good results, the glucose-isomerases which are usually used require the presence of magnesium ions and optionally a small amount of cobalt ions in the glucose syrup to be isomerized. It is therefore necessary for the levulose syrup obtained to be subjected to a deionization treatment by means of ion exchangers, which increases the cost of the levulose.

G. Lartigau, A. Bouniot and M. Guerineau; U.S. Patent 4,264,732; April 28, 1981; assigned to Rhone-Poulenc Industries, France have found that by enclosing microorganism cells containing glucose-isomerase in small plates or filaments of cellulose esters, which contain, as filler, at least one sparingly water-soluble magnesium compound and optionally a sparingly water-soluble cobalt compound, it is possible to obtain compositions of intracellular glucose-isomerase which have an increased lifetime and require the addition of none, or only a small amount of magnesium ions to the glucose syrup to be isomerized.

The preferred microorganism for producing the enzyme is *Streptomyces phaeochromogenes.* The cellulose ester used to form the small plates or filaments is preferably the diacetate or an industrial mixture of the diacetate and the triacetate, such as that employed for the manufacture of collodion. The magnesium compound is preferably magnesium oxide, magnesium hydroxide, magnesium phosphate or basic magnesium carbonate or a mixture thereof, e.g., an industrial mixture such as magnesium hydroxycarbonate. Tribasic magnesium phosphate is the preferred filler. The cobalt compound, where present, is preferably a cobalt II compound such as cobalt II hydroxide carbonate, orthophosphate or oxalate.

The enzymatic composition can advantageously be prepared as follows: After having preferably subjected the cells to a heat treatment, which results in sterilization of the cellular mass and destruction of certain undesirable enzymes (such as proteases) which are less resistant to heat than glucose-isomerase, the cells are separated by centrifugation from the substrate in which they have developed. The slurry obtained is then dried until it has a dry material content of 20 to 95%, preferably about 30% by weight. The resulting slurry is intimately mixed with a filler comprising a sparingly soluble magnesium compound and optionally a sparingly soluble cobalt compound, the proportion of this total filler being from 5 to 100%, preferably about 50%, by weight of the total weight of the cells.

The cellulose ester (e.g., the abovementioned industrial mixture of the diacetate and the triacetate) is then added to the resulting mixture in an amount of 25 to 100%, preferably about 50%, of the weight of dry cells and in the form of a 5 to 20% w/v, preferably about 10% w/v, sol in a water-miscible organic solvent which is preferably acetone. If desired, the cellulose ester may be added before the filler.

A water/solvent mixture similar to that used to form the cellulose ester sol is then added to the resulting slurry which contains the cellulose ester, in order to give the final mixture the appropriate consistency for producing the shape which it is proposed to employ. The amount of solvent to be added can be from 1 to 10 times the weight of the dry cells employed.

The amount of water which has been mixed with the solvent beforehand is such that, allowing for the moisture in the cells, the water contained in the final mixture represents 0.6 to 8 times, preferably about 3 times the weight of cellulose ester employed; furthermore, the final weight ratio of water/solvent should be less than 0.2:1.

Frequently, where the method of preparation of the cells containing the enzyme gives rise to cell walls which allow strong diffusion of the enzyme (depending on the conditions under which the microorganisms are cultured, heat-treated or separated), it is preferable to subject the cells to a mild treatment with glutaraldehyde. This treatment can be carried out either at the end of the culture, or on the cell slurry, preferably with the cellulose ester but before introducing the filler (it then being necessary for the latter to be introduced last), or again, as will be seen below, after the mixture has been shaped. In the case where the glutaraldehyde treatment is carried out on the cells or on the slurry of cells and ester, 0.5 to 6%, preferably about 3%, of glutaraldehyde based on the weight of dry cells, is normally used. The mixture with glutaraldehyde is preferably intimate and it is preferably left to react for 10 minutes to 1 hour at 10° to 30°C.

Finally, the fluid paste obtained, which may have been homogenized, e.g., by means of a propeller stirrer or a turbine stirrer, contains the cells, the filler, the cellulose ester, water, the solvent and optionally glutaraldehyde. This paste can then be spread as a thin layer (1 to 3 mm thick) on a substrate and left to dry, for example, in air at a temperature of 15° to 30°C. After about one hour, when it is not yet brittle, it can be cut up into small units (plates) having a side length of a few millimeters, e.g., 1 to 10 mm. After detaching the plates from the substrate, drying is continued without allowing the brittle phase to be reached.

If desired, it is at this stage of the preparation that the glutaraldehyde treatment can be carried out by dipping the dried plates in an aqueous solution of glutaraldehyde. The product obtained can then be used for isomerizing glucose solutions.

Recovery of Fructose from Plant Parts Using Inulase

Fructose is a natural compound with a greater sweetening power than natural sugar, and can be tolerated better by diabetics. Various fructose polymers occur in nature, notably inulin and levan. Inulin is contained in various plants, mainly in tubers and roots of plants belonging to the composite family. Levan occurs in places such as grasses and grains and as a product of bacterial polymerization. The recovery of fructose from inulin by chemical or enzymatic hydrolysis is known. Inulin itself is recovered from plant parts by extraction.

It is an objective of *P.L. Kerkhoffs; U.S. Patent 4,277,563; July 7, 1981; assigned to Stamicarbon, BV, Netherlands* to provide a simpler and more direct process for the preparation of fructose from plants containing a fructose polymer. According to this process, fructose is obtained by contacting minced vegetable parts containing a fructose polymer with an aqueous medium containing an enzyme that hydrolyzes fructose polymers and, after removal of the remaining vegetable matter, isolating the fructose obtained by enzymatic conversion by known methods.

The advantages of this process are that an extraction step is avoided and that a pure fructose is obtained that can be crystallized more readily than the fructose obtained by chemical hydrolysis. Another important advantage is that the process can be conducted in a very weakly acid to neutral medium. As a result, the formation of by-products, such as the acid-catalyzed formation of fructose dianhydride and the base-catalyzed aldol condensation, both undesirable by-products, is suppressed almost completely. In the conventional inulin extraction with acid and hydrolysis significant amounts of by-product do form.

The recovery of fructose from plant material containing levan is effected by the enzyme levanase; fructose is recovered from plant material containing inulin with the enzyme inulase.

The plant part used as the starting material is one containing inulin, as inulin is contained in considerable amounts in the tubers or roots of plants that are easy to grow. Some suitable vegetable materials for use as starting material include dahlia tubers, members of the genus Cichoreum, parsnip, fleawort, costus roots and Jerusalem artichokes. The parts of the plants rich in inulin are minced into a pulpy mass. Equipment already known in the sugar and starch industry may be used for this purpose, such as slicers, cutters and ball mills.

Inulase is a well-known enzyme capable of breaking up (1-2) fructose compounds. Inulase can be obtained from vegetable material, for example, from Jerusalem artichokes, but is obtained in practice from cultures of microorganisms, such as, i.e., *Saccharomyces fragilis, Aspergillus niger,* and *Helminthosporium oryzae.* The enzyme may be used in the crude or pure form or as a cellular mass that exhibits inulase activity. For convenience, the activity of the enzyme preparations will be expressed in standard units (SU), one SU being the activity required to produce 1 micromol of fructose per minute from inulin. In most cases an amount of enzyme preparation corresponding to an activity of between 100 and 3,000 SU is used per gram of inulin contained in the vegetable matter.

The temperature and pH in the reaction are preferably chosen such that the enzyme has the optimum activity under the reaction conditions employed. The pH will preferably range between 4.5 and 6.5. The temperature will preferably be between 40° and 75°C. The reaction is preferably conducted for a time to substantially completely convert all of the inulin present in the plant parts, generally 1 to 24 hours.

Upon termination of the reaction, the remaining vegetable matter is separated from the aqueous fructose-containing solution by filtration or centrifugation. The aqueous solution of fructose can be processed further in the usual way to purify and recover the fructose. Any residual enzyme present can be removed by ultrafiltration or treatment with an ion exchanger. The fructose obtained according to this process is found to crystallize better than the fructose hydrolyzed and extracted with dilute acid.

Hydrolyzed Product from Whole Grain

E. Conrad; U.S. Patent 4,282,319; August 4, 1981; assigned to Kockums Construction AB, Sweden describes a process for preparing, in situ, enzymatically hydrolyzed protein and starch products from whole grain. The grain is thoroughly crushed and afterwards is subjected to a treatment essentially consisting of both the following steps.

(a) An enzymatic treatment in an aqueous medium with an endopeptidase so as to transform substantially all water-insoluble proteins present in the grain to water-soluble protein products, which thereafter are filtered and recovered from the crushed grain as a clear filtrate containing protein products containing about 50 to 60% peptides having at least 25 amino acid residues, 35 to 45% peptides having between about 5 to 20 amino acid residues and 4 to 8% peptides having up to 4 amino acid residues.

(b) The remaining crushed grain is subjected to another enzymatic treatment in an aqueous medium with at least one starch hydrolyzing enzyme so as to transform substantially all of the water-insoluble starch fraction in the grain to water-soluble, degraded products of starch, consisting of mono- and disaccharides. The starch hydrolyzing enzyme is amylase substantially free from other carbohydrate hydrolyzing enzymes.

Upon completion of process steps (a) and (b) any remaining water-insoluble husk components of the grain such as bran and water-insoluble starch components are separated.

The whole grain which can be used for the process may be maize, rye, barley, oats or rice, but is preferably wheat.

One advantageous way to use the process is to utilize whey or concentrated whey as the aqueous medium for the enzymatic treatment. In this case, a suitable lactase can be added with the endopeptidase to simultaneously transform the whey's lactose to glucose and galactose.

All the important components of the cereal raw material are recovered in the end products. The final syrup, which can be filtrated through a standard filtration procedure can be used either directly as a nutrient or in combination with other nutrients in drinks, breakfast flakes or food for children. A prepared syrup is very suitable for bread baking purposes, as it is similar to flour with regard to its constituents. The baked products obtained are positively affected with regard to color, taste and freshness. The syrup can also be used in beer production.

Purification of Unrefined Sugar Syrups

Sugar suitable for use in soft drinks must be strictly refined so that no impurities in it can affect the character of soft drinks using it or shorten their shelf life. The steps for such refining are many and expensive, involving washing, flocculation, decolorization and recrystallization. In countries where sugar refining capacity is low, only mill sugar is available to soft drink manufacturers, that is, the raw cane juice has been treated with lime juice and phosphates, evaporated, crystallized, centrifuged to remove most of the molasses, and spray-washed.

Before using such mill sugars for soft drinks it is necessary to remove from them the turbidity, color, flavor, and odor, and this must be done in the bottling plant itself, where various methods for in-plant purification of mill sugar are in use.

Purification processes involving entrapment of sugar impurities in a chemical floc and subsequent removal of the floc are in use in sugar mills and in refineries, but almost never in bottling plants.

The floc clarification process consists of adding to the dissolved, usually hot, sugar syrup small amounts of lime and phosphoric acid, or lime and soluble phosphate salts or aluminum sulfate. At about neutral pH the lime and phosphate or aluminum sulfate form an insoluble, calcium phosphate or aluminum hydroxide floc (primary floc) which contains insoluble matter, some of the colloids, and much of the color. The most efficient method for removing the primary calcium phosphate or aluminum hydroxide floc is by flotation with a gas. Usually, with either type of floc, a polyelectrolyte is added to form a more easily flotatable secondary floc.

Flotation of the secondary floc is accomplished by aeration, either by vigorous agitation or aeration with powerful, high shear centrifugal pumps equipped with air inlets. Mechanical aeration, although suitable for refineries, is less suited for bottling plants because of the capital expense of the extra equipment needed and the added utility costs. Further, the aeration step for an average batch may take 15 to 120 minutes to complete, depending on the size of the pump, thus interfering with plant schedules.

It would be of great benefit, therefore, to develop a process for floc flotation which would avoid the capital expense and power costs associated with conventional aeration, and which would overcome the seeming impossibility of using enzymatically generated oxygen gas bubbles to flotate sugar syrup flocs.

G.V. Gudnason and J.E. Stell; U.S. Patent 4,288,551; September 8, 1981; assigned to The Coca-Cola Company have found that the disadvantages associated with in-plant sugar flotation processes can be overcome by using an improved secondary floc flotation process by utilizing oxygen bubbles generated from hydrogen peroxide by catalase for secondary floc flotation instead of the conventional aeration procedure.

The process for purifying impure sugar syrups consists of adding to the syrup a suitable amount of a phosphate ion source (such as a soluble calcium phosphate or phosphoric acid), adding lime to form a primary floc, adding a suitable amount of hydrogen peroxide, adding small amounts of a catalase preparation, which immediately produces copious amounts of oxygen bubbles evenly dispersed throughout the syrup, and then adding a small amount of a polyelectrolyte to capture the oxygen bubbles in a secondary floc, thereby causing the secondary floc to rise to form a well-packed scum. The purified syrup may then be easily and rapidly filtered with or without addition of small amounts of carbon and filter aid.

With floc flotation either by air or by oxygen bubbles, much decolorization and removal of turbidity takes place. However, an advantage of the oxygen flotation method is that significantly more decolorization takes place in oxygen flotation than in air flotation. Small amounts of activated carbon may be added before the floc flotation to further reduce color, flavor, and aroma. The carbon will be contained in the flotated scum.

A further important advantage of the process is that complete aeration with oxygen bubbles takes only 1 to 2 minutes since all the bubbles are formed at once throughout the syrup, whereas mechanical aeration involves pumping the entire amount of syrup once or more, and may take from 15 to 20 minutes. A further advantage of the process is that the primary floc is not broken up after it is formed so that flotation can take place almost immediately after aeration, whereas during the turbulent pumping needed for mechanical aeration, the primary floc is completely broken up and must be given time to reform.

The patent contains a detailed description of the process and a schematic representation of a suitable apparatus for the floc flotation.

The polyelectrolyte used in the examples illustrating the process was Magnifloc 846A, a polyacrylamide polyelectrolyte (American Cyanamid) and added as a 0.1% aqueous solution, and all catalase is Takamine Catalase-L, a beef liver catalase in a stabilized form (Miles Laboratories, Inc.).

Thermally Stable Beta-Galactosidase for Hydrolysis of Lactose

Lactose is the sugar component of whey, the liquid which remains after the separation of solids from milk or cream during customary cheese-making processes, and which is often regarded as a waste material and presents a disposal problem. In view of its low sweetness and solubility and tendency to undesirable crystallization, lactose is unsatisfactory for use as a food sugar. Lactose, however, is a dimer of glucose and galactose and may be hydrolyzed to its separate sugar units which are then readily utilizable as food sugars. Straightforward acid hydrolysis of lactose may be used, though this is not usually satisfactory because side reactions take place giving rise to a product having undesirable flavors. Alternatively hydrolysis may be effected by use of a β-galactosidase enzyme, although the temperature optimums for known β-galactosidase enzymes, i.e., from about 30° to 40°C up to a maximum of about 50°C, permit microbial growth which seriously contaminates the product.

New β-galactosidase enzymes which may be used for hydrolysis of lactose while avoiding the problem of contamination by adventitious microbial growth are desirable.

Accordingly, *M.W. Griffiths, D.D. Muir and J.D. Phillips; U.S. Patent 4,332,895; June 1, 1982; assigned to National Research Development Corporation* have provided a β-galactosidase enzyme having thermal stability such that it has an activity half life of at least 1½ hours at 55°C, at least 1 hour at 60°C and at least 10 minutes at 65°C, as measured using ONPG (o-nitrophenyl-β-D-galactopyranoside) as substrate, which is derived from a bacterium of species *Bacillus stearothermophilus.*

Preferred strains of *B. stearothermophilus* for derivation of the new β-galactosidase, enzyme compositions and enzymically active whole cell preparations are represented by the related group of strains recently discovered by workers at the Hannah Research Institute (HRI), Ayr, Scotland. Cultures of these strains of *B. stearothermophilus* were deposited with the National Collection of Industrial Bacteria in Aberdeen, Scotland and are identified by the reference numbers NCIB 11407, 11408, 11409, 11410, 11411, 11412 and 11413, which are hereinafter referred to as the HRI A/S 1, 2, 3, 4, 5, 6 and 8 strains respectively.

The enzyme, enzyme compositions and enzymatically active whole cell preparations are derived by first cultivating a culture of a suitable strain of *B. stearothermophilus,* typically carried out at elevated temperature, preferably about 65°C, on a suitable culture medium, such as a basal salts culture medium, comprising lactose as sole carbon source. Any suitable cultivation regime may be employed including either batch or continuous culture. The purified enzyme or partially purified enzyme extracts and compositions comprising the enzyme are obtained by extraction of enzyme from the cells, e.g., by lysing, followed by purification procedures as required.

Purified enzyme or enzyme extracts may be immobilized on or with a suitable solid phase material such as an ion-exchange material, e.g., DEAE-cellulose or like material, to provide preferred immobilized enzyme composition products.

Whole cell preparations may be subjected to preparative treatment to render the cells permeable, e.g., with toluene, and thus enhance apparent enzyme activity for lactose hydrolysis. Permeable whole cell preparations may preferably be immobilized, for instance, in a gel matrix to provide an advantageously immobilized enzymatically active whole cell preparation. Particularly preferred, however, are permeable whole cell preparations immobilized with a suitable ion-exchange support material, including in particular DEAE-cellulose and like ion-exchange materials such as amino ethyl cellulose, DEAE-Sephadex and DEAE-Sepharose.

Preferred processes for production of immobilized enzyme composition products or especially immobilized whole cell products, immobilized with ion-exchange support materials, comprise prior treatment of the ion exchange material with glutaraldehyde or a similar linking reagent so as to covalently attach the enzyme or cells to the ion exchange support material. Such prior treatment, in particular with DEAE-cellulose and like ion-exchange support materials, advantageously gives products having greater retention of enzyme activity than products prepared without such prior treatment, and preferably also products of outstanding thermal stability. For example, whole cell products immobilized on DEAE-cellulose which has been pretreated with glutaraldehyde often have activity half lives of at least 7 days, in some cases about 15 days, at 60°C and pH 7.

The products of this process may be used for conventional methods of the hydrolysis of lactose to glucose and galactose.

ARTIFICIAL SWEETENERS

Addition Compounds of Dipeptide and Amino Acid Derivatives

Y. Nonaka, K. Kihara, K. Oyama, H. Satoh, S. Nishimura, Y. Isowa, M. Ohmori, K. Mori and T. Ichikawa; U.S. Patent 4,256,836; March 17, 1981; assigned to Toyo Soda Manufacturing Co., Ltd. and (Zaidanhojin) Sagami Chemical Research Center, both of Japan describe a process for preparing addition compounds of a depeptide derivative and an amino acid derivative. This is accomplished by reacting N-substituted monoaminodicarboxylic acids of general formula (2) with aminocarboxylic acid ester having the general formula (3) to produce the addition compounds of general formula (1). This reaction is carried out in an aqueous medium in the presence of a protease.

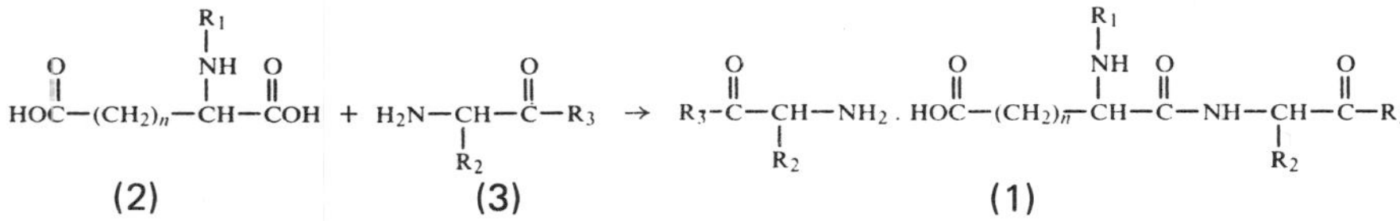

(2) (3) (1)

In the equation above, R_1 represents an aliphatic oxycarbonyl group, a benzyloxycarbonyl group which can have nuclear substituents, or a benzoyl, aromatic sulfonyl or aromatic sulfinyl group. R_2 represents a methyl, isopropyl, isobutyl, isoamyl or benzyl group; R_3 represents a lower alkoxy, benzyloxy or benzhydryloxy group and n is 1 or 2.

The N-substituted monoaminodicarboxylic acid having the formula (2) and the amino acid esters having formula (3) are in L-form or DL-form.

There is also provided a process for decomposing the addition compounds. The addition compounds are treated with an acidic solution for separating the solid component to obtain the corresponding dipeptide esters.

The process may be used specifically to produce α-L-aspartyl-L-phenylalanine alkyl esters for use as sweeteners. The methyl ester, for instance, has about 200 times the sweetness of sugar.

The proteases used are preferably metalloproteases which have a metal ion in the active center. Suitable metalloproteases are enzymes originating from microorganisms, such as neutral proteases from ray fungus, prolism, thermolysin, etc. It is also possible to use crude enzymes such as thermoase. In order to inhibit the action of esterase contained in the crude enzymes, it is preferable to use an enzyme inhibitor such as a potato inhibitor with the crude enzymes.

The pH range for the reaction is from 4 to 9, but a pH of 5 to 8 is preferable. The reaction is preferably carried out at a temperature of 20° to 50°C and is usually complete in from 30 minutes to 24 hours.

Suitable acids for decomposing the addition compounds are the customary organic or inorganic acids such as hydrochloric, sulfuric, phosphoric, formic, citric acids, etc.

Esterification of Alpha-L-Aspartyl-L-Phenylalanine with a Proteolytic Enzyme

A.A. Davino; U.S. Patent 4,293,648; October 6, 1981; assigned to G.D. Searle & Company has provided a process for the preparation of α-L-aspartyl-L-phenylalanine alkyl esters which are useful as sweetening agents. In particular, the dipeptide, α-L-aspartyl-L-phenylalanine is esterified by contacting the dipeptide with an alcohol in the presence of proteolytic enzyme having specific esterase activity in a water-alcohol solvent system in which the alcohol concentration is sufficient to reverse the hydrolytic action of the enzyme. Specific esterase activity refers to the proteolytic enzyme's ability to selectively alkylate the carboxyl group of the phenylalanine moiety. α-L-aspartyl-L-phenylalanine alkyl esters so produced are useful as sweetening agents as described in U.S. Patent 3,492,131.

Alternatively, N-protected-α-aspartyl-L-phenylalanine may be esterified by this process to give the corresponding N-protected-α-L-aspartyl-L-phenylalanine alkyl ester. The N-protecting group can be removed by conventional techniques to give α-L-aspartyl-L-phenylalanine alkyl ester. A preferred N-protecting group for use in the process is the carbobenzoxy group.

Preferred alcohols for use in the process are represented by the formula R—OH where R is lower alkyl having 1 to 7 carbon atoms. These lower alkyls may have straight or branched chains.

The esterase used in the process is a serine alkaline proteinase. These enzymes exhibit high esterase activity and selectively act on aromatic amino acids. Typical serine alkaline proteinases include subtilisin and alkaline proteinases from various strains of Bacillus, Aspergillus, Streptomyces, Penicillium, and Arthrobacter.

In an especially preferred embodiment the serine alkaline proteinase is subtilisin Carlsberg, a proteinase having specific esterase activity, and the alcohol is methanol which is present in a concentration of 60% by volume. The aqueous-alcohol mixture contains 5.0 mM calcium chloride and has a pH of 5.0, and the reaction is conducted at a temperature of 25°C.

The reaction proceeds smoothly under these conditions until completed. A preferred reaction period time is 1 to 90 hours. The reaction product is conveniently separated from the reaction system by standard chromatographic techniques and other methods recognized in the art.

PRODUCTION OF ETHANOL AND OTHER FUELS

ETHANOL PRODUCTION FROM STARCH

Using Manioc as Starting Material

With the ever-increasing depletion of economically recoverable petroleum reserves, the production of ethanol from vegetative sources as a partial or complete replacement for conventional fossil-based liquid fuels becomes more attractive. In some areas, the economic and technical feasibility of using a 90% unleaded gasoline-10% anhydrous ethanol blend ("gasohol") has shown encouraging results. According to a recent study, gasohol powered automobiles have averaged a 5% reduction in fuel compared to unleaded gasoline powered vehicles and have emitted one-third less carbon monoxide than the latter. In addition to offering promise as a practical and efficient fuel, biomass-derived ethanol in large quantities and at a competitive price has the potential in some areas for replacing certain petroleum-based chemical feedstocks. Thus, for example, ethanol can be catalytically dehydrated to ethylene, one of the most important of all chemical raw materials both in terms of quantity and versatility.

The various operations in processes for obtaining ethanol from such recurring sources as cellulose, cane sugar, amylaceous grains and tubers, e.g., the separation of starch granules from noncarbohydrate plant matter and other extraneous substances, the chemical and/or enzymatic hydrolysis of starch to fermentable sugar (liquefaction and saccharification), the fermentation of sugar to a dilute solution of ethanol ("beer") and the recovery of anhydrous ethanol by distillation, have been modified in numerous ways to achieve improvements in product yield, production rates and so forth.

For ethanol to realize its vast potential as a partial or total substitute for petroleum fuels or as a substitute chemical feedstock, it is necessary that the manufacturing process be as efficient in the use of energy as possible so as to maximize the energy return for the amount of ethanol produced and enhance the standing of the ethanol as an economically viable replacement for petroleum based raw materials. To date, however, relatively little concern has been given to the energy requirements for manufacturing ethanol from biomass and consequently, little effort has been made to minimize the thermal expenditure for carrying out any of the discrete operations involved in the manufacture of ethanol from vegetative sources.

The substitution of alcohol for at least a portion of petroleum based fuels is particularly critical for developing economies where proven domestic petroleum reserves are limited, such as in India and Brazil and these nations have therefore increasingly emphasized the production of alcohol from vegetative sources. The most common such operation employs cane sugar in a fermentation-distillation operation which conveniently utilizes the bagasse by-product as a fuel source. Cassava or manioc *(Manihot utilissima Pohl)* as a source of starch has also been considered for conversion into alcohol. However, since manioc lacks the equivalent of sugar cane's bagasse, the fuel for alcohol conversion must come from an external source. Thus, to make manioc root an economically attractive source of ethanol, it is essential to achieve rapid and high levels of conversion of the starch content to fermentable saccharide and of the fermentable saccharide to ethanol with high levels of thermal efficiency and at conservative plant investment and operating costs.

The composition of manioc is similar to other tropical starchy roots in that the bulk of the dried matter is carbohydrate, about 66 to 72% of which is starch in the form of granules of about 5 to 35 μ in dimension. Starch granules comprise amylose, a straight chain polymerized maltose and amylopectin, a branched chain polymerized maltose. However, cassava starch is distinguished from common sources of starch by its relatively low content, e.g., 17% of amylose as compared to potato starch (22%) and cornstarch (27%). Its corresponding relatively large percentage of branched chain amylopectin imparts different properties, requiring that its treatment in saccharification be somewhat unique.

Accordingly, there has heretofore existed a need for a process for hydrolyzing manioc root starch and starches from other sources at rapid and high levels of conversion without any significant degradation of the resulting saccharide and at only a modest expenditure of thermal energy and of utilizing the saccharide in a thermally efficient, rapid continuous fermentation process to provide industrial ethanol at competitive prices.

W.C. Muller and F.D. Miller; U.S. Patent 4,243,750; January 6, 1981; assigned to National Distillers and Chemical Corporation have developed such a process, in which an aqueous slurry of starch is first liquefied, i.e., converted to a pumpable partial hydrolysate, employing a liquefying agent selected from the group consisting of strong acid and liquefying enzyme.

The partial hydrolysate is thereafter saccharified in a primary saccharification vessel or vessels in the presence of a saccharifying enzyme for a period of time sufficient to convert from about 60 to 80 weight percent of the original starch present (calculated on a dry basis) to fermentable sugar with the remaining portion of the original starch being present in the form of partial hydrolysate. The fermentable sugar is introduced, with or without the partial hydrolysate therein having been previously further saccharified in a secondary saccharification vessel or vessels, into a series of fermentation vessels wherein fermentation of sugar to ethanol by yeast and the saccharification of partial hydrolysate, if present, to fermented sugar takes place and wherein the ethanol content of the fermentation medium in each fermentation vessel is progressively increased as the sugar content of the fermentation medium is consumed, there being at least two strains of yeast selected for the fermentation, one of which provides a high rate of ethanol production in a fermentation medium containing a relatively low concentration of ethanol and a relatively high concentration of fermentable sugar and the other of which provides a high rate of ethanol production in a fermentation medium containing a relatively high concentration of ethanol and a relatively low concentration of fermentable sugar.

Liquefaction is rapidly accomplished at elevated temperature and pressure employing a strong acid such as hydrochloric acid to effect partial hydrolysis, a liquefying enzyme such as alpha-amylase or a two-step procedure employing acid for initial liquefaction and liquefying enzyme for further liquefaction. Following the cooling and reduction in pressure of the liquefied starch, saccharification is carried out in the presence of a dextrogenic enzyme such as amyloglucosidase until conversion of starch to fermentable sugar has reached a level of from about 60 to 80 wt %.

The yeast selected for introduction into the fermentation vessel is one which provides high rates of ethanol production in the presence of relatively low concentrations of ethanol and relatively high concentrations of fermentable sugar. Yeasts which will perform in this manner can be selected employing known microbiological techniques. Thus, for example, several strains of yeast can be introduced into a laboratory of large scale fermentation vessel (e.g., a chemostat) in which initial ethanol, sugar and nutrient concentrations are noted and predetermined levels of temperature and pH are accurately maintained so as to simulate the conditions present in a commercial fermentation unit. As the different strains of yeast compete with one another for survival over a prolonged period which can be several weeks or even months, only one or a few strains will have survived, the surviving organisms being optimal producers of ethanol under the conditions selected for the operation of the fermentation unit.

Using the same procedure, the mutation of a single yeast organism to provide an optimal ethanol producer under the fermentation conditions selected can be induced. The foregoing screening procedure can also be used to evaluate and isolate selected strains of yeast produced by techniques of induced mutation, e.g., those employing ultraviolet radiation, gamma rays, etc., to accelerate the incidence of mutation. Other useful techniques for obtaining different strains of yeast for evaluation as ethanol producers under predetermined fermentation conditions include cross breeding of two different strains to yield a third and genetic engineering in which genetic materials from two different strains are recombined to form a completely new genetic "blueprint."

A yeast which has been found to provide especially good rates of ethanol production at relatively low concentrations of ethanol and relatively high concentrations of fermentable sugar is *Saccharomyces bayanus.* The yeast in the fermentation vessels is preferably present at a level of from about 3 to 6 weight percent of the fermentation medium (based on dry weight of yeast). Once continuous fermentation has started and a steady state has been achieved, there will be no need to add more yeast other than those amounts necessary to make up for cells which die.

Hydrolysis of an Aqueous Starch Slurry

While the enzyme hydrolysis of starch generally provides sugar which is of a superior quality compared to that obtained by acid hydrolysis, the somewhat longer conversion times and the relatively high cost of enzymes tend to militate against the use of enzyme hydrolysis methods where maximum rates and levels of sugar production, not product quality, is of foremost consideration.

Accordingly, there exists a need for a process for hydrolyzing an aqueous starch slurry to a solution of fermentable sugar starch at rapid and high levels of conversion while dispensing with or minimizing the need to neutralize acid present in the sugar solution.

W.C. Muller and F.D. Miller; U.S. Patent 4,266,027; May 5, 1981; assigned to National Distillers & Chemical Corporation have suggested such a process, in which an aqueous slurry of starch is first liquefied, i.e., converted to a pumpable partial hydrolysate, employing a liquefying agent selected from the group consisting of strong acid and liquefying enzyme and the partial hydrolysate is thereafter saccharified in the presence of an acidic cationic exchange resin to provide an aqueous solution of fermentable sugar containing little free acid. Any amount of acid which may be present is of a relatively low level of magnitude and its neutralization with base will result in the production of but a correspondingly small amount of salt.

A concentrated aqueous slurry of starch which contains from about 30 to 40 weight percent dry substance is delivered from a starch slurry tank to a steam jet mixer where it is combined with steam and thereafter passed through a starch liquefier. Prior to introduction into the steam jet mixer, the starch slurry is combined with a strong acid as partial, but preferably exclusive, liquefying agent to provide a pH of preferably from about 1.2 to 2.2. Suitable acids include nitric acid, sulfuric acid, hydrochloric acid and phosphoric acid.

It is further within the scope of this process to employ a liquefying enzyme, e.g., alpha-amylase, for the liquefaction step thereby dispensing with an acid neutralization procedure prior to the use of the product sugar liquor in ethanol fermentation.

The liquefied starch is conveyed to a saccharifying vessel containing an acidic cationic exchange resin for catalyzing the further hydrolysis of the starch to fermentable sugar. In the event enzyme is employed in the liquefying operation, the cation exchange resin can also be used for recovery of the enzyme.

Examples of suitable cationic exchangers are Amberlite IR-120 and Amberlyst 15 (Rohm and Haas Company), Chempro-20 andDuolite C-25 (Chemical Process Company), Dowex 50 (Dow Chemical Company), Nalcite HCR (National Aluminate Corporation) and Permutit Q (Permutit Company).

Following saccharification, the sterile sugar liquor is conveyed to a flash tank where steam is used to adiabatically cool the liquid mass, preferably to about 212°F. Alternatively, the vapors are directly contacted with a water jet condenser supplied with cold water with the liquid condensate passing to a sump sewer. When any acid has been used, either for accomplishing initial hydrolysis or for optimizing enzyme activity, it is necessary to add a neutralizing agent to the sugar liquor in at least a stoichiometric amount prior to subjecting the sugar to fermentation. Advantageously, the acid is neutralized with ammonia. The ammonium nitrate, sulfate, chloride and/or phosphate which is produced by neutralization of the acid is retained in the product fermentable sugar produced by the process and satisfies a nutritional need of the yeast used in the conversion of the sugar to ethanol. The neutralized sugar liquor is delivered to a storage vessel or, if desired, directly to a fermentation unit.

In another patent, *F.D. Miller and W.C. Muller; U.S. Patent 4,330,625; May 18, 1982; assigned to National Distillers & Chemical Corporation* adjust the process described in the previous patent to make it more suitable for use in converting starch derived from dry milled whole grain to ethanol in thermally efficient large-scale production. The process is modified in the following way. First, the aqueous slurry of the starch is subjected to a preliminary hydrolysis, catalyzed by a strong acid.

Typically, this first preliminary hydrolysis is conducted for a period of time which will yield a slurry containing preferably from about 16 to 20 dextrose equivalents. Under the conditions of acid hydrolysis, the accompanying water-insoluble protein and oil and the water-soluble components of the starch will remain largely unaffected. The partially hydrolyzed starch emerging from the liquefier is then conveyed to a first centrifuge, filter or other separating device where an aqueous partially hydrolyzed starch stream is recovered to undergo a further, final hydrolysis to fermentable sugar in the starch saccharifying unit and a stream of protein and oil is recovered as overflow to be washed with fresh water in an overflow stream washing unit.

The washed aqueous stream of protein and oil is then conveyed to an oil separating unit, in which the oil is removed in a known or conventional manner, e.g., extraction with a solvent such as n-hexane. The deoiled protein stream is then passed to a second centrifuge, filter or other separating device with the aqueous underflow being used as process water elsewhere in the system, e.g., for preparation of the starch slurry, and the protein which is recovered as overflow being recovered as a valuable by-product.

The aqueous partial starch hydrolysate fraction is thereafter subjected to a further, or final, acidic cationic exchange resin catalyzed hydrolysis to provide an aqueous solution of fermentable sugar containing little free acid. The amount of free acid which is present as a result of the preliminary hydrolysis step is of a relatively low level of magnitude and its neutralization with base will result in the production of but a correspondingly small amount of salt. Neutralization of the free acid present in the sugar liquor is again preferably performed with ammonia.

Employing the foregoing starch hydrolysis process, only minimal quantities of fresh water need be used to accomplish conversion of the starch to fermentable sugar thus reducing the amount of water which must be removed from product ethanol obtained from the fermentation of the sugar, and consequently, the amount of thermal energy which must be expended in the manufacture of the ethanol. Moreover, substantially all of the water-insoluble protein contained in the original starch can be recovered for other commercially valuable uses, notably animal feed, and due to the water recycle feature which is made possible by the process herein, a good portion of the water-soluble components of the starch are retained in the solution of product fermentable sugar and are therefore available for satisfying certain nutrient requirements of the yeast employed in the fermentation of the sugar to ethanol.

The process herein with appropriate modification is also applicable to the hydrolysis of starch contained in degerminated cereal grains, i.e., grains from which a portion of the oil has been removed, dehulled grains, and degerminated and dehulled grains.

Fermentable Sugar from Dry Milled Corn

Processes for the enzymatic hydrolysis of starch and the acid hydrolysis of starch to provide fermentable sugars are well known. Such processes applied to the starch-containing fractions from dry milling corn procedures are wasteful of the protein and edible oils from these fractions.

Accordingly, there has heretofore existed a need for a process for converting starch derived from dry milled corn to fermentable sugars while recovering substantially all of the protein and oil content of the starch component of the dry milled corn prior to the complete hydrolysis of the starch.

W.C. Muller and F.D. Miller; U.S. Patent 4,287,304; September 1, 1981; assigned to National Distillers & Chemical Corporation have therefore provided a process for converting starch derived from dry milled whole corn, which starch contains relatively substantial amounts of water-insoluble protein and oil and relatively small amounts of one or more water-soluble components selected from the group consisting of sugar, lipid, protein, vitamin and mineral, to fermentable sugar to provide substrate for the thermally efficient large-scale production of ethanol.

An aqueous slurry of the starch is subjected to a mild, i.e., thinning or liquefying hydrolysis to provide a sterile partial starch hydrolysate containing the water-insoluble protein and oil and the water-soluble components of the starch in a substantially unaltered condition. The slurry is then separated into an aqueous partial starch hydrolysate portion containing a part of the water-soluble components and a water-insoluble protein and oil portion containing the remaining part of the water-soluble components. The aqueous partial starch hydrolysate portion is subjected to further hydrolysis and the resulting aqueous solution of fermentable sugar together with part of the water-soluble component of the original starch feed is conveyed to a fermentation unit where conversion of the sugar to ethanol and further hydrolysis of any remaining partial starch hydrolysate to fermentable sugar takes place.

The water-insoluble protein and oil portion may be combined with water to dissolve the water-soluble components associated therewith with the resulting aqueous slurry thereafter being separated into a protein and oil portion substantially free of any of the water-soluble components of the original starch, and an aqueous portion containing water-soluble components. The protein and oil may be used directly in animal feed or, if desired, they may be separately recovered for individual use. The aqueous portion containing water-soluble components of the starch is advantageously recycled for use in mildly hydrolyzing another quantity of starch.

Employing the foregoing starch hydrolysis process, only minimal quantities of fresh water need be used to accomplish conversion of the starch to fermentable sugar thus reducing the amount of water which must be removed from product ethanol obtained from the fermentation of the sugar, and consequently, the amount of thermal energy which must be expended in the manufacture of the ethanol. Moreover, substantially all of the water-insoluble protein contained in the original starch can be recovered for other commercially valuable uses, notably animal feed, and due to the water recycle feature which is made possible by the process herein, a good portion of the water-soluble components of the starch are retained in the solution of product fermentable sugar and are therefore available for satisfying certain nutrient requirements of the yeast employed in the fermentation of the sugar to ethanol.

Acid Hydrolysis Followed by Continuous Fermentation

Another process has been developed by *W.C. Muller and F.D. Miller; U.S. Patent 4,291,124; September 22, 1981; assigned to National Distillers & Chemical Corporation* designed particularly for using manioc root starch and other carbohydrate polymers from vegetative sources for fuel for alcohol conversion.

In this process, an acidified aqueous slurry of carbohydrate polymer particles such as starch granules and/or cellulose chips, fibers, etc., is mixed with steam under pressure and conveyed through a conduit at substantially steady state temperature and pressure for a period sufficient to accomplish at least about 60% weight conversion, and preferably at least about 80% weight conversion, of the carbohydrate polymer to fermentable hydrolysate but without appreciable conversion of carbohydrate polymer to nonfermentable products.

The pressurized acidified carbohydrate slurry is then passed through an expansion valve or critical flow orifice resulting in an abrupt reduction of pressure on the product hydrolysate and a cessation of, or considerable reduction in, further hydrolyzing activity. This abrupt reduction in pressure serves to prevent or diminish the likelihood of any further reaction tending to produce unfermentable reversion or decomposition products and thus maximizes the yield of useful hydrolysate, i.e., fermentable sugar.

Following conversion of the carbohydrate polymer to fermentable sugar, the latter is introduced, with or without any partial hydrolysate which may be present therein having been previously further saccharified, into a series of fermentation vessels wherein fermentation of sugar to ethanol by yeast and the saccharification of partial hydrolysate, if present, to fermented sugar takes place. The ethanol content of the fermentation medium in each fermentation vessel is progressively increased as the sugar content of the fermentation medium is consumed, there being at least two strains of yeast selected for the fermentation, one of which provides a high rate of ethanol production in a fermentation medium containing a relatively low concentration of ethanol and a relatively high concentration of fermentable sugar and the other of which provides a high rate of ethanol production in a fermentation medium containing a relatively high concentration of ethanol and a low concentration of fermentable sugar.

Employing two or more organisms which maintain high rates of ethanol production in the presence of different concentrations of ethanol and fermentable sugar provides a faster, more efficient fermentation than that attainable employing a single strain of yeast in each fermentation vessel as is the current practice. In this way, this fermentation process is particularly well suited for the production of ethanol which is price competitive with ethanol produced from nonvegetative sources.

A yeast which has been found to provide especially good rates of ethanol production at relatively low concentrations of ethanol and relatively high concentrations of fermentable sugar is *Saccharomyces bayanus.* A strain of yeast which provides a high rate of ethanol production in a fermentation medium containing a relatively high concentration of ethanol and a relatively low concentration of sugar is *Saccharomyces cerevisiae.*

Continuous Fermentation Process Using Two Yeast Strains

In the various known processes for obtaining fermentable sugar from starch by hydrolysis of the latter, particularly those which employ acid as the hydrolyzing agent, some of the product fermentable sugar will undergo chemical modification to provide oligomers, for example, dimers and/or trimers, which are not readily converted to ethanol using yeasts commonly encountered in the brewing industry. The oligomers which have resisted conversion to ethanol and are therefore present in the product "beer" stream at the conclusion of fermentation are recovered in the distillation bottoms during the process of concentrating the ethanol. Until now, it has been necessary to rehydrolyze the recovered oligomers to fermentable sugar and recycle the sugar rehydrolysate to fermentation when maximum total utilization of raw material for the production of ethanol is desired.

However, the additional capital investment needed to provide a plant having the capability to recover, rehydrolyze and recycle the oligomers, and the increased operational complexity and consumption of energy which this capability necessarily entails, have to date militated against the general adoption of the foregoing procedures. Thus, an otherwise valuable raw material for ethanol production, saccharide oligomer, is either being routinely discarded or diverted to uses other than ethanol production.

W.C. Muller and F.D. Miller; U.S. Patent 4,315,987; February 16, 1982; assigned to National Distillers & Chemical Corporation have found, however, that the use of two or more yeast organisms, as described in the previous patent, results in a significantly greater degree of direct carbohydrate conversion to ethanol compared to that provided by the use of a single strain of yeast in the current practice, and makes the product much more price-competitive with ethanol which has been produced from nonvegetative sources.

The strain of yeast which provides a relatively high rate of conversion of fermentable sugar to ethanol is *Saccharomyces cerevisiae*, and that which provides a relatively high rate of conversion of fermentable sugar oligomer to ethanol is *Saccharomyces carlsbergensis* or *Saccharomyces cerevisiae* var. *ellipsoideus.*

Continuous Fermentation Process Minimizing Thermal Expenditure

In the continuous ethanol fermentation process such as that previously described, an important improvement has been made by *W.C. Muller and F.D. Miller; U.S. Patent 4,310,629; January 12, 1982; assigned to National Distillers & Chemical Corporation.*

This comprises separating effluent from the first fermentation vessel into a first yeast stream and a first substantially yeast-free aqueous ethanol stream; recycling as much of the first yeast stream as necessary to the first fermentation vessel to maintain a high level of yeast cells therein; introducing the remaining portion of the first yeast stream to the second fermentation vessel; introducing the first, substantially yeast-free aqueous ethanol stream into the second fermentation vessel; separating effluent from the second fermentation vessel into a second yeast stream and a second, substantially yeast-free aqueous ethanol stream; and recycling as much of the second yeast stream as necessary to the second fermentation vessel to maintain a high level of yeast cells therein.

By providing for yeast recycle which is totally independent of fermentation conditions, it is possible to readily and conveniently maintain constantly high levels of yeast in both fermentation vessels so as to obtain an optimum overall rate of ethanol production.

Fermentation Conversion of Granular Starch

The conventional fermentation for production of ethanol from starch or starchy substances includes a starch liquefaction and hydrolysis sequence to convert the starch into a glucose solution which becomes the growth medium for the yeast organism, *Saccharomyces cerevisiae.*

Efforts by the art to integrate low temperature enzymatic liquefaction of starch with conduct of the fermentation so as to achieve improvements in the fermentation process, e.g., improved fermentation efficiency, and/or reduced fermentation time, and/or fermentation with a fermentation broth containing more than 25% solids content (to reduce water and energy costs) have, however, been lacking.

The rationale of the process developed by *N.W. Lützen; U.S. Patent 4,316,956; February 23, 1982; assigned to Novo Industri A/S, Denmark* derives from a discovery that the fermentation may be carried out on a slurry of solid and completely ungelled starch, i.e., granular starch, dosed with alpha-amylase and glucoamylase. During the course of the fermentation the starch is enzymatically liquefied and saccharified into fermentable sugars and the sugars are fermented. Control over the fermentation rate is possible through variations in the starch concentration in the slurry, by preconditioning of the starch, and through variations in the concentration and proportions of alpha-amylase and glucoamylase in the slurry.

Briefly stated, the process involves use of a fermentation medium which contains suspended granular starch particles, alpha-amylase and a glucoamylase. The enzymes are relied upon to liquefy the solid starch particles and to saccharify the dissolved starch to the fermentables glucose and maltose during the course of the fermentation.

The proportions of granular starch in the medium and of each enzyme are set to provide a controlled release of fermentables to the yeast, allowing thereby control over the fermentation reactions. Characteristic of the process is recovery of enzymes from the fermentation medium for use anew. Any undissolved starch particles remaining in the fermentation broth at the time the fermentation is halted contain thereon considerable amounts of alpha-amylase. Removal of residual starch particles from the fermentation broth for a later refermentation is a preferred way to recover enzymes.

For fermenting high starch concentration slurries (35 to 40% the preferred range) the following type of operation may be used. Conveniently, the starch slurry from a wet corn milling operation, usually close to 40% starch by weight, may be fed directly into the following fermentation system.

The fresh granular starch, glucoamylase, alpha-amylase, brewer's yeast, and water containing the essential nutrients are charged into a fermenter. The fermented broth passes to a centrifugal separator wherein yeast and unconverted granular starch are separated out for recycle by way of a line back to the fermenter. Some of the yeast and starch is removed to a purge line. The centrifugal broth passes into a still for separation into the alcohol overhead, removed through one line, and still bottoms, removed by way of another line. Some of the stillage is purged through yet another line, while the balance is cooled in a heat exchanger before return to the fermenter. This mode can, of course, be constructed for continuous operation.

The enzymes used are, of course, the glucoamylase and either a fungal alpha-amylase from *Aspergillus oryzae* (such as the commercially available Fungamyl) or a Bacillus alpha-amylase from *B. licheniformis* (available as Termamyl) which is stable enough to withstand brief exposures to still pot temperatures. As much as 75% of the alpha-amylase and 15% of the glucoamylase values are recoverable in the starch recycle, making the process an economical one.

PROCESS IMPROVEMENTS

Decreasing Energy Demand and Water Consumption with Increasing Yield

The process of *B.I. Dahlberg, L.K.J. Ehnström and C.R. Keim; U.S. Patent 4,287,303; September 1, 1981; assigned to Alfa-Laval AB, Sweden* is one for the production of ethanol from a raw material flow consisting of a carbohydrate-containing substrate mixed with cellulose-containing fibers and/or other nonfermentable, solid material.

This raw material flow is separated into one flow rich in solid substance and one substrate flow free from solid substance which is fermented in one or several fermenters. The flow of fermentation liquor is then separated by centrifugal separation into at least one yeast concentrate flow and one yeast-free flow, of which the yeast concentrate flow is recirculated to the fermenter, while the yeast-free flow is separated into one flow enriched in ethanol and one residual flow, of which at least part is recirculated to the fermenter.

The objectives of Dahlberg et al, in this patent, were to minimize the energy demand and water consumption of the process, while increasing the sugar yield to be fermented into alcohol. These objectives are met by bringing the flow which is heavy in solids content into contact with at least part of the residual flow, the mixture thus obtained being separated partly into a flow of solid substance, from which the remaining substrate has substantially been removed, and partly into one flow, enriched in substrate, which flow is recirculated to the fermenter.

In one suitable embodiment of the method, the mixture is separated by at least one sieve, a flow rich in solids being fed to the sieve together with at least part of the residual flow. The solids being retained against the passage through the sieve by means well known in the art, producing a liquid fraction, enriched in substrate, which passes through the sieve and is recirculated to the fermenter.

The mixture can also be separated by at least one centrifugal separator provided with a rotor, journalled horizontally in bearings and a conveying screw, arranged coaxially within the rotor for discharge of a separated heavy phase into one solid substance fraction, which forms the heavy phase, and into one liquid fraction, enriched in substrate, which is recirculated to the fermenter.

Vacuum sieves can also be used for the separation into solid phase fraction and substrate-enriched liquid fraction.

It is especially advantageous to bring the flow which is rich in solids into contact with at least part of the residual flow by providing several separation means in series, the flow rich in solid substance and at least part of the residual flow being brought to move countercurrently.

Such countercurrent washing of substrate from the flow, rich in solid substance, can also be performed in different types of extractors, for instance in the form of columns, provided with so-called extraction trays.

In one especially suitable embodiment the yeast-free flow is separated by a distillative method into one flow enriched in ethanol and one residual flow, called slop, which is thus utilized for transferring the substrate that is adhering to the solid substance to the fermenter.

Minimizing Energy Requirements of Fermentation Process

Fermentation to produce ethyl alcohol is itself well known, but the existing processes require large amounts of heat and are slow and cumbersome such that, even using complex equipment to maximize efficiency, it is questionable as to whether the energy which can be obtained by combustion of the alcohol product will exceed the energy needed to produce and distill the alcohol.

The overall thrust of the process of *L.B. Crombie; U.S. Patent 4,306,023; December 15, 1981* is to simplify the needed construction and operation, to enable continuity of operation, and to minimize the energy needed to produce the alcohol.

This process of producing ethyl alcohol comprises separately supplying to a wort within a fermentation tank:

(1) a moist particulate feed containing a component convertible to a fermentable sugar and including at least about 20% by weight water, but insufficient water to provide a continuous aqueous phase in the particulate feed. This moist feed is sprayed into the fermentation tank by means of an auger feed with a restricted outlet, by which the moist feed is under pressure and at a temperature in excess of about 250°F, under which conditions rapid conversion to fermentable sugar is induced. The moist feed is maintained at a temperature of about 300° to 450°F and is sprayed into the wort; and

(2) an aqueous liquor which mixes with the converted feed and is added to the wort, which contains yeast or some other microorganism for converting the fermentable sugar to ethyl alcohol. This wort is maintained at an elevated temperature at which the yeast or other microorganism can reproduce and convert the fermentable sugar to ethyl alcohol.

A mixture of air and carbon dioxide gas exiting from the fermentation tank are bubbled through the wort to agitate it and to maximize the fermentation process. This process is carried out continuously, a portion of the wort being continuously or periodically removed from the fermentation tank, solids being removed from the portion of the wort which is removed to provide a beer, and the beer being distilled to remove alcohol and provide a sugar-containing aqueous liquor. The sugar-containing aqueous liquor is recycled to the fermentation tank to provide a portion of the aqueous liquor used in component (2) above.

The moist particulate feed (component 1) contains starch and is heated to a temperature of at least about 350°F for a period of at least 7 to 10 seconds. It may be ground corn.

The moist feed contains added acid to lower its pH and thereby minimize the time and temperature needed to produce fermentable sugars and to enhance fermentation conditions.

The temperature of the wort is prevented from becoming excessive by increasing the ratio of aqueous liquor to feed or by increasing the amount of fresh cold water in the aqueous liquid supplied to the fermentation tank.

Using Ethanol-Containing Beer as a Recycled Diluent in Hydrolysis

The manufacture of ethanol from cellulose-containing materials has been known for many years. Two reactions are needed. The first is the hydrolysis of cellulose into fermentable sugars, and this reaction is also referred to as "saccharification". The second reaction involves the conversion of the sugars to ethanol, and this is commonly done by a fermentation with yeast.

The commercial manufacture of ethanol from cellulose has lagged because of several major process problems involving saccharification, the reactivity of the raw materials and the energy required to purify the ethanol.

The objectives of *W.H. Hoge; U.S. Patent 4,321,328; March 23, 1982* are to provide a commercially practical process for producing ethanol from cellulose by various changes in the saccharification and fermentation procedures. The essential features of this process comprise the following:

(a) forming a slurry of defibered cellulosic material by mechanically defibering the cellulosic material with beer which is recycled from the fermentation reactor, this beer containing ethanol and saccharifying enzyme;

(b) subjecting the slurry mass to saccharification with recyled enzyme supplemented by newly added enzyme;

(c) subjecting the saccharified slurry mass, containing sugars, ethanol, saccharifying enzyme and unreactive materials, to fermentation by yeasts;

(d) recycling of the ethanol-containing beer from the fermentation vessel back to the saccharification reaction mass, which results in increasing the ethanol concentration in the feed liquid which is ultimately passed to a distillation column for the recovery of ethanol. The entire process is carried out under substantially atmospheric pressure conditions and without vacuum treatment.

Various microbiological agents and mixtures of such materials can be used as saccharifying agents to form sugars from incoming raw materials in accordance with this process. If the incoming cellulosic material contains almost no starchy material the saccharifying agent would be cellulase enzyme containing cellobiase activity. Further, where the incoming cellulosic material contains starchy impurities, e.g., such as in garbage, or if it contains a mixture of starch and cellulose, e.g., such as in manioc root, it is preferable to include small quantities of alpha-amylase and glucoamylase enzymes which are commonly used in the processing of cornstarches. Where the starting cellulosic material is woody material with a high lignin content, it is desirable to include lignin-attacking agents such as the pleiotropic mutants of wood-rotting fungus, *Polyporus adustus.* This feature is described in the *Canadian Journal of Microbiology,* 20, (1974) pages 371-378.

Various pretreatments of the cellulosic materials obviously may be incorporated into the cellulose preparation system for the purpose of making the cellulosic materials more reactive. Such pretreatments may include swelling with alkaline materials, acid treatments and irradiation.

Process and Special Apparatus for Fermentation of Corn, Wheat, Milo, etc.

J.E. Lionelle, J.A. Staffa and W.L. McCormick; U.S. Patent 4,347,321; August 31, 1982; assigned to Bio-Systems Research, Inc. have developed a process for producing alcohol from fermentable substances which is efficient and easy to carry out. The apparatus designed for the process is easy to operate and durable.

The apparatus in general consists of a steam boiler, a mash cooking, fermentation, and beer boiling vessel and a stripping vessel. The steam boiler is a conventional design and is fired by any convenient fuel such as gas, wood, coal, etc.

In a specific and preferred embodiment of the process, construction of all vessels, the condenser, the packed columns and all conduits and the like which are contacted by liquid or vapor therein, are fabricated of stainless steel. The cooking vessel has a capacity of 396 gallons and the stripping vessel has a capacity of 75 gallons. The stripping columns are each five feet long and eight inches in diameter and are packed with ½ inch porcelain balls.

Three hundred gallons of water are introduced into the cooking vessel. A gas fired boiler supplies steam to the cook vessel jacket to heat the water in it. One half pound of ground limestone is added through an access port to ensure a source of calcium for the proper action of the alpha-amylase enzyme. When the water temperature reaches about 110°F, 15 bushels of finely ground corn are added. Citric acid is added to adjust the pH to about 6.0. When the slurry is heated to 135°F, 8 ounces of alpha-amylase enzyme (Takatherm) is added, the amount being according to the manufacturer's recommendation.

The mash is brought nearly to a boil and an additional 200 ml of alpha-amylase enzyme is added. The mash is cooked at about 190°F until all the starch is broken down, generally about 30 minutes. A sample port is provided on the cooking vessel to permit sampling of the mash to detect the presence of starch by the iodine test. Cooking is terminated when the test indicates the absence of starch. Agitation is preferably employed during the entire mash cooking step, including preliminary heating.

Additional water is added to bring the total volume of the cooked mash to about 360 gallons. This addition of water will cool the cooked mash to about 155°F and further cooling is accomplished by running cooling water through the jacket. Two pounds of urea are added to insure a nitrogen source for the yeast. The pH is adjusted with citric acid to 4.0 to 4.5 and two pounds of glucoamylase (Diazyme L-100D) are added. When the temperature reaches 85°F, the vessel is inoculated with 1.0 lb of baker's yeast and the system is sealed and provided with a water trap gas vent. After about 1 minute of agitation, the agitation is turned off and fermentation is allowed to continue until CO_2 evolution ceases (generally about 20 to 48 hours). Temperature within the vessel is maintained at an appropriate temperature, depending upon the particular formulation, by the admission of cooling water and/or warm water into the jacket as fermentation progresses. Generally fermentation will be carried out at a temperature of preferably, 70° to 90°F. The beer thus produced has an alcoholic content of about 5 to 10%, 10-20 proof.

After fermentation is complete, steam is introduced to boil the beer present in the cooking vessel. Once equilibrium is reached and liquid alcohol product passes out through a conduit, the temperature of the stripping vessel is maintained at about 180° to 195°F. This temperature is readily maintained by the flow rate of coolant water through cooling conduits under steady conditions of heat input to the cooking vessel. When distillation proceeds to the point at which it is no longer possible to maintain the temperature of the stripping vessel at about 192°F, distillation is terminated. Yield is about 16 to 18 gallons of 192 proof alcohol and the amount of energy consumed is about 31,500 Btu per gallon of alcohol product. Distillation required about 2.5 to 3 hours, cooking requires about 3 to 6 hours, and fermentation requires about 20 to 36 hours. Average total time is 38 hours. Thus, yearly yield of ethanol from a plant of this size on an around-the-clock basis is over 4,000 gallons of 192 proof alcohol. A practical yield is thus about 2,500 gallons per year.

Various types of fermentable materials can be used in the process such as corn, wheat, milo and the like. It is preferred that the product has a proof of at least 190 and more preferably 192. The pH for mash cooking is preferably about 5.5 to 6.8 and for fermentation is preferably about 3.0 to 5.0. It will be apparent from the foregoing description that no external heat needs to be applied to the cooking vessel (or to the stripping columns). The absence of an external heat source for the stripping vessel is preferred to minimize energy requirements.

When fermentation is complete, the contents of the cooking vessel are removed through a discharge port. The solid residue contains about 28% protein and can be used as an animal feed. Some or all of the dregs in the stripping vessel, containing about 70% alcohol, are recycled to the cooking vessel to increase overall yield.

Continuous Fermentation Requiring Little or No Oxygen

J.D. Bu'Lock; U.S. Patent 4,357,424; November 2, 1982; assigned to Sim-Chem Limited, England describes a process for the continuous production of fermentation alcohol. The process requires little or no oxygen and maintains the treatment liquor in a uniformly mixed condition thus promoting the continuous growth of active biomass in the treatment system.

The apparatus generally comprises a reaction vessel of high aspect ratio connected near its upper end to a degassing vessel which is similar in cross-section to the vessel. The reaction vessel is equipped with a gas diffuser or sparge in the lower region of the vessel, and a gas input line beneath the sparge. The upper ends of the two vessels communicate with gas or vapor lines which join together and, via a pressure regulator, lead either to waste or to a compressor.

The output of the compressor is connected via an adjustable valve to the line feeding gas to the base of the reaction vessel. A feed line is also connected to the valve to permit the introduction into the feeding line of air or other treatment gas, whereby the resultant gas mixture is arranged to pass into the reaction vessel.

A settling vessel is connected to the degassing vessel at its lower end, and the lower end of the settling vessel is connected to the lower end region of the reaction vessel above the sparge. A line is provided for the removal from the settling vessel of evolved gases and/or vapors, and this is connected to the lines at the top of the reaction vessel.

The process can be carried out using the described apparatus in the following manner. In the instance of an ethanol/yeast fermentation, the process is initiated by filling the system with a process medium such as water and glucose, and this medium is inoculated with a yeast suspension. Initially, a supply of air alone is introduced into the gas input line for injection into the reaction vessel. Fermentation commences, and the culture is maintained in this manner until an adequate population of aerobically grown yeast has developed in the reaction vessel, at which time the valve is adjusted so that the air supply is largely replaced by carbon dioxide being recycled by the compressor from the upper ends of the three vessels.

A small proportion of air is advantageously retained, since the ethanol/yeast fermentation is, strictly, microaerobic rather than truly anaerobic. In cases where the fermentation is effected with a strictly anaerobic microorganism, a different procedure will be necessary in which the initial supply of gas is of some inert gas such as nitrogen or carbon dioxide, or a mixture of such, from an external source, and this will be replaced as soon as possible by the recycled gases generated in the fermentation process.

Once the system is running as described, then a continuous supply and removal of media at an appropriate rate may be established.

As will be seen, therefore, in operation of the process the introduction of substrate into the reaction vessel is continuous, and the liquid/biomass mixture passes progressively into the degassing vessel. In the degassing stage of the process, conditions exist which are less turbulent than those in the reaction vessel and there is present a relatively low availability of the original substrate, so that the production of carbon dioxide by the biomass is reduced and the conditions permit the residual gas to be given off thus promoting separation of the degassed biomass from the fermented liquor owing to the relative specific gravities thereof.

Thus a part or the whole of the descending degassed biomass, and the liquid phase surrounding same, passes into the settling vessel where the direction and magnitude of the liquor flow towards the outlet permits the suspended biomass to settle out and so return to the reaction vessel to assist in maintaining the fermentation process.

It will be seen that this process enables substantial productivity benefits to be achieved by the increased concentration of biomass in the fermentation system and this is attained by simple means in an enclosed system, without the use of pumps or centrifuges or other mechanical devices and without the dangers of excessive solids accumulation in so-called "dead spots" where liquid flow might otherwise be impeded. It is expected that considerable biomass concentrations can be achieved yet still maintained in a well mixed condition so far as the main body of the fermentation liquor is concerned.

PROCESSES USING SPECIAL MICROORGANISMS OR ENZYMES

Thermophilic Ethanol-Forming Anaerobe for Use on Cellulose

Relatively few anaerobic microorganisms have been isolated and characterized that grow on carbohydrates (are glycolytic) and yield ethanol under thermophilic and extreme thermophilic conditions. Representative examples of well-characterized glycolytic anaerobic bacteria that will grow in a nutrient-culture in the thermophilic to extreme thermophilic ranges belong to the genus Clostridium and include: *C. thermoaceticum, C. tartarivorum, C. thermosaccharolyticum, C. thermocellum, C. thermocellulaseum* and *C. thermohydrosulfuricum.*

These microorganisms are, of course, of interest for the possible anaerobic fermentation of various carbohydrates, such as saccharides, to assist in combination (mixed cultures) with other bacteria to efficiently break down cellulose, and for the direct production of ethanol and other products of fermentation under thermophilic conditions. Yeast (Saccharomyces species) fermentation of sugar, as is well known, ordinarily must be conducted at less than about 37°C under semiaerobic conditions to yield ethanol. Further, the conditions must be carefully controlled to avoid contamination of harmful bacteria, fungi and molds.

In the isolation of thermophilic anaerobic bacteria, in particular, Clostridium species, from mud samples from the hot springs at various locations in Yellowstone National Park *L.G. Ljungdahl and J.K.W. Wiegel; U.S. Patent 4,292,407; September 29, 1981; assigned to the U.S. Department of Energy* have found two strains of a new anaerobic thermophilic glycolytic species. In the isolation, purification and characterization of this newly discovered species, which is not a Clostridium, it has been found that the new species is an efficient producer of ethanol from various carbohydrates, in particular, the most common mono- and disaccharides.

The newly discovered thermophilic anaerobe is a new species of a new genus isolated in a biologically pure culture and designated herein as *Thermoanaerobacter ethanolicus.* A representative strain of this new microorganism in a biologically pure subculture, designated JW 200, has been deposited in the strain collection of the American Type Culture Collection under the accession number ATCC 31550.

The isolated strain of *T. ethanolicus* has been cultured in aqueous nutrient medium under anaerobic, thermophilic conditions to produce a recoverable quantity of ethanol.

The microorganism described here was discovered in and isolated from mud samples of hot springs in Yellowstone National Park. One strain JW-201 was isolated from an acidic spring, the Dragon Mouth, with a pH of about 5.5 and the second strain JW 200 from an alkaline spring, White Creek, with a pH of about 8.8. The strains are very similar and were discovered in association with the anaerobic thermophilic Clostridia strains hereinabove mentioned.

Although the new microorganism strains share the ability to ferment certain carbohydrates at thermophilic temperatures with the Clostridia mentioned above, they do not form spores and are therefore excluded from the genus Clostridium. Other characteristics exclude them from characterized and known genera. In view of the morphology and fermentation characteristic, hereinafter described, these new strains are deemed a new genus and species designated *Thermoanaerobacter ethanolicus,* ATCC 31550 being representative of these strains.

Isolation of the new strains from the mud samples was accomplished using the anaerobic technique according to Hungate as modified by Bryant and Robinson, which technique will be familiar to those skilled in the art. The medium use for isolation and enrichment cultures and to maintain the isolated strains has the following preferred composition: KH_2PO_4, 1.5 g/ℓ; $Na_2HPO_4 \cdot 12H_2O$, 4.2 g/ℓ; NH_4Cl, 0.5 g/ℓ; $MgCl_2$, 0.18 g/ℓ; yeast extract (Difco), 2.0 g/ℓ; glucose, 8.0 g/ℓ; and Wolfe's mineral solution, 5 ml. The medium is prepared under anaerobic conditions and must be stored under an atmosphere of an inert gas, such as nitrogen or argon. The pH of the medium is in the range of about 6.8 and 7.8, preferably 7.3, and is adjusted as required with a sterile, anaerobic NaOH or HCl solution. Stock cultures are maintained on the same medium solidified with 2% agar and stored at 4°C. Liquid medium cultures can be stored at -18°C after the addition of an equal volume of glycerol.

Although in the exemplary nutrient medium, glucose is the preferred carbohydrate substrate, other monosaccharides, such as, xylose, ribose, mannose, fructose and galactose, and disaccharides, such as sucrose, lactose, maltose and cellobiose can be used. Growth also occurs on pyruvate, pectin, and starch. It should be noted that both strains of *T. ethanolicus* require yeast extract for growth. Without yeast extract, no growth was obtained in subsequent subcultures. Although growth is much less than in the presence of glucose, yeast extract concentrations above 0.5% can serve as the only carbon, nitrogen and energy source.

The new strains of *Thermoanaerobacter ethanolicus* ATCC 31550 (JW 200) can be conveniently cultured using the same nutrient medium as used for isolation under anaerobic conditions at temperatures between about 36° and 78°C with the optimum temperature for growth being about 68°C. Doubling time at 68°C, is about 90 minutes. It is significant, however, that excellent growth can be also maintained at higher temperature, such as 72°C, doubling time at this temperature being about 120 minutes. The ability to maintain a significant growth at these extreme thermophilic conditions, is, of course, one of the distinguishing characteristics of *T. ethanolicus.* Such growth is not pH dependent in that growth occurs in the very wide pH range of from 4.5 to 9.8. For optimum growth, the pH of the medium should be between about 5.7 and 8.6, with the preferred pH being about 7.3.

Another distinguishing characteristic of *T. ethanolicus* is its ability under the abovedescribed conditions to ferment a wide variety of saccharides with a significant yield of ethanol—yields of ethanol as high as 1.8 mol of ethanol per mol of fermented glucose having been achieved at a temperature of 60°C and at a pH of 7.8 under anaerobic conditions. However, to maintain a process of ethanol fermentation, when concentrations of ethanol reach about 5%, the ethanol should be removed by conventional distillation techniques. Such a distillation can be accomplished using reduced pressure (partial vacuum) even during fermentation in the presence of the microorganism *T. ethanolicus* since the organism will tolerate the boiling point of ethanol (78°C).

Using an Anaerobic Thermophilic Culture System

In a second patent by *L.G. Ljungdahl and J.K.W. Wiegel; U.S. Patent 4,292,406; September 29, 1981; assigned to the U.S. Department of Energy,* there is provided a mixed culture system of the microorganism *T. ethanolicus* described in the previous patent and the microorganism *C. thermocellum.* In a mixed nutrient culture medium that contains cellulose, these microorganisms have been coupled and cultivated to efficiently ferment the cellulose to produce recoverable quantities of ethanol. This fermentation is conducted under anaerobic, thermophilic conditions.

A strain of *C. thermocellum* ATCC 31549 (JW 20) has been isolated from a cotton bale from Louisiana. As with *T. ethanolicus*, this strain of *C. thermocellum* was isolated in biologically pure form using the Hungate technique as modified by Bryant and Robinson. The nutrient medium described in the previous patent that was used for *T. ethanolicus*, was used for isolation and enrichment of *C. thermocellum*, except that cellulose in the form of paper derived from wood pulp, (10.8 g/ℓ) was used instead of glucose for the substrate and the amount of yeast extract was increased to 5.0 g/ℓ. The medium is prepared under anaerobic conditions and must be stored under an atmosphere of an inert gas, such as nitrogen or argon. The pH of the medium is in the range of about 6.8 to 7.8, preferably 7.3.

Cellulose, the naturally occurring polymer of glucose, is available from a variety of sources, both in an untreated, impure, natural state, such as in the form of plant tissue, and in a hydrolyzed or treated form, such as paper prepared from wood pulp. Those skilled in the art will recognize that treatment of natural, impure cellulose is desirable for more efficient fermentation of the cellulose material. For this process, cellulose that has been treated physically and chemically to break down the existing lignin protective covering and expose the cellulose component is the preferred form of cellulose material for efficient fermentation. Previously treated waste cellulose material, such as paper, is particularly preferred.

C. thermocellum can be conveniently grown in the same substrate medium as used for isolation under anaerobic conditions at temperatures between 45° and 65°C with the optimum temperature for growth being about 60°C.

The same cellulose containing medium that is used for the isolation and cultivation of *C. thermocellum* is used for the coupled fermentation of cellulose using *T. ethanolicus* and *C. thermocellum* under anaerobic thermophilic conditions. Such direct fermentation of cellulose (paper) provides a significant yield of ethanol. As high as 1.46 mols of ethanol is produced per glucose unit of cellulose at a temperature of 60°C and at a pH of 7.5 under anaerobic conditions. Ethanol product from this fermentation can be recovered by conventional distillation techniques.

Use of Immobilized Enzymes

I. Chibata, J. Kato and M. Wada; U.S. Patent 4,350,765; September 21, 1982; assigned to Tanabe Seiyaku Co., Ltd., Japan have developed a process for producing ethanol by the conversion reaction of fermentative sugar using an immobilized enzyme. It was found that an immobilized microorganism exhibits superior activity over a long period of time. Ethanol can be produced in concentrations as high as 100 to 200 mg/ml within a relatively short time when the conversion is carried out in the following steps:

(1) contacting the immobilized microorganism with a culture broth containing a fermentative sugar and other nutrients necessary for growth of the microorganism;

(2) microbiologically converting the sugar into ethanol;

(3) adding an additional fresh culture broth containing sugar to the conversion reaction system; and

(4) separating the broth containing ethanol produced.

The initial concentration of the sugar should be 50 to 100 mg/ml; the sugar is converted into ethanol until its concentration is 20% or less of the initial concentration; and the additional sugar is added in the fresh culture broth in a concentration of not less than 100 mg/ml, until the ethanol is produced in a concentration of not less than 100 mg/ml.

Various members of the genera of Saccharomyces and Zymomonas may be used as the microorganism, such as brewer's yeast, distillery yeast, etc., e.g., *Saccharomyces cerevisiae* ATCC 4124 and IFO 2018, *Saccharomyces fermentati* IFO 0422, *Z. mobilis* IFO 13756 and *Z. anaerobea* ATCC 29501.

The immobilization of these microorganisms can be effected by known methods, for example, a sulfated polysaccharide-gel method (U.S. Patent 4,138,292), a polyacrylamide-gel method [*Advances in Applied Microbiology,* vol 22, p. 1 (1977)] and the like, or by the following process.

A small amount of microbial cells is mixed with a solution of a gel base material and the resulting mixture is gelatinized in pellets or film. The gels are then incubated in a nutrient culture broth suitable for growth of the microorganism at 15° to 45°C to obtain a desired immobilized microorganism which has a dense layer of the microbial cells formed within the supporting gel near the surface of the gel. As a gel base material, there can be used a known gel base material such as sulfated polysaccharide, polyacrylamides, sodium alginate, polyvinylalcohol, cellulose succinate, casein succinate and the like. Examples of sulfated polysaccharides which may be used are carrageenan, furcellaran and cellulose sulfate.

Use of an Immobilized Cell Reactor

In most parts of the world, large quantities of biomass exist which could serve as an energy mechanism or as raw materials for manufacturing chemicals. For example, recent studies indicate that recoverable agricultural residues in the United States alone total 300 million dry tons per year. In the United States, corn stover is one possible source of biomass. It accounts for about half of the total agricultural residue and about 70% of that is produced in the central states so that collection and transportation could be centralized. At present, these corn residues could supply all of the petrochemical needs of the United States with a conversion efficiency as low as 40%.

Hemicellulose and cellulose can be converted into energy or chemicals by direct combustion, pyrolysis or biological conversion. Biological conversion is preferred, however, because of its higher efficiency and the preservation of minerals and nutrients for return to the soil.

In order to bioconvert agricultural residues, however, it is necessary to hydrolyze the hemicellulose and cellulose into monomeric sugars, i.e., pentoses and hexoses, respectively. As is well known, the hydrolysis of hemicellulose and cellulose is catalyzed by enzymes or by mineral acids. The simple sugars can then be converted into alcohols, acids, aldehydes or gases depending on the microorganism selected. While a variety of chemicals can be made from the pentoses and hexoses, the focus of the process of *J.L. Gaddy and O.C. Sitton; U.S. Patent 4,355,108; October 19, 1982; assigned to The Curators of the University of Missouri* is on the production of ethanol because of present interest in gasohol and because it is a desirable starting material for chemical synthesis.

Up until now, there have been two ways to convert xylose and/or glucose into ethanol: batch fermentation and continuous stirred tank fermentation. Both of these ways have objectionably slow reaction rates and are quite susceptible to inhibition from materials in the substrate or in the product. In view of the above, there is a need for a stable, high productivity reactor for converting glucose and/or xylose into ethanol, these simple sugars being produced in abundance by hydrolyzing cellulose and hemicellulose containing materials such as corn residues.

Immobilized cell reactors in accordance with this process are prepared by packing a column with an improved packing material with a fixed film of yeast organisms attached thereto. The improved packing material is prepared by coating the packing material with a coat of proteinaceous material, treating the proteinaceous coat with a polyfunctional reagent capable of reacting with at least one amino acid group making up the proteinaceous material, inoculating the treated coat with a yeast organism genetically stable for the production of ethanol and then incubating. It is important that the proteinaceous coat be formed of a polymeric material having reactive groups for reaction with the abovementioned polyfunctional reagent. It is also important that the yeast organisms be genetically stable for the production of ethanol and that the reagent which immobilizes them, not affect their ability to reproduce normally. Suitable organisms for fermenting glucose include *Saccharomyces cerevisiae* and suitable organisms for fermenting xylose include *Fusarium oxysporum* and *Candida utilis.*

It is preferred that the proteinaceous coat be gelatin and that the polyfunctional reagent be glutaraldehyde or diisocyanates. Of these, glutaraldehyde is preferred since it reacts faster and also tends to harden the gelatin film which prevents it from being washed away. The solid support is preferably porous or otherwise provided with a large surface area. Porous ceramic solid supports such as Raschig rings are particularly preferred.

With the yeast immobilized, it is possible to get much higher cell densities per packed volume than is possible with a batch reactor or with a continuous stirred tank reactor and to operate at dilution rates that exceed the maximum specific growth rate of the microorganism whereas with stirred tank reactors, where the organisms are not immobilized, this would result in washout. The higher cell densities and faster dilution rates produce higher productivities than are possible in conventional reactors. Immobilizing the yeast also has a beneficial effect on the stability of the system since the thickness of the film buffers upsets in pH and flow rates and mitigates inhibition by toxic materials in the substrate or in the products.

In a two-step process for hydrolyzing corn stover and other agricultural residues containing significant amounts of cellulose and hemicellulose ground corn stover is reacted in a prehydrolysis tank with a dilute mineral acid for such time as to hydrolyze the pentosans to xylose. During the prehydrolysis step some of the hexosans are also hydrolyzed but most are not. Suitable prehydrolysis conditions are provided with 4.4% sulfuric acid at 100°C for 50 minutes. The mixture is then passed through a separator such as a filter or a centrifuge. The filtrate comprises about 0.80% xylose and about 0.20% glucose and is processed through an acid recovery unit such as a dialysis unit where the acid is recovered for recycle. The resultant effluent from the acid recovery unit makes up the prehydrolyzate or first-stage hydrolyzate.

The solids recovered at the separator are dried and treated in an impregnator with concentrated mineral acid. This mixture is then fed into a second hydrolysis tank where it is refluxed for such time as to hydrolyze the hexosans to glucose. For this purpose, it has been found effective to mix the dried solids with 85% sulfuric acid, to which dilution water is then added such that the concentration of the sulfuric acid is diluted to 8%, followed by reacting at 110°C for 10 minutes. The mixture is then passed through a second separator such as a filter or a centrifuge. This filtrate comprises about 4.0% glucose and, like the first filtrate, is processed through the acid recovery unit. The effluent from the acid recovery unit makes up the hydrolyzate or second-stage hydrolyzate.

The prehydrolyzate and hydrolyzate produced as described are cooled by heat exchange with the reactor effluent streams. They are further cooled in heat exchangers to the optimal temperature required by the organisms for growth and are then converted, after essential vitamins and other nutrients have been added, into ethanol in the first and second immobilized cell reactors, respectively. The cells are harvested by centrifugation or the like and the ethanol is separated by distillation. The cells are stored for several days during which autolysis occurs. Essential minerals and media supplements can be added to the vitamin rich lysed cells and mixed with the incoming media or the cells can simply be returned to the soil along with the lignin collected at the separator.

OTHER ETHANOL-PRODUCING PROCESSES

Using Two Enzymatic Processes for Garbage-Containing Waste

In this manner, the fermentable carbohydrate contained in garbage can be turned with a better yield into an energy source of methane bacteria and carbon source. This process avoids a voluminous generation of hydrogen gas as was encountered in the conventional liquefaction process (acidic fermentation) and also permits the reduction of the required volume of neutralizer down to 10 to 20% of the conventional process.

The mechanism of anaerobic digestion requires two reactions—the liquefaction of volatile fatty acids by use of septic bacteria, and the conversion of the fatty acids to methane. These processes are usually accomplished by both bacteria groups in the same tank and the treatment takes an extended period of time, as much as 30 to 50 days.

M. Ishida, R. Haga and Y. Odawara; U.S. Patent 4,288,550; September 8, 1981; assigned to The Agency of Industrial Science and Technology, Japan have designed a process for a microbiological treatment of garbage and garbage-containing wastes which is efficient and economical. Its characteristic feature is the use of alcohol fermentation in lieu of the liquefaction process used previously. The fermentable carbohydrates in the waste are first converted into ethanol by yeasts, and the fermented product containing ethanol is then converted into methane by use of methane bacteria without any step of sterilization.

The organic wastes digestion process can be explained as follows. First, the garbage or garbage contained wastes which are stored in a tank are pulverized by a crusher. Next, water is added to the crushed garbage slurry if required to make the solids concentration in the slurry to 5 to 20%. Next, the slurry is charged into an alcohol fermentation tank and treated under a mix-cultured state of alcohol fermentation yeast, amylase producing yeasts, lipase producing yeasts and protease producing yeasts. Yeasts having strong alcohol fermentation activity such as those of the Saccharomyces genus *(S. cerevisiae, S. carlsbergensis)*, the Schizosaccharomyces genus *(Schizosaccharomyces pombe)*, Schwaniomyces genus, Torulopsis genus *(T. dattila)*, Brettanomyces genus, Candida genus *(C. Krusei)*, etc., are used as the alcohol fermentative yeast. As for amylase producing yeasts, *Endomycopsis fibuliger, Schizosaccharomyces pombe, Saccharomyces diastatics*, etc., are used. The protease producing yeasts *Candida lipolitica, Candida parapsilosis*, etc., are used, and the lipase producing yeasts, *Candida cylindracea, Candida lipolitica, Trichosporon pullulens*, etc., are used.

The slurry is kept stirred at a certain fixed temperature for several days, under anaerobic conditions. By this fermentation treatment, the carbohydrate contained in garbage is turned into alcohol, and the contained fat and protein are turned into lower molecular weight substances. The temperature employed for fermentation is in the range of 20° to 40°C. In case the pH decreases in the course of fermentation, it is necessary to add neutralizer so as to adjust it to the range of pH 4 to 6.5. The gas generated from the alcohol fermentation tank contains over 95% carbon dioxide, plus nitrogen, hydrogen and a trace of hydrogen sulfide. This gas is stored in a gas storage tank, after eliminating hydrogen sulfide therefrom at a desulfurizer.

The slurry from the alcohol fermentation is charged into a gasification tank and the low molecular substances, mainly ethanol, are converted into methane and carbon dioxide, by coming into contact with gasification bacteria (methane bacteria). For this gasification treatment to progress with satisfactory efficiency, it is required to keep up the temperature at the range of 20° to 75°C under an anaerobic condition and adjust pH to 7 to 8. As for gasification bacteria, the conventional ones such as Methanosarcina genus, Methanococcus genus, Methanobacterium genus, etc., can be used. The main ingredients of the generated gas are methane 60 to 85% and carbon dioxide 15 to 40%, and further small amounts of nitrogen, hydrogen and hydrogen sulfide are also involved.

The generating gas is then stored in a gas storage tank after hydrogen sulfide is removed by the desulfurizer. Following gasification the slurry is lead into a solids/liquid separation tank for the final treatment where it is separated to water and a digestive sludge.

Integrated Process for Producing Ethanol from Sugar Cane

According to the process by *F.W. Hayes; U.S. Patent 4,326,036; April 20, 1982;* ethanol is produced from sugar cane economically and in bulk and at high purity by a method which involves total utilization of the sugar cane, partly as a fuel to provide the heat requirements of the process, but mainly as a source of fermentable material.

The harvested sugar cane with or without the foliage, but generally without, the foliage being customarily burned off in situ in known manner prior to harvesting of the cane, is first chopped and shredded to provide a digestion mass of juice and fiber. In a first continuous digestion stage the digestion mass is subjected to the hydrolytic action of hemicellulase-producing organisms and/or isolated hemicellulase enzymes to break down at least some of the hemicellulose content of the cane into fermentable sugars, for example, pentoses.

Following the first digestion, the fibrous residue is separated from excess liquid, for example, by passage through a continuously operating centrifuge, and the liquid containing mainly sucrose, glucose and fructose from the original cane juice, pentoses produced by the hemicellulase digestion, and some initial fermentation products produced in a manner to be described, are passed to fermentation, while the fiber residue is passed to a second continuous digestion stage. In the second stage digestion, the fiber residue is subjected to the combined action of a cellulase enzyme (or whole cells of the organism) and an ethanol-producing fermentation culture, preferably a thermophilic organism, to break down a limited portion of the cellulose content of the fiber and initiate fermentation of the resulting glucose.

Following the second digestion, the fibrous residue is separated from the excess liquid, for example, by passage through a second continuously running centrifuge, and the liquid, containing glucose, fructose, some residual sucrose from the original cane and some initial fermentation products, is passed to fermentation proper in admixture with the liquid separated from the first digestion stage.

The fibrous residue or bagasse from the second digestion stage, the amount of which can be closely controlled by varying the reaction time intensity in digestion stages one and two, is now passed to a bagasse-burning boiler plant to provide at least some and preferably all the heat requirement of the process. Because of the high moisture content of the fibrous residue from the second centrifugal liquid extraction stage, this residue may be passed through a three-roller mill to obtain a further liquid extract which can be recycled to the first digestion stage, before the remaining bagasse is passed to combustion.

An important point is that the organisms used in the digestion stages, as a mixed culture, work in symbiosis, with an enhanced effect as compared with the same organisms in single culture.

A significant feature of this process is the double digestion of the sugar cane fiber, which not only contributes to the amount of fermentable material recovered from the cane, but enables complete utilization of the cane with close tailoring of the amount of bagasse passed to the boilers to the total heat energy requirements of the process, which in any case are lower than in normal raw sugar manufacture due to the fact that fermentation is performed on the juices as extracted, without evaporation and concentration. It is this combination of features which contributes significantly to the economic success of the process.

Following extraction, the two liquid extracts are combined, sterilized and fermented using a continuous fermentation technique, and incorporating membrane separation of the product ethanol as formed.

The hemicellulase and cellulase enzymes to be used in the digesters and the ethanol-producing microorganism to be used in conjunction with the cellulase enzyme in the second digestion stage, as well as in the final fermentation stage, will be a matter of choice and suitable enzymes and enzyme sources will be apparent to those skilled in the art.

As a guide, suitable hemicellulase and cellulase enzymes may be selected from *Trichoderma viride, Aspergillus wentii, Thielaviopsis paradoxa,* and *Thielatia terrestris* including mutants and strains thereof, while suitable ethanol-producing organisms include *Saccharomyces cerevisiae, Saccharomyces uvarum,* Thermoactinomyces sp., Zymomonas and *Bacillus stearothermophilus* and mutants and strains thereof, and especially thermophilic strains.

Preconditioning of Acid Hydrolysates from Lignocellulosic Materials

Traditionally, used wood, paper and agricultural by-products, such as sawdust, woodwaste, corncobs, straw, sugar cane bagasse, newspaper and the like have been regarded essentially as waste materials, and have been disposed of through incineration or by other, similarly unproductive, means. It is well known that the lignocellulosic constituents of such materials can be hydrolyzed to produce more valuable products which in turn can be converted into additional and different valuable products; however, such operations are in limited use, due largely to the relatively low returns on investment which they have been capable of generating. The capital expenditures required to design and construct the facilities for carrying out such recovery operations tend to be significant, thus demanding that relatively high conversion rates be attainable in order to justify the expense involved.

The objectives of *M.D. Faber, R.H. Ernst and P.H. Lefebvre; U.S. Patent 4,342,831; August 3, 1982; assigned to American Can Company* are to provide a process for fermentation of sugars present in acid hydrolyzates derived from lignocellulosic materials in which reaction times are relatively short, in which fermentation may be effected at relatively high sugar concentrations and in which control mechanisms are established which permit predictability, reproduction of results with consistency and production of end products of high value, such as ethyl alcohol.

The foregoing objects are attained in a method for preconditioning acid hydrolyzates, derived from lignocellulosic materials, to negate the effect of substances tending to inhibit the fermentation of such hydrolyzates and to a process for the production of ethyl alcohol from glucose contained in such preconditioned acid hydrolyzates. The hydrolyzate is preconditioned to remove and/or reduce or otherwise negate the effect of inhibitory substances to a level whereby the hydrolyzate may be readily fermented to ethyl alcohol in substantially theoretical yield. The preconditioning method broadly comprises the steps of:

(1) subjecting the hydrolyzate to steam to remove furfural and other steam-volatile substances therefrom;

(2) adding sufficient calcium oxide to the steam-stripped hydrolyzate, at room temperature, to adjust the pH to between about 10 and 10.5, maintaining the resulting mixture at that pH for about 1 to 3 hours and separating the hydrolyzate from the resultant precipitate;

(3) adding sufficient amounts of a mineral acid to adjust the pH of the hydrolyzate to about 5 to 7; and

(4) adjusting the concentration of the hydrolyzate to a glucose concentration of less than about 150 g/ℓ to provide a solution fermentable to ethyl alcohol.

The fermentation process broadly comprises the steps of:

(1) preconditioning an acid hydrolyzate to negate the effect of substances tending to inhibit the fermentation thereof by subjecting the hydrolyzate to the preconditioning method already described;

(2) inoculating the preconditioned hydrolyzate with yeast inoculum comprising from about 0.7 to 7 dry weight percent of yeast cells per 100 g/ℓ of glucose in the hydrolyzate;

(3) fermenting the inoculated hydrolyzate at a pH of 5 to 7 for a period sufficient to convert glucose to ethyl alcohol; and

(4) recovering ethyl alcohol from the fermentation mixture.

In a preferred embodiment, yeast cells are recovered, reconcentrated and recycled to a subsequent fermentation medium comprising preconditioned, concentrated hydrolyzate.

The yeast strains used for the fermentation may be various members of the genus Saccharomyces, such as *S. uvarum, S. cerevisiae* (bakers' yeast especially preferred) and also *Candida utilis.* It was also found that the use of phosphoric acid as the neutralizing acid had a definite positive effect on the rate of fermentation of the concentrated hydrolyzates.

Simultaneous Production of Ethanol and Fructose Polymers from Sucrose

Because fructose is sweeter than either glucose or sucrose, much effort has gone into developing processes for producing syrups in which more than 50% of the carbohydrate is fructose.

R.E. Heady; U.S. Patent 4,356,262; October 26, 1982; assigned to CPC International Inc. have provided for the first time a process for the simultaneous production of ethyl alcohol and fructose polymers. The process involves contacting a sucrose-containing substrate with a mixture of a fructosyl transferase enzyme and a yeast preparation. Purification of the reaction product by removal of cellular debris (e.g., yeast cells) and ethanol yields a syrup containing the fructose polymers. This syrup is useful as a specialty carbohydrate for sweetener and other applications. It may also be hydrolyzed to yield a syrup whose principal sugar is fructose. The fructose content of the sugars in these syrups is generally higher than 66% by weight and ranges up to about 75% and even higher, depending upon the composition of the sucrose substrate and the reaction conditions employed.

This process is unique in its simplicity. This is readily apparent when it is compared with prior art processes for production of high fructose syrups from corn syrup or sucrose. This process requires no separation from glucose. Rather the glucose is converted by fermentation to an easily separated by-product, ethyl alcohol, by means of a yeast preparation that does not ferment the fructose polymers present. Furthermore, the conversion to fructose polymers and the fermentation of the glucose both occur in one reaction mixture without the isolation of intermediates.

Sucrose-containing substrate may be any substrate in which sucrose is the predominant sugar. It includes molasses, turbinados, meladura, mixtures of sucrose and invert sugars, mixtures of sucrose and fructose-bearing syrup as well as purified sucrose.

The fructosyl transferase enzyme refers to any enzyme that catalyzes transfructosylation such as the enzyme preparation derived from *Pullularia pullulans,* ATCC 9348 (synonymous with *Aureobasidium pullulans*). Preferably, the fructosyl transferase enzyme preparation of this process contains the fructosyl transferase enzyme in a purified form, that is, separated from the fermentation culture medium in which it was produced.

The fermentation process is carried out using aqueous solutions of the substrate. Substrate concentrations from as low as about 10% (w/v) may be employed. However, it is preferred to use as concentrated solutions as practical, preferably ranging from 30 to 50% (w/v), so that there will be less need to evaporate water from the final product. The reactions are carried out at 24° to 32°C, with the pH of the system preferably from 5.0 to 5.6.

The concentration of yeast cells used to carry out the fermentation may vary over a wide range. However, it is convenient to employ about 1 g of wet cells for every 10 to 20 ml of 35% (w/v) substrate. Wet yeast cells obtained by centrifugation contain about 70 to 76% moisture. The amount of fructosyl transferase enzyme used may also vary widely. A practical rate of reaction is observed when from 10 to 30 fructosyl transferase enzyme units are used per gram of sucrose in the substrate.

The sucrose-containing substrate may be treated concurrently with a fructosyl transferase enzyme and a suitable yeast preparation to carry out this process. Alternatively, the sucrose-containing substrate may be treated first with a fructosyl transferase enzmye preparation at a suitable temperature, preferably 50° to 60°C, for from 3 to 6 hours before the mixture of substrate and fructosyl transferase enzyme is allowed to undergo fermentation with the yeast.

Any conventional means, such as centrifugation or filtration, may be used to remove the yeast cells from the reaction mixture. Recovery of alcohol is most conveniently accomplished by distillation from the fermentation mixture.

If high fructose syrup is desired as a product, the fructose polymers may be hydrolyzed. Hydrolyzing agents and conditions of hydrolysis must be chosen so that the fructose is not destroyed. The reaction may be catalyzed by an acid or an acidic resin. Alternatively, the hydrolysis may be accomplished by means of enzymes such as those contained in commercially available invertase enzyme preparations.

PRODUCTION OF OTHER FUELS

Methane Production by Anaerobic Digestion

Methane production of anaerobic digestion has been widely practiced, particularly with respect to digestion of sewage sludge organic waste. In recent times, the worldwide energy shortage has furthered consideration and improvement of such nonfossil sources of energy.

D.L. Klass and S. Ghosh; U.S. Patent 4,316,961; February 23, 1982; assigned to United Gas Pipe Line Company describe a process for improved methane production from and beneficiation of anaerobic digestion of plant material and/or organic waste comprising anaerobic digestion of plant material and/or organic waste of normally low biodegradability in the presence of extract of different plant material. The extract is present in about 10 to 90 volume percent of the digester contents. The process may be carried out under mesophilic or thermophilic temperatures for detention times in excess of about four days. Under steady state anaerobic digestion, the plant material and/or organic waste of normally low biodegradability in the presence of the extract of different plant material results in synergistic action providing higher methane yields and production rates than those that result from the anaerobic digestion of the individual feed components separately.

The plant material used in the process may be of terrestrial or aquatic origin. It is particularly preferred that the plant material be a mixture of terrestrial and aquatic plant materials.

Terrestrial plants include warm season grasses, such as Bermuda grass and elephant grass; cool season grasses, such as Kentucky blue grass and Merion blue grass; reedy plants, such as bamboo, rice, cattails, herbaceous plants, such as kudzu and maize; deciduous trees, such as eucalyptus and poplar; and coniferous trees, such as white and red pines. Exemplary aquatic plants include water hyacinth, duck weed, algae, sea kelp and sargassum.

The term organic waste as used here means all types of organic refuse including sewage sludge, animal waste, municipal waste, industrial waste, forestry waste, agricultural waste, and the like.

An important aspect of this process is the anaerobic digestion of biomass consisting of plant and/or organic waste materials of normally low biodegradability in the presence of an extract of different plant material. The plant and/or organic waste material and extract may be premixed and slurried prior to introduction into the digester or the individual feed materials may be separately introduced into the digester and mixed within the digester. The important aspect is that the mixture of biomass material of normally low biodegradability and extract be together in the active digestion zone. Feeding and associated wasting may be continuous or intermittent.

Any active methane producing mesophilic or thermophilic anaerobic digestion system may be used. Exemplary methane-producing organisms suitable for use in this process include members of Methanobacterium, Methanococcus and Methanosarcina, specific members being *Methanobacterium formicicum, Methanosarcina barkerii, Methanobacterium omelianskii, Methanococcus vannielii, Methanobacterium sohngenii, Methanosarcina methanica, Methanococcus mazei, Methanobacterium suboxydans* and *Methanobacterium propionicum*. It is usually preferred to use mixed cultures to obtain the most complete fermentation action. Nutritional balance and pH adjustments may be made to the digester system as is known to the art to optimize methane production from the culture used.

In a further patent, *S. Ghosh and D.L. Klass; U.S. Patent 4,329,428; May 11, 1982; assigned to United Gas Pipe Line Company* describe a process which provides improved methane production resulting in higher yield and higher production rates by anaerobic digestion of a mixture of plant material and organic waste. The process is suitable for production of synthetic natural gas and through anaerobic digestion of a mixture of plant material and organic waste allows better matching of organic waste and plant material feed supplied for year round operation. The process results in digester effluent which is easily dewatered and has a low concentration of soluble organics, providing easy disposal and recycling to the digester. It may be used for methane production from plant material which is, by itself, recalcitrant to anaerobic digestion.

The methane production occurs by anaerobic digestion in the presence of acid-forming bacteria and methane-producing organisms at temperatures about 20° to 70°C for detention times about 5 to 30 days.

The mixture preferably comprises about 30 to 70 weight percent on a dry solids basis of organic waste and about 30 to 70 weight percent on a dry solids basis of plant material and the anaerobic digestion is carried out under mesophilic temperatures of about 45° to 75°C for detention times of about 8 to 30 days.

The organic waste comprises municipal solid waste and the plant material comprises a mixture of terrestrial plant material such as Bermuda grass and aquatic plant material such as water hyacinth.

Cellulose Fermented with a Combination of Clostridium Organisms

L.H. Grove; U.S. Patent 4,326,032; April 20, 1982 has found that a deliberate combination of microorganisms of the genus Clostridium (or the enzymes produced thereby) can be utilized to convert cellulose directly to a liquid organic fuel of relatively high energy (e.g., above 4 kg-cal/g when anhydrous). It has been found that at least two Clostridium species (or their enzymes) should be used in combination to achieve this result. Both species are preferably somewhat aerotolerant, and both should be carbohydrase-producing. One of the two species should be capable of producing cellulase, and other species preferably produce cellobiase, glucosidase, and/or glucase. Typical of such combinations would be *Clostridium cellobioparum* with a saccharase-producing organism such as *Clostridium acetobutylicum.*

The deliberate combination of Clostridium organisms (or their enzymes) is effective for a variety of substrates, including commonly occurring cellulosic materials (agricultural wastes, municipal sewage, waste paper, and other inexpensive sources of cellulose and hemicelluloses). Although ratios of the deliberately combined organism populations used can range from about 1:9 to 9:1 with respect to the two Clostridium species, it is preferred to maximize alcohol/ketone production and minimize carboxylic acid production, which objective appears to be obtained most effectively with at least 40% (by weight or activity or population) of the cellulase system or cellulase-producing organism, more preferably 50 to 80%.

Fermentation can proceed under normal ambient or moderately elevated temperatures, e.g., 30° to 40°C. The preferred pH in the fermentation zone is 5.8 to 6.4. The fuel obtained from this process is nonpotable, relatively high energy, volatile (typically boiling within the range of 20° to 200°C), and typically high in antiknock properties. High yields of this fuel are obtained in an efficient manner. Indeed, the efficiencies and the economies of the process are sufficient to permit small scale, low volume production, as low as a few gallons per week for the small farmer.

To sum up the key aspects of the process: (a) Water, a cellulose- or hemicellulose-containing particulate mass, and a fermentation agent are blended to form a fermentation medium. The fermentation agent comprises the combination of Clostridium organisms described previously (or their enzymes).

(b) The particulate mass is permitted to ferment until fermentation products are produced.

(c) The resulting nonpotable hydrocarbon oxidate fuel is then recovered from the fermentation medium by conventional means. The most commonly used conventional means is distillation; however, it is also known to utilize the principles of fractional crystallization, solvent extraction with gasoline, etc.

Hybrid Biothermal Liquefaction Process

D.P. Chynoweth and P.B. Tarman; U.S. Patent 4,334,026; June 8, 1982; assigned to Institute of Gas Technology describe a hybrid biothermal liquefaction process for improved production of alcohol-containing liquid fuels.

The process comprises adding organic carbonaceous feed to a fermentation reactor, fermenting the organic carbonaceous feed under thermophilic or mesophilic conditions in an active alcohol-producing liquid culture, separating alcohol containing liquid fuel product and fermentation residue and introducing the fermentation residue into a thermochemical converter which may be a thermal gasifier or thermal liquefier, or both.

At least a substantial portion of the carbonaceous material in the fermentation residue is converted in the liquefier or gasifier under elevated temperature conditions to thermal conversion products used directly or converted by catalytic synthesis to alcohol-containing liquid fuel and thermal residue. At least a portion of the thermal conversion products or their derivatives may be passed to the fermentation reactor to beneficiate the fermentation process. In one embodiment, at least a portion of the thermal residue may be returned to the fermentation reactor to provide inorganic nutrients for the fermentation. In another embodiment, ammonia may be added to the fermentation reactor, the ammonia being a thermal converter product or produced from product gases of the thermal converter.

The process provides for the catalytic synthesis of alcohols from the gases produced by thermochemical gasification or liquefaction of the fermentation residue from the fermentation reactor providing higher alcohol production per unit of feed than conventional fermentation processes. In another embodiment, the low Btu gases produced by thermochemical gasification or thermochemical liquefaction of the fermentation residue may be used as fuel gas.

The fermentation of the organic carbonaceous feed may be carried out at temperatures of generally about 20° to 40°C; retention times usually about ½ to 8 days; and loading rates; pretreatment of feed; fermentation reactor mixing, recycling, batch and continuous processes as known to the art for fermentation. Pretreatment of the feed by methods such as enzymatic or acid hydrolysis may be necessary to produce sugars for fermentation.

Any active alcohol producing fermentation system may be used, such as *Saccharomyces cerevisiae,* standard yeast fermentation or *Clostridium acetobutylicum* and *Clostridium thermosaccharolyticum.*

Recovery of Water-Soluble Oxygenated Hydrocarbons from Fermentation Process

In a typical ethanol fermentation process, an aqueous solution of fermentable substrate, i.e., dextrose, and a quantity of fermenting microorganisms, i.e., yeast cells, are introduced into one or the first of a series of fermentation vessels wherein the sugar is metabolically converted by the yeast into product ethanol and carbon dioxide gas. Since the metabolic evolution of ethanol is exothermic, provision is made for the cooling of the fermentation medium to maintain a range of temperature conducive to high levels of ethanol production, i.e., from about 68° to 104°F, and preferably from about 68° to 99°F. The resulting dilute solution of ethanol (so-called "beer") which can contain up to about 12 weight percent ethanol is thereafter subjected to distillation if the ethanol is to be recovered in a more concentrated form.

Additional ethanol can be recovered from the gases, largely carbon dioxide containing relatively minor amounts of ethanol, which are evolved during fermentation by scrubbing the gases with water. The carbon dioxide once freed of ethanol is then discharged to the atmosphere. The foregoing description with appropriate changes in fermentable substrate and fermenting microorganism is generally applicable to the production of other water-soluble oxygenated hydrocarbons such as butanol and acetone, the by-product carbon dioxide also being vented to the atmosphere. In some known fermentation processes, the by-product carbon dioxide is returned to the fermentation vessel in order to maintain agitation which will prevent the fermenting microorganisms and other insolubles from settling.

It has been found by *W.C. Muller and F.D. Miller; U.S. Patent 4,336,335; January 22, 1982; assigned to National Distillers & Chemical Corporation* that ethanol and other water-soluble oxygenated hydrocarbons which can be volatilized at temperatures lower than the temperature of the fermentation process by which they are produced can be more efficiently recovered from the fermentation vessel with a significant saving of process energy by pressure within the fermentation vessel, scrubbing the carbon dioxide gas which evolves during fermentation with water to recover substantially all of the oxygenated hydrocarbon present therein as a dilute aqueous solution, heating the scrubbed carbon dioxide gas to a temperature above the temperature of the fermentation medium and passing the heated gas through at least a portion of fermentation medium to vaporize and carry off a substantial portion of the product water-soluble oxygenated hydrocarbon dissolved therein. As the heated carbon dioxide gas flows through the cooler fermentation medium, the heat transferred to the fermentation medium results in the volatilization of dissolved water-soluble oxygenated hydrocarbon. At the same time, volatilization of the oxygenated hydrocarbon produces an evaporative cooling effect which serves to maintain the temperature of the fermentation medium within a range favoring maximum ethanol production.

After recovery from the carbon dioxide gas in a scrubbing unit, the product oxygenated hydrocarbon present in dilute aqueous solution can be concentrated to any desired level employing any known technique such as distillation and advantageously is recovered by the distillation method which is disclosed herein and which represents a further aspect of this process. Optionally, the by-product carbon dioxide gas can be pressurized, preferably to at least 5 pounds above the pressure within the fermentation vessel, to facilitate its passage through the fermentation medium.

POLYSACCHARIDE SYNTHESIS

PRODUCTION OF XANTHAN GUM

Fermentation of inoculated medium with Xanthomonas organisms for 36 to 72 hours under aerobic conditions results in the formation of xanthan gum which is separated from other components of the medium by precipitation with acetone or methanol in a known manner. Because of the time required to ferment each batch, the low biopolymer content of the fermented medium and the processing steps required for recovery and purification of the product, xanthan is relatively expensive. Earlier work has indicated that heteropolysaccharides produced by the action of Xanthomonas bacteria on carbohydrate media have potential application as film-forming agents, as thickeners, for body-building agents, and edible products, cosmetic preparations, pharmaceutical vehicles, oil field drilling fluids, fracturing liquid, and emulsifying, stabilizing and sizing agents. Heteropolysaccharides, particularly xanthan gum, have significant potential as mobility control agents in micellar polymer flooding. Xanthan gum has excellent viscosifying properties at low concentration. It is resistant to shear degradation and exhibits only minimal losses in viscosity as a function of temperature, pH, and ionic strength. For these reasons, it is an attractive alternative to synthetic polyacrylamides for enhanced oil recovery operations.

However, in order for xanthan gum to be used in enhanced oil recovery operations as a mobility control agent, the cost must be sufficiently low to make such operations economical. It has been shown that the economics of xanthan gum fermentation are sensitive, at least in part, to the yield of xanthan produced in relationship to the amount of glucose consumed. Therefore, any process improvements which enhance xanthan yield will improve the overall economics. For example, in a normal fermentation process increasing the xanthan yield from 60% to 80% can lower the per pound price of xanthan by as much as 10%.

Yield Improvement by Addition of Cholate or Deoxycholate

W.P. Weisrock; U.S. Patent 4,301,247; November 17, 1981; assigned to Standard Oil Company (Indiana) describes a method for improving the efficiency of the process for production of heteropolysaccharides, such as xanthan gum, by action of the bacteria of the genus Xanthomonas on suitable nutrient media. The improvement consists of culturing the microorganism in the presence of a sufficient amount of an additive compound selected from a group consisting of deoxycholic acid, cholic acid, salts thereof, and mixtures thereof, whereby the yield of the heteropolysaccharide produced is increased.

The additive compound should be added in amounts sufficient to cause increased yields in both modes of operations.

Preferably, in the batch mode, the concentration of either of deoxycholate ion or cholate ion should be at least 200 ppm in the nutrient medium, but less than an amount which would be uneconomical or toxic to the microorganisms used. Most preferably, the concentration should range from 400 to 700 ppm.

Preferably, in a continuous operation mode, the concentration of either deoxycholate ion or cholate ion is at least 50 ppm in the medium. Most preferably, the concentration ranges from 300 to 500 ppm.

Any form of deoxycholate or cholate is useful in this process. Preferably, forms of deoxycholic acid, cholic acid, salts thereof, and mixtures thereof, as well as various bile extracts, i.e., ox bile and beef bile are useful. Due to the cost of purified deoxycholate and cholate compounds, use of crude extracts of bile are preferable.

Use of a Stimulatory Organic Acid

The process developed by *A.L. Demain and P. Souw; U.S. Patent 4,245,046; January 13, 1981; assigned to Massachusetts Institute of Technology* for the production of xanthan gum by the fermentation of *Xanthomonas campestris* NRRL B-1459 is based on the finding that the process is stimulated by an organic acid selected from the group consisting of pyruvic acid, α-keto-glutaric acid, succinic acid or mixtures thereof.

In accordance with this process, a nutrient medium is used containing a carbon source comprising at least one sugar and a source of citric acid such as citric acid, citric acid monohydrate or a water-soluble citric acid salt. The use of the citric acid source is preferred since increased production is obtained therewith. In addition, the nutrient medium contains the usual nutrient additives. To the nutrient medium is added a source of pyruvic acid, α-keto-glutaric acid and/or succinic acid as a stimulator for the conversion of the carbon sources to xanthan gum. The preferred stimulatory organic acid is succinic acid or a source thereof, particularly in amounts of between about 0.6 and 1.0 w/v % based upon the weight of the sodium salt. The nutrient medium is adjusted to a pH of between about 6.8 and 7.2, autoclaved, cooled and then mixed with a culture of the bacterium *Xanthomonas campestris* strain NRRL B-1459. The resultant composition then is cultured at a conventional temperature such as between about 24° and 26°C with continuous agitation and aeration for about 72 to 96 hours.

The crude xanthan gum then is isolated by first diluting the fermentation medium to less than 100 cp with 33 volume percent ethanol. The bacterial cells then are removed from the fermentation medium by centrifugation. Removal of the cells can be either preceded by or followed by the introduction of a water-soluble alcohol such as methanol, ethanol, isopropanol or the like which is utilized subsequently to effect precipitation of the xanthan gum from the fermentation medium. Precipitation is effected by the combination of the alcohol and a salt such as potassium chloride, which is added in an amount of about 1 w/v % based upon the volume of water. Subsequent to or concomitant with the introduction of the precipitating salt, the precipitating 95% ethanol is added to the fermentation medium in a concentration effective to promote precipitation of the xanthan gum. A gelatinous flocculant precipitate of low density is separated by centrifugation and is purified such as by slurrying it in 70% ethanol, redissolving it with 33% ethanol, diluting to less than 50 cp, followed by precipitation with the salt and the alcohol. The resultant precipitate then can be further purified as above and dried in a conventional manner.

High Phosphate Process

The preparation and uses of xanthan gum are well known to those skilled in the field of heteropolysaccharides. While aqueous compositions of xanthan gum have many desirable properties, such compositions have a chunky or nonuniform flow.

The objective of *R.A. Empey and J.G. Dominik; U.S. Patent 4,263,399; April 21, 1981; assigned to Merck & Co., Inc.* was to provide a low-calcium, smoothly flowing xanthan gum without imparting high shear to the media during fermentation. They found there is a correlation between the calcium content of xanthan gum and the flow characteristics of aqueous compositions containing xanthan gum. The aqueous compositions include solutions of xanthan gum as well as o/w emulsions. In general, aqueous compositions of xanthan gum containing not more than about 0.04% by wt of calcium and fermented under high shear conditions have desirable flow properties, and aqueous compositions of xanthan gum containing not more than about 0.02% by wt of calcium, and fermented under high shear conditions have best flow properties.

The low calcium xanthan gum of this process may be prepared by a heteropolysaccharide-producing bacterium, *Xanthomonas campestris* by the whole culture fermentation of a medium comprising a fermentable carbohydrate, a nitrogen source, and appropriate other nutrients.

The bacterium is grown in a medium which is substantially free of calcium ions. By substantially free is meant up to about 4 ppm of calcium ion per each 1% of xanthan gum concentration in the completed fermentation broth, and preferably up to about 2 ppm of calcium per each 1% of xanthan gum concentration in the completed fermentation broth. Thus, if the xanthan gum is to be produced at a final concentration of about 2.1 to 2.3%, the total calcium ion content of the completed fermentation broth should not exceed about 9 ppm and preferably should not exceed about 5 ppm. To obtain such a low calcium medium the calcium content of the water in the fermentation medium may be reduced to the appropriate level by any means such as by chemical means, e.g., ion-exchange treatment, or by distillation, or by the use of soft water.

As commercial sources of organic nitrogen contain appreciable amounts of calcium ion, it is important that the nitrogen source of the process be a material which is substantially free of calcium ions. An example of such a nutrient material is Promosoy 100 (Central Soya), a soy protein concentrate. Use of this material at 500 ppm imparts 1 to 2 ppm calcium to the medium.

Increasing Yield by Gradual Addition of Assimilable Carbon Source

W.C. Wernau; U.S. Patent 4,282,321; August 4, 1981; assigned to Pfizer, Inc. has found that the gradual addition of a source of assimilable carbon, preferably glucose, to the aqueous nutrient medium during the course of a Xanthomonas fermentation results in substantially increased xanthan yields.

Conventional media for the production of xanthan (Xanthomonas polymer) contain a suitable carbohydrate in the aqueous nutrient media at a concentration from about 1 to 5% w/v. Suitable carbohydrates include, for example, glucose, sucrose, maltose, fructose, lactose, processed inverted beet molasses, invert sugar, high quality filtered thinned starch or mixtures of these carbohydrates. The preferred source of assimilable carbon is glucose.

The use of concentrations of glucose greater than 5% w/v in a typical batch oxidative Xanthomonas fermentation leads to excessive inhibition of Xanthomonas growth and premature cessation of the fermentation. It has been found that this problem can be obviated by the "feeding" of glucose during the course of the Xanthomonas fermentation. The gradual addition of glucose to the fermentation medium initially low or free of glucose not only allows a final glucose addition up to 7% w/v but results in substantially increased xanthan concentration in the final fermentation broth (5% w/v and 70% yield based on total glucose present in the second state inoculum and amount added to the final fermentor). This represents an increase of greater than 60% over a conventional batch process.

The addition of glucose solution (approximately 15 to 50% w/v) is started immediately after inoculation of the medium (selected from any of those described in the literature for production of xanthan) with an appropriate bacteria of the genus Xanthomonas, such as *Xanthomonas campestris* NRRL B-1459.

The glucose is fed at an exponentially increasing rate from 0 to 24 hours after inoculation and at a constant rate from about 24 to 120 hours. Some sugar accumulation (8 g/ℓ) occurs during the early stages (peaks at about 48 hours) and drops to nondetectable levels (about 72 hours). Other nutrients may be fed with the source of assimilable carbon without changing the essential nature of the process.

Air is introduced into the production fermentor via conventional means. Oxygen demand of the fermentation can be matched to equipment limitations of mixing and oxygen transfer in order to reduce toxic acid by-product accumulation. This can be effected by reducing the feed rate of assimilable carbon source until oxygen demand more suitably matches oxygen transfer capacity of the system.

The hydrophilic colloids produced by Xanthomonas species are polysaccharides which have found wide food and industrial applications. Of special interest is the increasing focus on the use of Xanthomonas polymers in displacement of oil from partially depleted reservoirs.

Xanthomonas Bipolymer for Use in Displacement of Oil from Partially Depleted Reservoirs

In further work on xanthan production, particularly for the purpose of using it to displace oil from partially depleted reservoirs, *W.C. Wernau; U.S. Patent 4,296,203; October 20, 1981; assigned to Pfizer Inc.* has prepared a pyruvate-free Xanthomonas colloid-containing fermentation broth suitable for the preparation of mobility control solutions used in oil recovery which comprises aerobically fermenting a mutant strain of the genus Xanthomonas in an aqueous nutrient medium. The pyruvate-free xanthan and the deacetylated form of this new biopolymer provide mobility control solutions which are especially useful for enhanced oil recovery where high brine applications are involved. The mobility control solutions produced in accordance with this process are employed in oil recovery in the same manner as previously known mobility control solutions.

A number of processes have been developed in recent years to recover further quantities of oil from already worked reservoirs by the use of mobility control solutions which enhance oil displacement by increasing the viscosity or permeability of the displacing fluid. Of interest are those enhanced recovery processes employing polymer flooding with a polysaccharide or polyacrylamide to increase the viscosity of the displacing fluid. Polyacrylamides have been found to suffer such deficiencies as viscosity loss in brines and severe shear sensitivity. Since, as was well documented in the prior art, xanthan gum is insensitive to salts (does not precipitate or lose viscosity under normal conditions), is shear stable, thermostable and viscosity stable over a wide pH range, xanthan gum is a good displacing agent. Moreover, the gum is poorly adsorbed on the elements of the porous rock formations and it gives viscosities useful in enhanced oil recovery (5 to 90 cp units at 1.32 sec^{-1} shear rate) at low concentrations (100 to 3,000 ppm).

A troublesome problem encountered in some oil fields with brines containing high salt concentrations or areas where brine (especially brine high in calcium content) is used as a diluent in preparing xanthan mobility control solutions is the tendency of the xanthan to precipitate out of solution or flocculate. The particulate matter soon plugs the oil-bearing formation at the site of injection. In addition, the desired viscosity is lost from solution.

The reduction of the ionic character of the pyruvate-free xanthan minimizes its compatibility with calcium and other ions. Additional reduction in the ionic nature of the xanthan can be accomplished by deacetylation of the pyruvate-free xanthan.

The process for preparing the deacetylated pyruvate-free xanthan-containing fermentation broth suitable for the preparation of mobility control solutions used in oil recovery comprises (a) fermenting by aerobic propagation a pyruvate-free xanthan producing strain of *Xanthomonas campestris* ATCC 31313 in an aqueous nutrient medium whose ingredients comprise a carbohydrate, a source of assimilable nitrogen and trace elements and continuing the fermentation until at least about 100 ppm of pyruvate-free xanthan is present in the broth, and (b) adjusting the pH of the whole or filtered broth to at least about 9, allowing the resultant broth to stand at room temperature for at least about 10 minutes and optionally neutralizing.

Improving Specific Productivity During Continuous Fermentation

In further work on methods of improving the efficiency of the process for making xanthan gum by action of Xanthomonas on suitable nutrient media, *W.P. Weisrock; U.S. Patent 4,311,796; January 19, 1982; assigned to Standard Oil Company (Indiana)* found that an improvement in specific productivity can be made by increasing the average cell concentration through the stepwise increase of growth limiting nutrients in the medium. The expression "specific productivity," for the purpose of this process is a measurement of the amount of product formed by a given quantity of cells in a given unit of time, e.g., generally expressed as gram of product per gram of cells per hour.

By growing a species of the genus Xanthomonas, for example, *Xanthomonas campestris*, in continuous culture in a medium containing glucose, mineral salts, and NH_4Cl and either glutamate or glutamate plus yeast extract, the specific productivity can be improved by first operating (under nitrogen-limited conditions) at a cell concentration of about 2 g/ℓ and then raising the cell concentration up to 4 to 5 g/ℓ. (The cell concentration is given in terms of dry weight.)

The increase in cell concentration is obtained by increasing the concentration of nitrogen, the limiting nutrient in the medium. Of course, other elements in place of nitrogen as the limiting nutrient, such as, for example, phosphorus, sulfur or potassium, may be used. The amount of limiting nutrient may vary from about 0.1 to 15%, preferably 0.3 to 10% of the desired cell concentration at steady state although concentration levels of from about 0.001 to 0.07% are useful. Where nitrogen is the limiting nutrient, it can be supplied by NH_4Cl, glutamate, yeast autolysate, yeast extract or various combinations thereof.

Semicontinuous Production Using *Xanthomonas campestris*

Two patents by *W.P. Weisrock; U.S. Patents 4,328,308 and 4,328,310; May 4, 1982; both assigned to Standard Oil Company (Indiana)* describe the production of xanthan gum by a semicontinuous process.

Most, if not all, of the installed plant capacity presently in use for the manufacture of xanthan by fermentation methods is restricted to use of the batch fermentation process. In using this particular technique, one of the chief disadvantages is the lag time required to prepare adequate quantities of inoculum (seed) for each batch run. For the quantities of nutrient medium involved in industrial operations, as much as four days time is needed to provide enough inoculum. After the fermentation has been completed and the product xanthan separated, the spent mash in the fermenter must be withdrawn and the fermenter cleaned out and sterilized before a new charge of sterile medium can be introduced.

The efficiency of conventional batch-type plants can be substantially improved by employing semicontinuous fermentation. This process can be effectively conducted in existing batch-type plants with a minimum of modification. In semicontinuous fermentation, the fermenter is filled to the desired volume with a suitable sterile growth medium. A culture of the desired microorganism is then introduced into the growth medium and growth and/or product formation are allowed to occur under known conditions.

When the fermentation is complete, a volume of culture broth, amounting to from 25 to 90% of the total volume, is withdrawn and the desired culture or product is recovered. Thereafter, a volume of sterilized fresh medium is introduced into the fermenter generally equivalent to the volume withdrawn. The microorganisms remaining in the fermenter resume growth and/or product formation when placed in contact with the sterile medium. After fermentation is complete, the cycle is repeated. This process is usually referred to as "serial culture" or "serial transfer." The procedure is considered to be semicontinuous in that no steady-state condition is reached as in continuous fermentation but the need for fermenter cleanup and preparation of fresh culture inoculum is avoided as is required in the batch process.

However, in the production of xanthan by semicontinuous fermentation using the common strains of *Xanthomonas campestris*, such as NRRL B-1459 which is generally used, degeneration of the cultures occur after just a few serial transfers. The result of this degeneration is a loss of xanthan-producing ability, and the appearance of atypical bacterial variants.

It has been found, however, that successful production of xanthan in good yield can be obtained by the use of degenerative resistant *Xanthomonas campestris* organisms identified as *Xanthomonas campestris* XCP-1 ATCC 31600, XCP-19 ATCC 31601, and P-107 ATCC 31602.

Subcultures of these living organisms can be obtained upon request from the permanent collection of the American Type Culture Collection. The accession numbers in this repository for *Xanthomonas campestris* XCP-1, XCP-19 and P-107 are given above.

The medium employed may be an inexpensive minimal medium consisting primarily of essential inorganic salts, glucose, and NH_4Cl. The medium may or may not also contain a yeast extract or yeast autolysate as a supplemental nitrogen source. In general, it may be said that any nutrient medium containing essential salts and assimilable sources of carbon and nitrogen may be employed.

The operating conditions employed in this semicontinuous fermentation process may vary widely. In general, the following conditions are preferred: agitation = 500 to 1,000 rpm; air rate = 0.5 to 1.0 v/v/min; temperature = 25° to 30°C; dissolved oxygen = 20 to 60% saturation.

By a Process of Emulsion Fermentation

Fermentations to produce fermentation polysaccharides such as xanthan gum are usually carried out in aerated, deep vessels fitted with stirrer assemblies to provide effective mixing and aeration of the fermenting reaction mass. Aeration and mixing are required to assure air and nutrient exchange between the growing cells and the reaction medium. In polysaccharide fermentations, viscosity of the reaction mass increases with product formation since the product is soluble in the aqueous reaction medium. This viscosity increase reduces the efficiency of stirring and aeration and ultimately limits the amount of polymer which can be produced.

L.G. Maury; U.S. Patent 4,352,882; October 5, 1982; assigned to Kelco Biospecialties Limited, England has found that fermentations to produce polysaccharide gums can be carried to significantly higher gum concentrations if the aqueous culture medium is dispersed in a water-immiscible oil which is a nonsolvent for the polysaccharide gum and the fermentation is effected within that dispersion. This process provides a method of conducting a fermentation reaction wherein an aqueous culture medium comprising a carbohydrate source and a nitrogen source is inoculated with a polysaccharide gum producing microorganism and the medium is mechanically agitated and aerated under conditions to effect fermentation thereof, the culture medium being dispersed in about 20 to 80% of its weight of a water-insoluble oil in which the resultant polysaccharide is also insoluble. Preferred oil concentration is 40 to 80%.

Substantially any water-immiscible, water-dispersible organic oil can be employed as the oil phase. Preferred oils are the higher boiling liquid hydrocarbons of the type having about 8 and more carbon atoms, boiling at about 100°C and higher. These are usually supplied in complex mixtures such as paraffin oil, mineral oil, odorless mineral spirits, deodorized kerosene, or narrow boiling aliphatic hydrocarbons, although pure hydrocarbons such as n-octane per se will work equally well. Vegetable oils such as corn oil, peanut oil, soybean oil and safflower oil can also be used. Certain halogenated hydrocarbons have also been found useful.

Dispersion and stabilization of the aqueous phase of the reaction mass in the oil phase are further aided by the presence of an emulsifier. Preferred are nonionic emulsifiers of HLB ranging from about 12 to 18. Typical of such emulsifiers are the ethoxylated fatty acids and ethoxylated glycerol, glycol and sorbitol fatty acid esters.

The microorganism used to effect the fermentation was *Xanthomonas campestris.*

PRODUCTION OF OTHER POLYSACCHARIDES

Heteropolysaccharide S-21

It is known that heteropolysaccharides can be produced by certain microorganisms. Some of such heteropolysaccharides function as hydrophilic colloids and because of their viscosity properties and rheology have been used as thickening agents for aqueous systems. Xanthan gum, i.e., is a heteropolysaccharide. Illustrative of prior art heteropolysaccharides, their preparation and uses are U.S. Patents 3,020,207; 3,256,271; 3,894,976; 3,915,800 and 3,894,976.

As with other fields of technology, research has continued with the objective of discovering new heteropolysaccharides having useful properties as thickening, suspending and/or stabilizing agents.

K. S. Kang and G.T. Veeder; U.S. Patent 4,247,639; January 27, 1981; assigned to Merck & Co., Inc. have found that a high viscosity anionic heteropolysaccharide composed of about 33% mannose, 29% glucose, 21% galactose and about 17% glucuronic acid and also containing about 5.7% acetyl and about 4.9% pyruvate is obtained by an aerobic fermentation of an organism isolated from a soil sample from the Canal Zone. This heteropolysaccharide has desirable thickening, suspending and/or stabilizing properties in aqueous systems, and is referred to as Heteropolysaccharide S-21.

This heteropolysaccharide is a high molecular weight polysaccharide containing primarily carbohydrate residues and a minor amount of protein. It is sometimes referred to as a "gum."

The bacterium employed in the process, identified as Strain tTR-45, is a mutant of *Klebsiella pneumoniae* Strain S-21 that was isolated from the rhizosphere soil of a plant of the genus *Aechmea epiphytes* belonging to the pineapple family. The soil sample was obtained in the Canal Zone. Strain tTR-45 requires thymine for growth at 37°C but does not require thymine for growth at 30°C. A deposit of Strain tTR-45 was made with the American Type Culture Collection under Accession No. ATCC 31314. The culture is available to the public without restriction.

This organism requires a fermentation medium that supplies a carbon source, a phosphorus source, a nitrogen source, a magnesium source and an iron source. The carbon source typically is hydrolyzed starch with a D.E. range of 12 to 31. The starch can be hydrolyzed with commercially available α-amylases. The phosphorus source may be either Na_2HPO_4, NaH_2PO_4, K_2HPO_4 or KH_2PO_4 or a mixture thereof. The concentration may range from about 0.025 to 0.5%. The magnesium source may be supplied with magnesium chloride or magnesium sulfate in concentrations of from about 0.005 to 0.02%. The nitrogen source may be sodium or potassium nitrate, ammonium nitrate, ammonium sulfate or ammonium chloride, as well as organic sources such as soy peptone Type T (Sheffield Chemical), Promosoy 100 (Central Soya Chemurgy Division), NZ-amine Type A (Sheffield), or Ferm Amine Type IV (Sheffield). The medium may contain either inorganic or organic nitrogen or mixtures thereof. The concentration of inorganic nitrogen in the medium may range from about 0.045 to 0.2% and with the organic nitrogen from about 0.01 to 0.1%. The iron may be supplied to the fermentation as $FeCl_3$ or $FeSO_4$ at levels of 1 to 10 ppm.

The pH of this fermentation preferably is maintained between about 6.3 and 7.7, and the temperature between about 28° and 33°C for maximum polysaccharide production.

The fermentation time is typically from about 48 to 60 hours when proper conditions of medium, temperature, pH and other fermentation parameters are met.

The heteropolysaccharide S-21 may be used as a thickening, suspending, or stabilizing agent.

Heteropolysaccharide S-119

An exopolysaccharide containing D-glucose, D-galactose, pyruvic acid, and O-acetyl groups in the approximate proportions 6:1:1:1.5 is described in *Carbohydrate Research*, 26 (1973) 409-419. These organisms are described as *Agrobacterium tumefaciens* A-8 and A-10.

K.S. Kang, G.T. Veeder and P.J. Mirrasoul; U.S. Patent 4,269,939; May 26, 1981; assigned to Merck & Co., Inc. have found that a variant strain of *A. radiobacter*, ATCC 31643, produces a water-soluble heteropolysaccharide of composition similar to that described for *A. tumefaciens* A-8 and A-10 when incubated in a selected nutrient medium. An unrestricted deposit of this previously undescribed organism was made with the American Type Culture Collection on May 12, 1980, under Accession No. ATCC 31643. The organism was isolated from a soil sample obtained in Kahuka, Hawaii. The organism was picked as a gummy colony after five days' incubation at 30°C from an E-1 agar plate with 1% D.E. corn syrup as the carbon source. The isolate was then pure cultured on nutrient agar.

The nutrient medium for growing the ATCC 31643 (hereafter termed S-119) comprises 5.0% glucose, 0.05% K_2HPO_4, 0.20% enzymatic digest of soybean meal, 0.15% NH_4NO_3, 0.05% $MgSO_4 \cdot 7H_2O$, 1 ppm Fe^{2+}, and 1 ppm Mn^{2+}. The pH for the culture may range from 6.5 to 7.2, and the temperature of the medium is 30°C. The nutrient medium should be substantially free of Ca^{2+}.

Heteropolysaccharide S-119 has been proven an effective antimigrant in pad-dyeing systems.

Heteropolysaccharide S-10

The heteropolysaccharide developed by *K.S. Kang, G.T. Veeder, III and D.D. Richey; U.S. Patent 4,286,059; August 25, 1981; assigned to Merck & Co., Inc.* may be referred to as Heteropolysaccharide S-10. It is prepared by fermentation of a suitable nutrient medium with a strain of *Klebsiella pneumoniae* which does not grow at 37°C in the absence of iron. A deposit of this organism was made with the American Type Culture Collection on August 11, 1971 under Accession No. ATCC 21711. At the time of deposit, based upon classification tests in the 7th Edition of *Bergey's Manual of Determinative Bacteriology* (1959), the organism was identified as *Erwinia tahitica*. When the 8th Edition of this work appeared in 1974, however, a change in classification tests resulted in a change of identification of the organism from *Erwinia tahitica* to *Klebsiella pneumoniae*. A restricted deposit of a doubleblocked mutant of this organism which requires iron, insufficient of which is available in the human body, for growth at 37°C has been deposited with the American Type Culture Collection on July 25, 1977, under Accession No. ATCC 31311.

The bacterium identified as ATCC 21711 is a nonvirulent strain of *Klebsiella pneumoniae* isolated from a Tahitian soil sample. *Klebsiella pneumoniae* is an ubiquitous organism found in water, soil, and all manner of vegetable matter. In addition, it is found in large numbers in industrial settings such as cooling tower water and paper mill effluents. In none of the situations is there evidence for *Klebsiella pneumoniae* causing any health problem. Soil isolates such as this are generally considered to be safe.

Heteropolysaccharide-10, which contains about 3% protein and about 97% carbohydrate, the carbohydrate portion of which contains about 19% of a uronic acid, about 39% glucose, about 29% galactose and about 13% fucose, is compatible with Methylene Blue chloride dye. The process for preparing it comprises cultivating under submerged aerobic conditions at a temperature of about 28° to 35°C a heteropolysaccharide-producing strain of *Klebsiella pneumoniae* or a heteropolysaccharide-producing mutant thereof in an aqueous nutrient medium containing as a source of carbon hydrolyzed starch having a D.E. of about 20 to 35, a source of nitrogen, a source of magnesium and a source of phosphorus, until substantial viscosity has been imparted to the medium, and recovering the heteropolysaccharide.

The fermentation is carried out for from about 45 to 60 hours, and the carbon source is a cornstarch hydrolyzate. The recovery of heteropolysaccharide is effected by precipitation with lower alkanol, such as isopropanol.

Heteropolysaccharide-10 imparts viscosity to an aqueous medium when dissolved in water in low concentration. Because of this, its sensitivity to shear, its pseudoplasticity, its stability with salts, and because of its overall rheology, Heteropolysaccharide-10 is useful as a thickening, suspending and stabilizing agent in aqueous systems. More specifically, it is useful as an additive to textile printing pastes or in formulating low drift aqueous herbicidal compositions. It is also of value as a thickening or suspending agent in salad dressings, in forming thickened puddings and as a thickener in adhesive compositions.

Heteropolysaccharide-10 is of particular value as an additive in aqueous paints because of its ability to improve the flow and leveling of such paints, and because of its pseudoplasticity. It is also useful as a fluid loss control agent in drilling muds, completion fluids and similar aqueous media from which fluid losses to subterranean strata have to be controlled.

Heteropolysaccharide S-156

In similar work by *G.T. Veeder and K.S. Kang; U.S. Patent 4,298,691; Nov. 3, 1981; assigned to Merck & Co., Inc.* it has been found that a variant strain of *K. pneumoniae*, ATCC 31646, produces a water-soluble heteropolysaccharide of composition similar to that described for *K. pneumoniae* K-63 when incubated in a selected nutrient medium. An unrestricted deposit of this previously undescribed organism was made with the American Type Culture Collection on May 12, 1980, under Accession No. ATCC 31646. It is listed with the ATCC as a Class 2 agent.

When this organism is grown in a low nitrate medium, it produces a polysaccharide of high viscosity. The nutrient medium should contain no more than about 0.06% (w/v) ammonium nitrate. It comprises also 3% hydrolyzed starch (D.E. 16 to 20), 0.05% K_2HPO_4, 0.05% enzymatic digest of soybean meal, 0.01% $MgSO_4 \cdot 7H_2O$ and 1 ppm ferrous ion.

On completion of the fermentation, the aqueous nutrient medium at a pH of 4.5 to 7 is heated to a temperature of 70° to 100°C for about 2 to 5 minutes in the presence of from 0.5 to 15% by weight of dialdehyde (weight based on the dry weight of Heteropolysaccharide S-156 in the medium) to recover the Heteropolysaccharide S-156.

Heteropolysaccharide S-156 is particularly useful as a thickening agent for water-based paints.

Heteropolysaccharide S-53

K.S. Kang and G.T. Veeder; U.S. Patent 4,311,795; January 19, 1982; assigned to Merck & Co., Inc. have also developed a polysaccharide referred to as S-53. It may be prepared by fermentation of a suitable nutrient medium with a new strain of *Klebsiella pneumoniae*, ATCC 31488.

Various classification keys for the genus Klebsiella and the culture descriptions of *Klebsiella pneumoniae* species and strains are found in the 7th and 8th Edition of *Bergey's Manual*.

The heteropolysaccharide S-53 is produced during the aerobic fermentation of suitable aqueous nutrient media under controlled conditions via the inoculation with the organism of *Klebsiella pneumoniae* ATCC No. 31488. The media are usual media, containing sources of carbon, nitrogen and inorganic salts.

The preferred carbon source is hydrolyzed starch, prepared by reacting approximately 30% of starch which can be of different sources such as corn, tapioca, or potato, etc., with 1-2% of commercially available α-amylase at approximately 70° to 80°C for approximately 30 minutes. The exact quantity of the carbohydrate source or sources utilized in the medium depend in part upon the other ingredients of the medium but, in general, the amount of carbohydrate usually varies between about 2 and 4% by weight of the medium.

In general, many proteinaceous materials may be used as nitrogen sources in the fermentation process. Suitable nitrogen sources include, for example, soy peptone, yeast hydrolysates, primary yeast, soybean meal, cottonseed flour, hydrolysates of casein, corn steep liquor, distiller's solubles or tomato paste and the like. The sources of nitrogen, either alone or in combination, are used in amounts ranging from about 0.05 to 0.2% by weight of the aqueous medium.

Among the nutrient inorganic salts which can be incorporated in the culture media are the customary salts capable of yielding sodium, potassium, ammonium, calcium, phosphate, sulfate, chloride, carbonate, nitrate and like ions. A medium containing $NaNO_3$ definitely produces a higher yield than a medium to which $NaNO_3$ is not added. Also included are trace metals such as iron and magnesium.

The fermentation is carried out at temperatures ranging from about 25° to 35°C; however, for optimum results it is preferable to conduct the fermentation at temperatures of from about 28° to 32°C. The pH of the nutrient media for growing the *Klebsiella pneumoniae* ATCC No. 31488 culture and producing the polysaccharide S-53 can vary from about 5.5 to 7.5.

Although the polysaccharide S-53 is produced by both surface and submerged culture, it is preferred to carry out the fermentation in the submerged state.

For large-scale work, it is preferable to conduct the fermentation in suitable tanks provided with an agitator and a means of aerating the fermentation medium. According to this method, the nutrient medium is made up in the tank and sterilized by heating at temperatures of up to about 121°C. Upon cooling, the sterilized medium is inoculated with a previously grown seed of the producing culture, and the fermentation is permitted to proceed for a period of time as, for example, from 2 to 4 days while agitating and/or aerating the nutrient medium and maintaining the temperature at about 30°C. This method of producing the S-53 is particularly suited for the preparation of large quantities.

The product is recovered from the fermentation medium by precipitation with a suitable alcohol, such as isopropanol.

The polysaccharide S-53 imparts viscosity to an aqueous medium when dissolved in water in low concentrations. Because of this, its sensitivity to shear and overall rheology, it is useful as a thickening, suspending and stabilizing agent in aqueous systems; for example, as an additive to textile printing pastes or in formulating low drift aqueous herbicide compositions, salad dressings, thickened puddings, and adhesive compositions. It is particularly useful in thickening latex semigloss and flat paints, or in other paint applications.

Conversion of Guar Gum to Gel-Forming Polysaccharides Using α-Galactosidase

Guar gum is 78 to 82% of the endosperm component of guar seed. Guar has been grown for centuries in India and Pakistan, where it is one of the crops used as a food for both humans and animals. In the United States, it is grown in north Texas and southern Oklahoma. Guar seeds contain approximately 14 to 17% hull, 35 to 42% endosperm, and 43 to 47% germ. They are commonly dry-milled to separate out the endosperm which is the industrial guar gum of commerce. Although guar gum is normally commercially used in its crude form, its principal component and the component giving its industrial value is the polysaccharide guaran. Guaran readily dissolves in water to form highly viscous solutions even at low concentrations of gum. The solutions remain stable because molecular segments of guaran cannot bind to each other when they collide in solution. This occurs essentially since the manopyranosyl chains are separated from each other by the derivatizing α-D-galactopyranosyl side groups.

According to *R.L. Whistler; U.S. Patent 4,332,894; June 1, 1982; assigned to Purdue Research Foundation* guar gum can be modified through the use of enzymes, for example, a commercially available α-D-galactosidase enzyme, to remove the α-D-galactopyranosyl side groups. In modification of guar gum, by controlling the length of time that the guar gum is exposed to the α-D-galactosidase enzyme, and, therefore, by controlling the extent of the removal of the α-D-galactopyranosyl side groups, the length of denuded mannan chain exposed by the enzyme's activity on the guar gum can be controlled. Control of the length of denuded mannan chain exposed provides a direct control on the amount of intermolecular association that adjacent guaran molecules in solution experience. To control this amount of intermolecular association is to control the relative strength or weakness of a gel produced from guar gum by the activity of the α-D-galactosidase enzyme.

In an illustrative preparation technique, guar gum is treated with a commercially available D-galactosidase enzyme, and a 1% by weight solution of the treated guaran or guar gum in water is permitted to stand for 30 minutes. At the end of 30 minutes, a viscosity measurement is conducted. A reduction, and specifically a large reduction, in the viscosity of the solution indicates the presence of mannosidase enzymes in the starting enzyme treatment. The mannosidase enzymes have cleaved, or broken, the mannan backbone or main chain of the guaran and rendered the treated solution useless. On the other hand, gel formation in the 1% solution after standing for 30 minutes will be indicated by an increase in the viscosity of the 1% solution. The promotion of gel-forming characteristics in the 1% solution, and the relative "strength" of the resultant gel, can be made by any of a number of methods for measuring gel strength known in the industry.

Another method for measuring the success of the treatment with D-galactosidase enzyme is the change in viscosity of the treated guaran or guar gum when a standard concentration of the treated guaran or guar gum is mixed with a standard concentration of a synergistically active gum such as xanthan gum of carrageenan. The synergistic increase in viscosity of a mixture of these two normally water-soluble polysaccharides will be indicative of the degree of removal of D-galactopyranosyl side groups by D-galactosidase enzyme.

Production of a Gelable Curdlan-Type Exopolysaccharide

H.G. Lawford; U.S. Patent 4,355,106; October 19, 1982; assigned to George Weston Limited, Canada has developed a microbial process whereby a gelable extracellular microbial polysaccharide, also known as an exopolysaccharide, can be produced continuously by means of a stable, curdlan-producing strain of microorganism.

Exopolysaccharides have found diverse applications in industry, principally the food industry and the petroleum industry. In the food industry they replace or extend natural plant gums as stabilizers, emulsifiers and thickeners. In the petroleum industry they have been used in both drilling and oil recovery operations.

Many different microorganisms are known which produce exopolysaccharides. The chemical composition and structure of the exopolysaccharide are relatively species specific, i.e., they are dependent upon the strain of microorganisms used in its production.

In this two-stage continuous process for the production of a gelable curdlan-type exopolysaccharide (1) a stable, curdlan-producing strain of microorganism is grown aerobically under conditions of continuous culture at pH 6 to 8 and at a temperature of from 22° to 37°C in an agitated culture medium containing assimilable carbon, nutrients, and inorganic salts and also containing assimilable nitrogen in an amount so limited that the effluent from the first stage contains substantially no inorganic nitrogen; and (2) the effluent is continuously introduced into a constant volume fermenter wherein it is mixed with a nitrogen-free carbohydrate. The resultant mixture is aerated and mixed at pH 5.5 to 6.5 and a temperature of from 25° to 35°C, the volume and dilution rate in the fermenter being selected so that the residence time of the microorganism in the fermenter does not exceed an equivalent to the maximum length of time during which the activity of a batch culture of the microorganism with respect to product synthesis is maximal, and subsequently isolating the product.

The preferred stable, curdlan-producing strains of microorganisms for use in the process are *Alcaligenes faecalis* var. *myxogenes* ATCC 31749 and 31750. The growth limiting nutrient in the culture medium in the first stage of this synthesis is nitrogen. The amount of assimilable nitrogen in the culture medium defines the steady-state level of biomass being continuously produced in the first stage and the rate at which the medium is pumped into the first constant-volume fermenter establishes the growth rate of the microorganism or a rate at which biomass is generated in the system. The biomass effluent from the first fermenter contains a negligible amount of nitrogen so that growth is not possible in the second-stage fermenter which is also maintained at a constant volume. Tests have shown that this use of a two-stage process with a medium of defined composition permits continuous production of curdlan-type polysaccharide in good yield.

Heteropolysaccharide Biopolymer PS-87

Certain polysaccharides can be obtained by microbial biosynthesis utilizing specific strains or species of bacteria. Some of these polysaccharides have been employed as thickeners or suspending agents, particularly in water-based systems such as foods, cosmetics and pharmaceuticals. Generally, however, it has been found that microbial polysaccharides have certain limitations when employed in such products, in terms of their ability to function as thickeners or suspending agents. Some polysaccharides are, for example, unstable, particularly when the products containing them are subjected to shear, for example, when filling or dispensing through a narrow orifice. Limitations such as these can be partly overcome by increasing the concentration of the microbial polysaccharide in the product, but this can change the character of the product in other respects and can substantially increase raw material costs. Research has accordingly been directed to finding new polysaccharides that would be of value as improved thickeners or suspending agents, and which do not suffer from these problems.

R.B. Cox and D.C. Steer; U.S. Patent 4,357,423; November 2, 1982; assigned to Lever Brothers Company have found that a new heteropolysaccharide, referred to as Biopolymer PS-87, can be obtained by fermentation of a nutrient medium with a strain of the species of the bacterium known as *Bacillus polymyxa*. This heteropolysaccharide, which is useful as a thickener or suspending agent in foods, cosmetics and pharmaceuticals, has unexpectedly superior pseudoplastic properties. It comprises glucose, galactose, mannose and glucuronic acid. A 1% by weight solution of Biopolymer PS-87 has pseudoplastic properties, a consistency at 20°C of at least 150 poises and a yield stress value at 20°C of at least 30 dynes/cm^2.

The microorganism which is preferably employed in the production of Biopolymer PS-87 is a specific strain or mutant of the species *Bacillus polymyxa* and has Accession No. NCIB 11429. The organism was isolated from seawater by plating out onto simple molasses plus mineral salts agar medium and incubating at 30°C. Those colonies which developed a mucoid or sticky appearance were streaked onto new plates of the same medium in order to obtain pure isolates of the organism. The pure cultures were cultivated in liquid molasses-containing medium in shaken flasks at 30°C to confirm a satisfactory growth pattern with an accompanying increase in viscosity indicative of polysaccharide production.

In addition to employing the particular strain of *Bacillus polymyxa* for the production of Biopolymer PS-87 it is also possible to employ genetically modified bacteria which have been adapted to synthesis Biopolymer PS-87. The basic organisms from which the modified organisms can be derived include the bacteria *E. coli, Pseudomonas* sp., *Klebsiella* sp. and *Bacillus* sp. The genetically modified microorganisms are made by incorporating into basic microorganisms genetic information carrying the genes of the Biopolymer PS-87 synthesis mechanism.

One such method comprises the steps of: (a) producing from the basic microorganism a mutant deficient in genetic material specifying Biopolymer PS-87 synthesis; (b) preparing a plasmid hybrid consisting of plasmid DNA covalently joined to DNA specifying Biopolymer PS-87 biosynthesis; (c) introducing the plasmid hybrid into the Biopolymer PS-87 synthesis deficient basic microorganism; (d) culturing the resultant microorganism in conditions favoring growth by the Biopolymer PS-87 synthesis mechanism; and (e) selecting one or more clones of microorganisms growing by the Biopolymer PS-87 synthesis mechanism.

Another such method comprises the steps of: (a) producing from the basic microorganism a mutant deficient in genetic material specifying Biopolymer PS-87 synthesis; (b) identifying a phage DNA or a temperature phage for the basic microorganism; (c) introducing into the phage or phage DNA a piece of DNA specifying Biopolymer PS-87 synthesis, thereby to produce a phage hybrid; (d) lysogenizing the Biopolymer PS-87 synthesis deficient basic microorganism with the phage hybrid; (e) culturing the resultant microorganism in conditions favoring growth by the Biopolymer PS-87 synthesis mechanism; and (f) selecting one or more clones of microorganisms exhibiting the Biopolymer PS-87 synthesis mechanism.

According to a preferred embodiment, a process for the production of Biopolymer PS-87 comprises the steps of: (1) cultivating *Bacillus polymyxa* NCIB 11429 under submerged aerobic conditions in an aqueous nutrient medium having a pH of from 4.5 to 7.5, until substantial accumulation of Biopolymer PS-87 has occurred; (2) heating the culture medium to a temperature of at least 50°C at a pH value of at least 8; (3) subsequently separating coagulated cells and cell debris from the culture medium to provide a visually clear solution of Biopolymer PS-87. The medium will preferably contain a trace amount of from about 0.5 to 5 mg/ℓ of biotin. This can be provided as biotin itself or as a source of biotin.

MISCELLANEOUS PROCESSES AND PRODUCTS

WASTE MATERIAL TREATMENTS

Process for Decolorizing Pulp and Paper Mill Wastewater

The purpose of the process of *J.E. Blair and L.T. Davis; U.S. Patent 4,266,035; May 5, 1981; assigned to Sybron Corporation* is to provide a biological process for treatment of pulp and paper mill wastewater effluent to not only remove biodegradable organic matter therefrom but to specifically reduce or decolorize pulp and paper mill wastewater.

This objective is accomplished by treating wastewater effluent from a pulp or paper mill with a microorganism of the strain *Pseudomonas aeruginosa* 4-5-14.

The mutant *Pseudomonas aeruginosa* 4-5-14 (hereinafter mutant strain) was produced by mutation of a parent strain of *Pseudomonas aeruginosa* isolated from the soil surrounding a pulp and paper mill wastewater lagoon at a large Kraft paper mill located in Franklin, VA. On the basis of its morphological, cultural, and physiological characteristics, the strain has been identified as a member of the species *Pseudomonas aeruginosa* and has been designated herein as *Pseudomonas aeruginosa* 4-5-14. A culture of the strain has been deposited in the American Type Culture Collection and has received an accession number, ATCC-31482.

This mutant strain can be cultured in wastewater from a pulp or paper mill either using a batch process, a semicontinuous process or a continuous process, and is cultured for a time sufficient to degrade the colorant materials present in the wastewater and remove them or break them down into components capable of being degraded by other organisms normally found in biological wastewater treatment systems.

The mutant strain can be employed in ion exchange resin treatment systems, in trickling filter systems, in carbon adsorption systems, in activated sludge treatment systems, in outdoor lagoons or pools, etc. Basically, all that is necessary is for the microorganism to be placed in a situation of contact with the wastewater effluent from a pulp or paper mill. In order to degrade the material present in the wastewater, the organism can be cultured at conditions of about 15° to about 40°C, preferably about 18° to about 37°C. Desirably, the pH is maintained in a range of about 6.0 to about 8.5, preferably 7.0 to 8.0. Control of the pH can be by monitoring of the system and an addition of appropriate pH adjusting materials to achieve this pH range.

The culturing is conducted basically under aerobic conditions of a dissolved oxygen concentration of about 2 ppm or more, preferably about 5 ppm or more. These conditions can be simply achieved in any manner conventional in the art and appropriate to the treatment system design being employed. For example, air can be bubbled into the system, the system can be agitated, a trickling system can be employed, etc.

The wastewater to be subjected to treatment by this process may contain sufficient nitrogen and phosphorus for culturing without the need for any additional source of nitrogen or phosphorus being added. However, in the event the wastewater is deficient in these two components, suitable available nitrogen and phosphorus can be added.

Fermentation of Plant Materials with *Cyathus stercoreus*

The process developed by *D.T. Wicklow and R.W. Detroy; U.S. Patent 4,275,167; June 23, 1981; assigned to U.S. Secretary of Agriculture* relates to the biological modification of the lignocellulosic components of gramineous agricultural plant materials for the purpose of making the cellulose more available for use in a variety of applications. For instance, the digestibility of lignocellulosic grasses by cattle is limited to only that portion of the cellulose which is not physically bound by the lignin component. Also, with the current trend toward the production of alcohol as a renewable energy resource, the feasibility of employing lignocellulosic residues as sources for fermentable sugars hinges on the amount of free cellulose which can be enzymatically hydrolyzed to a glucose. Typically, up to about 50% of the cellulose in natural, grassy plant tissue is rendered unavailable for such uses by the lignin associated therewith.

Wicklow and Detroy have developed a microbiological procedure for treating gramineous plant materials, which is characterized by the surprising result of preferentially degrading the lignin over the cellulose to an extent heretofore not observed for any known biological process. The procedure comprises inoculating the plant material with the fungal microorganism *Cyathus stercoreus,* fermenting the inoculated material under conditions favorable for the growth of the microorganism, and recovering the resultant free cellulose-enriched material from the fermentation.

Suitable lignocellulosic substrates which can be preferentially degraded by this process include fresh plant materials of grassy species belonging to the family Gramineae. The term "fresh" is intended to include recently harvested as well as dried materials, but excludes plant materials previously digested by animals. Of particular interest are fresh gramineous agricultural residues; that is, the portions of grain-bearing grassy plants which remain after harvesting the seed. Illustrative of such residues are wheat straw, oat straw, rice straw, corn stalks, corn husks, and the like. Due to the unique combination of chemical substructures characteristic of the natural lignins in grasses, the process is not expected to have comparable utility when applied to other than the grassy plants.

The microorganism, *C. stercoreus* (Bird's Nest Fungus), is one of many well-known fungal colonists of grassland, ruminant and horse dung. It characteristically appears late in the decomposition sequence and forms basidiocarps on the surface of the dung. *C. stercoreus* is unique in its ability to degrade gramineous lignin at a rate on the order of twice that for degrading the associated cellulose and is able to free at least five times more cellulose for conversion to glucose than is available in unfermented wheat straw.

C. stercoreus can be cultivated on gramineous substrates under aerobic conditions at temperatures in the range of about 20° to 30°C and in the presence of a sufficient amount of moisture. It is necessary that water be present at a level of at least about 100% based on the dry weight of the substrate, and generally should be in the range of about 100 to 400%. The water may either be that which is naturally present in the material, or it may include water added to bring the moisture content up to the desired level. Other conditions for cultivation are not particularly critical.

The amount of free cellulose available for cellulase hydrolysis to glucose in the fermented substrate is typically five to six times, and even as much as eight times, the amount found in the starting material. This represents an approximate 40 to 60% reduction in the level of lignin accompanied by a mere 20 to 25% loss of the original cellulose content.

The available cellulose-enriched plant material can be fed directly to ruminants, though it may be desirable to sterilize it by any conventional means. Alternatively, the fermented substrate can be subjected to enzymatic hydrolysis with cellulase in order to provide a fermentable source of glucose for use in the production of alcohol. Other end uses for the enriched substrate would be obvious to a person skilled in the art.

Removing Oleaginous Material from Wastewater by Combination of Microorganisms

The objective of the process developed by *P.W. Spraker; U.S. Patent 4,288,545; September 8, 1981; assigned to Sybron Corporation* is to provide a biological treatment for removal of oleaginous materials from industrial, domestic and wastewater, making it suitable for pollution-free discharge. This is accomplished by treating the wastewater with a synergistic microbial combination comprising:

(a) a microorganism of the strain *Pseudomonas aeruginosa* mutant $SGRR_2$; and

(b) at least one of:

  (1) a microorganism of the genus Bacillus; and

  (2) a microorganism of the genus Pseudomonas other than the strain *Pseudomonas aeruginosa* mutant $SGRR_2$.

The mutant of the species *Pseudomonas aeruginosa* $SGRR_2$ (ATCC accession number 31480) was produced by mutation of a parent strain of *P. aeruginosa* designated *P. aeruginosa* HCP (hereinafter parent strain) isolated from the soil in Salem, Virginia, and having the ATCC accession number 31479.

The *P. aeruginosa* $SGRR_2$ (hereinafter mutant strain) has been found, when used in combination with microorganisms of the genus Bacillus and microorganisms of the genus Pseudomonas other than this mutant strain, to be capable of degrading oleaginous materials.

The mutant was tested against oleaginous materials and found to be substantially nonactive against oleaginous materials in wastewater when used alone, as was the parent strain *P. aeruginosa* HCP, but synergistically active in degrading oleaginous materials when used in combination with the parent strain. Further, it was found that when the $SGRR_2$ mutant strain was used in combination with microorganisms of the genus Bacillus, such as *Bacillus subtilis,* also substantially not active alone against oleaginous materials, the combination was synergistically active against oleaginous materials.

The microbial combination of (a) the *P. aeruginosa* $SGRR_2$ and (b) the other organism of the genus Bacillus or the genus Pseudomonas (other than *P. aeruginosa* $SGRR_2$) can be employed in cell count proportions ranging from about 1:99 to about 99:1 of (a) to (b) to achieve the objects of the process.

The microbial combination employed can be cultured in wastewater containing the oleaginous material either using a batch process, a semicontinuous process or a continuous process, and such a microorganism combination is cultured for a time sufficient to degrade the oleaginous materials present in the wastewater and remove them or break them down into components capable of being degraded by other organisms normally found in biological wastewater treatment systems.

The microbial combination can be used in ion exchange resin treatment systems, in trickling filter systems, in activated sludge treatment systems, in outdoor lagoons or pools, etc. Basically, all that is necessary is for the microbial combination to be placed in a situation of contact with wastewater containing the oleaginous material. In order to degrade the oleaginous material present in the wastewater, the microbial combination can be cultured at conditions of about 15° to about 42°C, preferably about 20° to about 38°C. Desirably, the pH is maintained in a range of about 5.5 to about 8.5, preferably 6.5 to 8.0. Control of the pH can be by monitoring of the system and an addition of appropriate pH adjusting materials to achieve this pH range.

The culturing is conducted basically under aerobic conditions of a dissolved oxygen concentration of about 2 ppm or more, preferably about 5 ppm or more. These conditions can be simply achieved in any manner conventional in the art and appropriate to the treatment system design being employed.

Suitable examples of other organisms of the genus Pseudomonas, other than the *P. aeruginosa* $SGRR_2$ mutant strain, which can be employed in the combination include those of the alcaligenes group, such as *P. alcaligenes,* and those of the fluorescent group, such as *P. fluorescens, P. putida,* and other *P. aeruginosa* strains.

Suitable microorganisms of the genus Bacillus which can be employed in combination with the mutant strain include *B. subtilis, B. licheniformis, B. cereus, B. thuringiensis, B. megaterium, B. circulans, B. coagulans, B. brevis, B. sphaericus, B. fastidiosus,* etc.

The microbial combination utilized in the process results in the ability to obtain approximately a 75-fold increase over that obtained with the use of Bacillus strains alone, and approximately a 4-fold increase over the results obtained where the *P. aeruginosa* SGRR_2 mutant strain is employed alone or where the parent *P. aeruginosa* HCP, from which the mutant strain was developed, is used alone.

Removal of Surface Active Agents and Detergents from Waste

Wastewaters containing organic and inorganic materials as pollutants are unsuitable for reuse and undesirable for release into the biosphere due to problems of pollution which results when they are discharged untreated. To remove, or at least minimize, this difficulty, domestic, municipal and industrial wastewaters are conventionally processed in biological treatment systems, for example, aerated lagoons or activated sludge systems, for removal of biodegradable organic matter prior to reuse or discharge to receiving bodies of water.

While the biological processes occurring during such a biological treatment provide the ability to produce effluent with lower biochemical oxygen demand (BOD) and low chemical oxygen demand (COD), unfortunately, removal of materials such as anionic and nonionic surface active agents, detergents and like materials using conventionally employed biological treatment systems has not met with a large amount of success. Even when the biological treatment system is capable of degrading materials such as anionic and nonionic surface active agents, detergents and like materials, the degradation process is often too slow or insufficient resulting in a concentration build-up of these materials or a carry through the system of these materials undigested.

The objective of *J.E. Blair and L.T. Davis; U.S. Patent 4,317,885; March 2, 1982; assigned to Sybron Corporation* is to provide a process whereby anionic and nonionic surface active agents, detergents, and like materials present in domestic, municipal and industrial wastewaters can be removed.

It has been accomplished by providing a process for treating wastewater containing anionic and/or nonionic surface active agents, detergents and like materials, which comprises treating wastewater containing synthetic nonionic surface active agents, detergents and like materials with a microorganism of the strain *P. fluorescens* 3P (ATCC 31483).

The parent of this Pseudomonas strain was produced by selection of strains of microorganisms from a wastewater lagoon at a large textile manufacturing plant in South Carolina. The mutant strain was made by subjecting the parent strain to mutagenesis using 0.02% sodium nitrite at a pH of 6.5 to 6.8. The selected mutant strain can be cultured in wastewater from any type of industrial plant containing anionic and/or nonionic surface active agents, detergents and like materials either using a batch process, a semicontinuous process or a continuous process, and is cultured for a time sufficient to degrade the anionic and/or nonionic surface active agents, detergents and like material present in the wastewater and remove them or break them down into components capable of being degraded by other organisms normally found in biological wastewater treatment systems.

The mutant strain can be employed in ion exchange resin treatment systems, in trickling filter systems, in carbon adsorption systems, in activated sludge treatment systems, in outdoor lagoons or pools, etc. Basically, all that is necessary is for the microorganism to be placed in a situation of contact with the wastewater effluent. In order to degrade the material present in the wastewater, the organisms can be cultured under conditions preferably about 20° to about 38°C. Desirably, the pH is preferably 6.5 to 8.0.

The culturing is conducted basically under aerobic conditions of a dissolved oxygen concentration of about 2 ppm or more, preferably about 5 ppm or more. These conditions can be simply achieved in any manner conventional in the art and appropriate in the treatment system design being employed.

In the above manner, anionic and/or nonionic surface active agents, detergents and like materials which have been previously considered in the art to be difficultly degradable or nonbiodegradable, as well as other organic compounds which might be present in wastewater systems, can be advantageously treated to provide treated wastewater suitable for discharge after any additional conventional processing such as settling, chlorination, etc., into rivers and streams.

Producing an Agglutinating Substance Using a Dematium Strain

S. Shinohara; U.S. Patent 4,258,132; March 24, 1981 describes a process for producing an agglutinating agent by the culture of a microorganism belonging to the Dematium genus. A process is also described for using this microorganism for the agglutination of proteins, organic and inorganic substances, minerals and living germs suspended, dispersed or floating in a liquid.

The microorganism used is the fungus Dematium (Dematiaceae) FERM-P 4257 (ATCC 20524). The fungus is cultivated at an acidic pH in a carbon source such as a carbohydrate until the agglutinating substance is produced and then the agglutinating substance is separated from the culture medium. The carbohydrate should be 0.5 to 5 wt % of the culture medium.

The fact that the described substance can be obtained in a maximum yield in the medium of a very low carbon source concentration of around 1% indicates that wastes from agricultural industries, stock raising industries and food processing industries having a carbon source (glucose, sucrose, etc.) content as low as about 1% are suitable as the medium for the culture of the fungus. Accordingly, there is provided an effective process for the treatment of those wastes. If this fungus is cultured in the presence of a carbon source-producing fungus, the intended agglutination-active substance can be obtained in a high yield. Thus, the process has a great industrial and economical value.

Tests showed that this agglutinating substance containing an inorganic ion (aluminum ion for acidic solutions and calcium ion for weakly alkaline solutions) has the property of completely agglutinating and precipitating organic substances, inorganic substances and living germs in the form of dispersion, suspension or colloid in water or floating in water if the substance is added in even a very small amount (0.1-3.0 ppm based on the liquid). It may safely be said that the agglutinating effect of this substance is far stronger than that of commercially available agglutinating agents (inorganic and organic). Another merit of this agglutinating substance is that it does not pose the problem of secondary environmental pollution, since it is a metabolic product of a microorganism. Agglutination conditions are as follows:

(1) Optimum pH range is from acidic to faintly acidic range. Under alkaline condition, the agglutinating power is not exhibited.

(2) Reaction temperature ranges from ambient temperature to an elevated temperature. The temperature has no influence on the agglutinating power.

(3) Slow stirring is required after the addition of the agglutinating substance.

(4) Amount of the substance used as agglutinating agent is 0.1-3.0 ppm. The amount is fixed irrespective of substances to be agglutinated with few marked exceptions.

Example 1: By adding 6-9 ppm of this agglutinating agent to the wastewater from a noodle-making device in the presence of 2 ppm of aluminum sulfate, the chemical oxygen demand (COD) was reduced by 80% and percent transmission was increased to about 88%. It was found that the agglutination can be completed in several minutes and the agglutinated substances can be filtered out very easily.

Example 2: The urine and wastewater from a swinery are added to the agglutinating agent and then aluminum sulfate. White fluffy flocks are formed immediately. The flocks are precipitated very easily to leave transparent supernatant liquid. Percent transmission thereof is up to 99.0%. Decoloring rate is 95.0% and, simultaneously, COD can be reduced by 40%. The flocks thus precipitated can be filtered out very easily.

The above results indicate that the agglutinating substance exhibits a remarkable effect on liquids and wastewaters containing microorganisms, particularly bacteria and that the agglutinating substance can be utilized advantageously for the treatment of wastewaters by sedimentation.

MISCELLANEOUS PROCESSES

Isopenicillin N and Its Derivatives with a Cell-Free Extract of *Cephalosporium acremonium*

A.L. Demain, T. Konomi and J.E. Baldwin; U.S. Patent 4,248,966; February 3, 1981; assigned to Massachusetts Institute of Technology describe procedures for producing isopenicillin N [formula (1) below] and many of its derivatives to the exclusion of penicillin N. The mechanism of the reaction, broadly stated, is that the tripeptide δ-(L-α-aminoadipyl)-L-cysteinyl-D-valine [hereafter known as LLD; see formula (2) below] undergoes ring formation resulting in the production of isopenicillin N provided that certain enzymes in the cell-free extract are inactivated so that the conversion does not go beyond the isopenicillin stage.

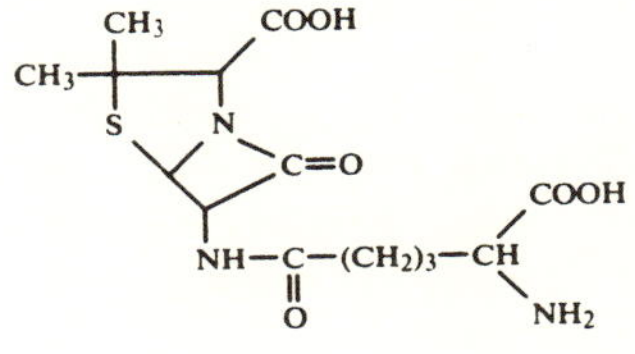

(1)

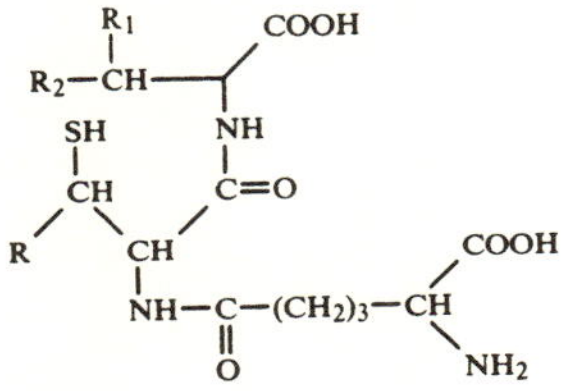

(2)

In formula (2) above, R, R_1, and R_2 are hydrogen, methyl, ethyl, propyl, or isopropyl or halogenated analogs of the foregoing radicals.

In addition to the foregoing, the cell-free system can catalyze the reactions when tripeptides containing certain synthetic β-substituted derivatives of L-cysteine and of D-valine and certain analogs of α-amino adipic acid are used in place of LLD. Thus, the process is capable of producing the isopenicillin derivatives which are entirely new.

Thus in addition to LLD containing α-amino adipic acid, α-AAA analogs can be used as starting materials to synthesize isopenicillin derivatives. Such analogs include: L-2-bromoadipic acid, L-2-azidoadipic acid, L-2,5-diaminoadipic acid, adipic acid, L-glutamic acid, L-α-aminopimelic acid, L-S-carboxymethylcysteine, L-3,3-dimethyl-S-carboxymethylcysteine and L-5-methyl-S-carboxymethylcysteine.

It has also been discovered that providing aeration by shaking and providing energy by the addition of adenosine triphosphate (ATP), and especially ATP plus an ATP regeneration system, increases the amount of isopenicillin produced. The ATP regenerating system comprises a phosphate donor and a phosphotransferase enzyme. The preferred donor is phosphoenolpyruvate and its corresponding phosphotransferase enzyme, pyruvate kinase.

In accordance with the process, a starting material comprising LLD or its derivatives is intimately contacted with a cell-free extract, made from *Cephalosporium acremonium,* in the presence of ATP and is thereby transformed by enzymes in the extract to an isopenicillin derivative. *C. acremonium* is a well known microorganism, and several strains are available from the American Type Culture Collection under numbers such as ATCC 20339, ATCC 14553 and ATCC 35255.

The preferred method of preparing the cell-free extract comprises lysing a protoplast pellet made from whole cells obtained from 40 to 70 hr mycelia and treated with, e.g., Cytophaga lytic enzyme L_1 preparation and Zymolyase-5000. After treatment with the enzymes, the protoplast pellet suspension is centrifuged and gently homogenized. A second centrifugation enables separation of a supernatant liquid extract which may be used to produce the isopenicillins after late enzyme(s) has been inactivated. Thus, if a suitable starting material is mixed with this cell-free extract and ATP, an isopenicillin derivative-rich solution results. (As used herein "late enzymes" refer to enzymes that catalyze reactions in the latter part of a biosynthetic pathway.)

Deacetoxycephalosporin C

Isopenicillin N is a water-soluble, beta-lactam antibiotic shown in the last patent as formula (1).

The aminoadipyl side chain is in the L-configuration in isopenicillin N. Penicillin N, also an effective antibiotic, has a structure identical to isopenicillin N except that the aminoadipyl side chain is in the D-configuration. Penicillin N and isopenicillin N have a number of properties in common but differ in their antimicrobial activity toward certain classes of microorganisms.

Deacetoxycephalosporin C (hereafter referred to as DACPC) is useful as an antibiotic as such or as a starting compound for the production of cephalosporin antibiotics, such as cephalexin. DACPC is shown by the following formula:

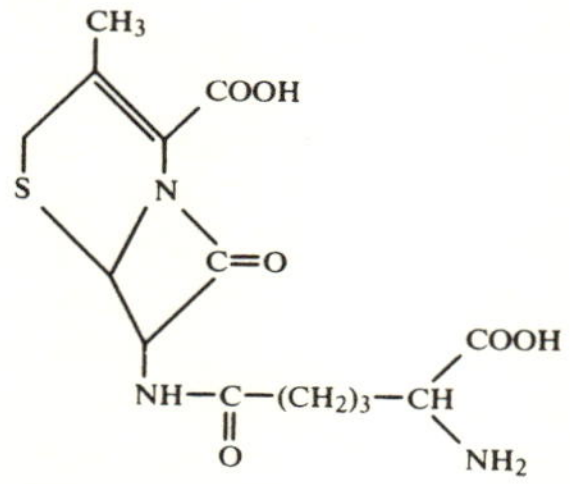

A.L. Demain, T. Konomi and J.E. Baldwin; U.S. Patent 4,307,192; December 22, 1981; assigned to Massachusetts Institute of Technology have found that certain cell-free extracts of *Cephalosporium acremonium* contains a racemase agent or agents capable of converting isopenicillin N to penicillin N. In accordance with this process, cell-free extracts of *C. acremonium* are prepared and used to catalyze reactions of isopenicillin N, 5-substituted derivatives of isopenicillin N and LLD much in the same manner as described in the previous patent. However, care is taken to preserve the racemase activity so that the catalyzed reaction of LLD (or its precursors) goes beyond the isopenicillin stage, through penicillin N, to form DACPC or a 6-substituted derivative thereof.

DACPC may be converted to cephalosporin C by standard techniques involving the hydroxylation of DACPC to deacetylcephalosporin C, followed by acetylation of the latter to cephalosporin C [e.g. Y. Fujisawa et al, *Agr. Biol. Chem.*, 39 (1975), pp 2049-2055].

The agent responsible for the inversion by which an isopenicillin is converted to its penicillin isomer has been found to be quite labile, and is rendered inoperative by conventional extract preparation procedures. It is accordingly critical to prepare the extract in a manner such that the operability of the racemase is preserved. The preparation of cell-free extracts of *C. acremonium* for use here is designed to preserve the racemase agent or agents and differs from the preparation described in the previous patent in the following particulars:

(1) fresh extracts, i.e., extracts which have never been frozen are used;

(2) it is preferable to avoid treatment of the extract such as homogenization; and

(3) when a phosphotransferase enzyme is employed in the synthesis, it is preferable that it be free of salt, e.g. free of $(NH_4)_2SO_4$.

The process, then, for making DACPC or a 6-substituted derivative thereof is as follows: The starting materials LLD or one of its derivatives or isopenicillin N or one of its derivatives plus a cell-free extract of *C. acremonium* are contacted in the reaction zone, sucrose and trace concentrations of KCl and $MgSO_4$ being included in the reaction system and the system being buffered to a pH of about 7.2 and ATP is provided to the reaction system as an air energy source. The reaction components are shaken vigorously to promote oxygen transfer and the reaction is allowed to continue for a sufficient amount of time to produce the cephalosporin compound.

Extraction of Uranium from Sea Water Using Mutant Microorganisms

It has been suggested that monocellular green algae be irradiated by x-ray and then placed in a uranium-rich nutrient solution until colonies form, whereupon the microorganisms particularly suitable for enriching uranium can be extracted.

M. Paschke, K. Wagener and M. Wald; U.S. Patent 4,263,403; April 21, 1981; assigned to Kernforschungsanlage Jülich GmbH, Germany have found, however, that not only are monocellular green algae suitable microorganisms for extracting uranium from the sea water, but that all microorganisms are suitable for this purpose which form threads or colonies, are adapted to be cultivated, and can easily be filtered, and with which the ratio of surface to volume is great. They have taken microorganisms having these required characteristics, dosed them with about 50 kilo-roentgens of x-ray for about an hour, selected the particularly suitable strains for enriching uranium, cultivating colonies of these strains to obtain a cell matrix, placing this matrix in a filter cage and passing uranium-containing sea water through the matrix.

An advantageous step in this method consists in that prior to the irradiation, individual cells are formed from the microorganisms which are used for the formation of the matrix. This is carried out by varying or changing outer conditions, in other words, by a change in the solution which contains the microorganisms or by a change in the osmolarity of the solution. The formation of individual cells can also be effected for instance by a change in the temperature of the culture or by the addition of enzymes. It has been found that by the formation of individual cells, the effectiveness of the matrix is increased.

Blue algae of the type oscillatoria have proved to be very suitable microorganisms because their thread-like character and the interweaving of the threads with each other permit the utilization of filter cages with a mesh of more than 50 μm. Blue algae frequently have a length of a plurality of mm but have only a diameter of one μm, and thus have a great surface with regard to its volume. By utilizing blue algae of the mentioned type, the extraction of uranium from the sea water is considerably more economical with utilization of monocellular green algae.

A further very suitable microorganism for carrying out this process is the mushroom *Aspergillus niger.* With submerged aerobic culture, ball-shaped sod (resin) forms which is distinguished by an extremely large surface.

This mushroom which normally grows in sweet water nutrient solution was first adapted to the osmolarity of the sea water by means of $NaNO_3$ occurring in ordinary nutrient solution. The cultures adapted in this manner were subsequently further cultivated in a nutrient solution which in sea water contained the normal nutrient salt concentration. As sea water there was used Helgoland deep water. If the thus cultivated mushrooms are introduced into synthetic sea water which contains 6 ppb uranium, it is possible after 24 hours to prove a uranium content which amounts to from 20 to 40 mg/kg in the dry mass of mushroom.

Preparation of Mutant Microorganisms for Production of Single Cell Protein

The staggering food deficit, particularly the problem of protein deficiency now faced by many nations, and the possibility of its spread, even to the industrial nations, has stimulated efforts to produce microbial, single-cell protein by fermentation processes. Processes for the biosynthesis of protein of high metabolic quality from various carbon energy sources have become well known over the last two decades.

Whereas much has been accomplished, and various microorganisms, inclusive particularly of the bacterium Cellulomonas (ATCC 21399) used alone or in symbiotic association with other microorganisms, have been successfully, reproducibly cultivated on various carbon or energy substrates, particularly cellulose, much might yet be done to provide greater yields of biomass at lesser cost, produce certain valuable by-products, particularly certain essential amino acids, and in general to accomplish such results with less complex growth mediums.

A particular object of *V.R. Srinivasan and Y.-C. Choi; U.S. Patent 4,278,766; July 14, 1981; assigned to Louisiana State University Foundation* is to provide a process for the production of mutants of Cellulomonas ATCC 21399 which can be cultivated on a carbon substrate, or carbohydrate, but particularly cellulose, as a source of comestible, digestible protein.

A yet more specific object is to provide a new process wherein such mutants provide protein of high nutrient value by the growth thereof in a simple medium which contains cellulose, but requires no yeast extract.

Yet another object is to provide a process of such character wherein the mutants during growth excrete certain amino acids, L-glutamic acid, L-lysine, and the like within the culture broth, thus providing major products due to the nature of the cellular biomass and, as well, additional very valuable products which can be extracted or otherwise separated from the culture broth.

These and other objects are achieved by a process for the production or derivation by spontaneous or induced mutations from a parent microorganism of the genus Cellulomonas (ATCC 21399) of selected microbial mutants which, unlike the parent microorganism, have the ability to excrete certain amino acids, notably L-glutamic acid or L-lysine, or both.

These mutant strains of Cellulomonas ATCC 21399 have been given the ATCC accession numbers 31230, 31231, and 31232.

The mutants are formed by mutation of the microorganism of the genus Cellulomonas (ATCC 21399), either by allowing the microorganism to undergo spontaneous mutation or by inducing mutations by the use of radiation or chemical mutagens, and the mutants are selected on the basis of their ability to excrete L-glutamic acid or L-lysine or both.

In a preferred method of forming desirable mutants, mutant microorganisms are formed from Cellulomonas (ATCC 21399) by chemical or UV-irradiation induced mutations in a dilute aqueous solution, or both containing the required nutrients at temperatures preferably from about 30° to about 35°C, at pH preferably from about 4 to about 6, during an incubation period ranging preferably from about 0.5 to about 2 hours. The desired strains are found by a replica plating technique wherein colonies, after contact with a mutagen, are transferred to a culture medium, e.g., an agar plate, and identified for their ability to excrete L-glutamic acid or L-lysine. The preferred strains are those which are capable of producing between about 0.1 to about 2 g/ℓ of L-glutamic acid with a 1 to 2% carbon source, or between about 0.1 to about 1 g/ℓ of L-lysine from a 1 to 2% carbon source.

The mutants can be cultivated on various energy or carbon substrates, alone or in compatible association with other microorganisms. Cellulose is a particularly preferred carbon source.

In the preferred embodiment, cellulose from any suitable source is first delignified, preferably by treatment with an alkali, preferably sodium hydroxide.

The alkali metal hydroxide is present in concentration ranging preferably from about 2 to about 20%, based on the weight of the total solution. Suitably, the physical size of the cellulose, if large, is reduced by cutting, sawing, grinding or crushing to assure intimate contact with the alkali. Generally, the cellulose is simply immersed in a solution of the alkali. Generally, the alkali treatment is conducted for periods ranging from about 0.1 hr to about 20 hr.

After contact between the cellulose and alkali, the treated cellulose is separated from the alkali by standard procedures known in the art, e.g., filtration, decantation, centrifugation, screening, passage between squeeze rolls, and the like. The alkali treated cellulose, generally after washing, is then placed in an oxidation, or circulating air oven and heated, preferably at temperatures ranging from about 25° to about 100°C. The time period that the treated cellulose remains in the oven is not critical, periods generally ranging from about 0.1 hr to about 20 hr.

To initiate fermentation, the oxidized, alkali-treated cellulose is next removed from the oven, treated ex situ to a pH ranging between 5 and 9, e.g., a pH of 7, and then charged into a fermentor.

The temperature of the broth is then adjusted to that desired for the most rapid growth of the microorganism, suitably from about 20° to about 40°C. The fermentation medium is then inoculated with the mutant microorganism, or microorganisms, to be cultivated, and harvested. Within the fermentor, comprising a closed vessel, a draft tube and air lift are provided to maintain vigorous agitation and suitable aerobic conditions. The pH of the medium is maintained while providing optimum growth temperatures for the microorganisms to be harvested. The medium is a simple one, containing glucose as the carbon source, and containing a mixture of salts inclusive of trace salts and specific vitamins.

It should be noted that the mutant microorganism, unlike the parent ATCC 21399, can be grown in a medium which does not include yeast extract.

Dissolving Collagen-Containing Tissues with Acid Proteases

R. Monsheimer and E. Pfleiderer; U.S. Patent 4,293,647; October 6, 1981; assigned to Röhm GmbH, Germany have developed a method for dissolving collagen-containing tissues such as wastes from leather processing, for example skin scraps, limed splits, machine fleshings, and the like, by enzymatic hydrolysis. The liquid hydrolyzates so obtained may thereafter be biologically decomposed or, optionally, may be further employed as industrially utilizable products.

It has been found that the collagen-containing waste materials from leather preparation can—because of their protein components, and particularly of those proteins having a collagen structure—be dissolved by enzymatic hydrolysis using one or more proteases, if such proteases are employed whose normal region of use lies in the acid pH range (preferably between 3 and 6) and if the enzymatic reaction is carried out in the acid pH range. The hydrolysis of collagen-containing wastes from leather preparation by means of acid proteases in the acid pH range has proved particularly advantageous if carried out in the presence of urea.

The use of acid proteases in combination is a special embodiment of this process. Of these, the combinations of acid proteases of plant origin with fungus enzymes are preferred, for example the combination of papain with proteases from Aspergillus species (*Asp. oryzae, Asp. saitoi, Asp. parasiticus*). It is advantageous also to use pepsin together with acid proteases of plant origin, such as papain, bromelain, or ficin, or with acid fungus proteases, for example those derived from Aspergillus species. The amount of enzyme to be employed depends on the kind and activity of the enzyme. In general, an amount of enzyme having an activity of 100 to 4000 mU_{Hb}, preferably 200 to 2000 mU_{Hb}, at pH 7.5, is employed per gram of dry subtrate. One U_{Hb} corresponds to the amount of enzyme which catalyzes the release from hemoglobin of fragments, soluble in trichloroacetic acid, which are equivalent to one micromol of tyrosine per minute at 35°C (measured at 280 nm). 1 $mU_{Hb} = 10^{-3}$ U_{Hb}.

The process offers the possibility of biologically decomposing the hydrolyzate of collagen-containing waste materials from leather preparation and of releasing it as an unobjectionable wastewater in a preflooder or into the public sewer system.

Single Cell Microorganisms Grown as Protein Source

Unicellular microorganisms (single cell protein) commonly referred to as SCP, offer a potential of being a vast cheap source of food protein. For this reason much research effort has been devoted to finding and/or preparing inexpensive, nontoxic, and abundant nutrients for the growth of such organisms.

J.G. Schulz and P.M. Bunting; U.S. Patent 4,302,539; November 24, 1981; assigned to Gulf Research & Development Company have found that an oxidation product of lignite is a useful carbon source in an aqueous mineral salts medium for the growth of microorganisms to manufacture protein. The lignite oxidation product is the acetone-insoluble, base-insoluble, solid fraction obtained from the low-temperature nitration and oxidation of lignite.

In the process of growing microorganisms on fossil fuel, the sole carbon source is an oxidation-nitration product of lignite which is manufactured by subjecting the slurry of lignite in aqueous nitric acid to a temperature of from 15° to 200°C for about 0.5 to about 15 hours, separating the solids, extracting the resulting solids with acetone, recovering the acetone-insoluble fraction, and dissolving the acetone-insoluble solid in base and neutralizing the solution to between pH 6.5 and pH 8.0.

Examples of representative genera of organisms which can be grown include the following organisms: Pseudomonas, Xanthomonas, Alcaligenes, Arthrobacter, Bacillus, Micrococcus, Azotobacter, Clostridium, Candida, Torulopsis, Rhodotorula, Debaryomyces, Endomycopsis, Hansenula, Penicillium, Aspergillus, Fusarium, and Geotrichum. A number of other microbial genera may also be cultivated on the coal acids.

Culture of Cocoa Bean Cells

J.E. Kinsella and C.H. Tsai; U.S. Patent 4,306,022; December 15, 1981; assigned to Cornell Research Foundation, Inc. have devised a means of culturing cocoa bean cells to produce triglycerides. The cells are grown by suspension cell culture in the presence of a growth-supporting medium for a time and at a temperature sufficient to propagate them and to produce triglyceride metabolites which are then separated from the cell culture.

The lipids formed in the process comprise triglycerides having an SUS configuration, i.e. having a saturated fatty acid in the SN-1 and SN-3 positions with an unsaturated fatty acid in the SN-2 position. As in cocoa butter obtained from the plant the fatty acids principally comprise palmitic acid and stearic acid, while the unsaturated acids comprise oleic acid and linoleic acid, the oleic acid present being in quantities substantially larger than linoleic acid.

The lipids are isolated from the cell culture by techniques known to the lipid chemist, e.g., extraction with warm organic solvent such as hexane followed by evaporation of the solvent. The lipids isolated from these cell cultures may be useful as cocoa butter extenders, food additives, in cosmetic preparations, and as lubricants.

The cocoa bean cotyledon is first grown on a solid or semisolid agar-based medium to form callus and the resultant callus is subjected to the suspension cell culture. The cell growth-supporting media used for the process are those known in the art to support soybean, tissue or cell cultures. They contain all the known nutrients required for plant cell growth.

The requirements vary with different plant species but generally sucrose or glucose, inorganic nitrogen (salts), some amino acids, the major and minor (trace) mineral elements, vitamins, and some auxins are included. The appropriate medium has to be determined for each plant species but some universal media are now available. In some instances media in which other cells have been grown contain stimulatory compounds which aid the establishment of primary cultures.

Preparation of Biologically Active Polymers such as Peptides

The area of peptide synthesis has received considerable attention in recent years. A significant problem has existed in synthetically achieving a high molecular weight, pure polypeptide wherein the amino acid sequence of the peptide actually prepared corresponds to that sought. To approach realization of the synthesis with the desired purity has heretofore been quite laborious.

The synthetic preparation of a polypeptide is a multistage molecular transformation procedure whereby a desired product is constructed by sequential chemical reactions of a precursor and an added compound, with the precursor at any given stage being the chemically reacted, reactable precursor from the preceding stage. Thus, the procedure is reiterative.

In this process a first amino acid is reacted with a second to form a dipeptide, and the peptide so formed at this stage is separated from the unreacted acids and a third amino acid is then reacted with the dipeptide to form a tripeptide. The procedure is reiterated until a polypeptide having the desired amino acid sequence, customarily termed the target peptide, is obtained. Sequence failure, whereby a portion of the elongated polypeptide chains have an improper amino acid sequence, can result from several causes, one being the failure to remove residual free acid from the reaction mixture prior to reaction with a subsequent amino acid, and the other being the incomplete reaction of all of the chains present with the amino acid added at each stage of the synthesis. In the preparation of low molecular weight polypeptides, the presence of chains containing different numbers of acids can be analytically ascertained and the desired peptide chains isolated. Conventional analytical techniques do not permit this to be done with respect to the higher molecular weight varieties, however, because the difference in molecular weight between properly and improperly synthesized chains is simply too small to be detectable.

G.P. Royer; U.S. Patent 4,315,074; February 9, 1982; assigned to Pierce Chemical Company has developed a process for making polypeptide which has the following advantages: (1) it provides a method which is susceptible to automation and can reliably be used to prepare pure, high molecular weight target peptides; (2) it provides a method for isolating an elongated polypeptide from its reaction environment so that the proper sequence can result on further reaction, which method is easy to accomplish with a minimum expenditure of time and minimum peptide loss; (3) it provides a facile way to minimizing the difficulties attendant on the reiterative preparation of polymers which result from sequence failure due to incomplete reaction by efficiently removing failed sequences from the growing chain population so that recovery of a desired pure polypeptide can be accomplished more conveniently; and (4) it provides a method for deblocking a peptide which is easy to accomplish and does not result in destruction of the chain being fashioned. In addition, the deblocking is accomplished in a manner which ensures the optical purity of the peptide being formed.

This process for synthesizing a peptide chain having a distinct sequence of amino acid segments comprises reacting, in an aqueous medium, a pure precursor containing a first amino acid segment of the peptide chain to be prepared having a free terminal carboxyl group or a free terminal amino group, with a second amino acid segment containing a free Nα-amino group and a blocked carboxyl group susceptible to enzymatic hydrolysis when the precursor has a free terminal carboxyl group or a free carboxyl group and a blocked Nα-amino group susceptible to enzymatic hydrolysis when the precursor has a free terminal amino group, deblocking the product peptide enzymatically, and then repeating the process of reaction and enzymatic deblocking until the desired peptide chain is prepared.

The precursor may conveniently contain a first amino acid segment covalently bonded to a polynucleotide handle or a polyethylene glycol handle.

The blocked group of the second amino acid segment may conveniently be an alkyl ester group.

To obtain high conversion, the elongation reaction can be accomplished with a large excess of the added sequencing segment, amounting to at least a 2:1 equivalent ratio, and preferably at least 5:1. However, the solution after reaction may nevertheless contain unreacted transformable precursor. In a reiterative procedure, the presence of such unreacted precursor can produce a sequence failure.

Therefore, a further aspect of the process centers on pruning. Pruning is selectively and chemically removing unreacted precursor from a reacted compound after the formation of the latter, thereby leaving a properly reacted compound which is free of material containing the base molecular structure of the unreacted precursor. Pruning is important in achieving ultimate separation and recovery of a pure target peptide.

A preferred method of enzymatic pruning is to pass the reaction solution through a column which contains a water-insoluble support material having immobilized on its surface an enzyme which selectively hydrolyzes substances either from the N-terminus or C-terminus. An aminopeptidase, such as aminopeptidase M or leucine aminopeptidase, is suitable for hydrolysis at the N-terminus. The hydrolysis is carried out at a temperature between 0° and 50°C and at a pH of 6.5 to 7.5. For hydrolysis directed at the C-terminus, a carboxypeptidase such as carboxypeptidase A, B, C or Y is useful. These Y and C enzymes, at pH 4 to 6, have been demonstrated as having nonspecific, C-terminal exopeptidase activity. A temperature between 0° and 60°C is employed.

Cyclodextrins Using Strains of Micrococcus

Cyclodextrins are crystalline dextrins which are cyclic oligosaccharides composed of 6, 7 or 8 glucose residues bound by α-1,4 bonds. They are monomolecular host molecules which can include in their molecular structure "guest compounds" which may be various kinds of organic compounds. This inclusion property has been widely used in the fields of medicine, agricultural chemicals, foods, cosmetics, etc.

The process of preparing cyclodextrins from the treatment of starch by enzymatic cultivation has been reported—all the microorganisms for producing suitable enzymes for the fermentation belonging to the genus Bacillus, plus one report using *Klebsiella pneumoniae M5.*

Y. Yagi, K. Kouno and T. Inui; U.S. Patent 4,317,881; March 2, 1982; assigned to Sanraku-Ocean Co., Ltd., Japan have isolated from the soil two new microorganisms belonging to the Micrococcus genus which produce the enzyme necessary for cyclodextrin production. They therefore provide a process for producing cyclodextrins which comprises cultivating a microorganism belonging to genus Micrococcus capable of producing cyclodextrin glycosyltransferase in a nutrient medium and treating starch or degraded starch with a cultured medium obtained by this cultivation, medium filtrate, concentrate thereof or an enzyme preparation obtained from various steps of purification, to give cyclodextrins. Furthermore, the process also provides a means for desirably regulating a ratio of α- and β-cyclodextrins in the enzyme reaction product.

The two microorganisms which produce the cyclodextrin glycosyltransferase enzyme have been named *Micrococcus varians* M 849 (FERM-P 4912) and *Micrococcus luteus* B 645 (FERM-P 4913).

Production of the enzyme is accomplished by cultivation of one of these microorganisms by an aerobic shaking culture in a conventional medium at a temperature of about 30° to 37°C and pH of 7 to 10.5 for from 24 to 96 hours. The enzyme is then recovered and purified by a known process.

The enzyme acts on various kinds of starch, for example, potato starch, corn starch and dextrin to make their reactivity to iodine negative and produce a large amount of cyclodextrins without showing substantial increase of reducing activity. In addition to the cyclodextrins, maltotetraose, maltotriose, maltose and glucose are produced as the final reaction products. When treating starch with this enzyme, a ratio of α-, β- and γ-cyclodextrins in the reaction product depends upon the substrate concentration, amount of enzyme used and pH of the reaction mixture among various reaction conditions. For example, under conditions where the substrate concentration is less than 2% at pH 4.5 to 6.5 with 10 U/g of starch, α-cyclodextrin production is larger than β-cyclodextrin. But, at pH 7 to 10, the dominant reaction product is β-cyclodextrin without dependency on substrate concentration and enzyme amount used. In both cases γ-cyclodextrin production is very small.

Production of Cellulose Fibers for Paper Manufacture by Using Microorganisms

Vast quantities of pulp wood are required for the production of paper and paper products. The harvesting of wood, transportation and processing to minute cellulosic fibers suitable for paper making are expensive in energy and resources.

R.L. Mynatt; U.S. Patent 4,320,198; March 16, 1982 has developed an apparatus and process for the production of polysaccharide (cellulose) fibers for use in paper manufacturing. Rather than from wood, the fibers are harvested already in minute size as the products of continuously cultivated bacterial or fungal microorganisms grown in a controlled manner. A suitable microorganism such as *Sphaerotilus natans* is grown on a pitted metallic plate by supplying a suitable flowing nutrient substrate. With abundant pellicle growth the nutrient flow is temporarily halted and a blade passed over the plate harvests the pellicle growth and deposits the harvest products onto a sluice conveyor. Upon retraction of the blade the nutrient substrate flow is restored until the harvest cycle is again repeated. The harvest products are then further processed as necessary to remove any undesirable noncellulosic materials. The particular processing, if necessary, depends upon the particular microorganism used.

An example of a suitable microorganism is the iron bacteria, *Sphaerotilus natans.* Cell growth of *Sphaerotilus natans* occurs in long filamentous polysaccharide (cellulose) sheets which as colonies exhibit well pronounced bulking qualities. The conditions for optimal growth of this organism permits in straightforward fashion continuous cultivation with multiple harvests from the original inoculation of the pitted plate.

Using continuous cultivation, conditions can be maintained which will prevent the culture from becoming contaminated by other bacteria. *Sphaerotilus natans* has low nutrient requirements, good resistance to extreme temperature variations and good tolerance for a high rate of nutrient substrate flow over the plate. A suitable nutrient for continuous cultivation of this microorganism would consist of 1 g/ℓ of water each of bacto-peptone and bacto-dextrose, 1.2 g/ℓ of water of magnesium sulfate, 0.05 g/ℓ of water of calcium chloride, 0.001 g/ℓ ferric chloride, and 12.5 g/ℓ of bacto-agar. The pH should be 7.0, the temperature about 25°C, and the flow rate 1.49 ft/sec.

Preparation of Spherical-Shaped Mycelial Pellets

Microorganisms have long been employed in the biocatalytic conversion of organic compounds. Exemplary of such conversions, to mention but a few, include conversion of simple sugars to useful products such as alcohols and organic acids; production of enzymes; synthesis of antibiotics and isomerization of sugars.

The use of mycelial fungi in biocatalytic systems has been limited due to the difficulty in handling such masses. While supported mycelia would be desirable, no satisfactory method has yet been developed and methods to date require preparation of the support surface in order to attach the active agent.

When mycelial microorganisms, such as fungi, grow in a liquid medium, their mycelia form a loose cotton-like mass. As used herein, mycelial microorganisms are defined as those living microorganisms, the vegetative portion of which forms filamentous hyphae. Such mycelial microorganisms include various fungi (including some yeasts), bacteria (e.g., Actinomyces) and algae such as blue-green algae. Of particular interest are mycelial fungi, especially from the genus Rhizopus and Mucor.

Some mycelial fungi can form pellets when they are grown in a circular action shaking incubator. The formation of pellet increases the cell density and facilitates the separation of the cell mass from the liquid products. Unfortunately not every fungus is able to form such mycelial pellets in a circular action shaking incubator.

L.F. Chen, C.S. Gong and G.T. Tsao; U.S. Patent 4,321,327; March 23, 1982; assigned to Purdue Research Foundation have provided a method for the preparation of mycelial microorganism pellets from all types of mycelial microorganisms, particularly fungi, with a rigid physical support core. Those fungi which are incapable of forming mycelial pellets in conventional shaking cultures can form mycelial pellets by the present method.

The pellets thus prepared may if desired possess a rigid physical support core, and thus exhibit good flow properties when they are used in a continuous fermentation reactor. These supported mycelial pellets grow into round pellets with a porous webbed layer of mycelia generally having a thickness of from about 0.1 mm to about 5 mm (preferably 1 to 3 mm), the webbed layer thereby forming a spherical encasement of structural integrity about the rigid core. The distance between the support core and the mycelial layer will vary from 0 to 1 cm (preferably about 5 mm) depending upon the initial spore concentration and incubation time. The pellets with such structural mycelial layers allow easy diffusion of nutrients into and passage of products out of the pellets.

The supported pellets may be prepared by various methods, one being based on the making of cellulose beads as described in U.S. Patent 4,090,022. In this case the steps of making mycelial pellets are as follows:

(a) providing a porous core support (e.g., prepare cellulose or cellulose derivative beads);

(b) absorb the fungal spores with these beads; and

(c) incubating the beads with agitation (e.g., in a circular action shaking incubator).

The physical and mechanical properties of the mycelial pellets can be improved by cross-linking the mycelium or impregnating the mycelial pellet with cellulose or cellulose derivatives or other inexpensive polymers. Suitable cross-linking agents which could be used include among others glutaraldehyde and diisocyanate.

One may also prepare biologically active mycelial pellets with agar, in accordance with a further embodiment. Agar is a carbohydrate and its solution melts at 90°C, solidifying at about 45°C. Because it is inert to most microorganisms, agar is used to solidify liquid culture media. These characteristics of agar can be utilized as a supporting core material to trap spores of mycelial microorganisms and produce mycelial pellets.

In general agar may be used as a supporting material in preparing mycelial pellets by:

(a) dissolving agar in hot water to form an agar solution;

(b) cooling the agar solution to a temperature of greater than 45°C, but below the temperature at which the spores added in step (c) would be killed (generally from between about 45° and 80°C, preferably about 50°C):

(c) mixing the agar solution of step (b) with spores of a mycelial microorganism and dispersing the resulting mixture in the form of droplets into cool air or an aqueous precipitation solution at a temperature below 45°C whereby the agar is precipitated in the form of uniform beads containing the spores;

(d) separating the agar beads from the precipitation solution if used, or alternatively, one may employ the same solution for both precipitation and incubation; and

(e) incubating the agar beads with agitation (e.g. in a circular action shaking incubator) for a period of time sufficient to produce a round pellet characterized by a rigid round agar core surrounded by porous webbed layer of mycelial having structural integrity of a microorganism.

Mycelia used for making the pellets were fungi of the genera Rhizopus, Mucor, Aspergillus, Penicillium or Trichoderma.

Mutant Strains of *Streptococcus mutans* for Controlling Dental Caries

A large body of evidence has implicated *Streptococcus mutans* as a principal pathogen in dental caries of both rodents and humans. The characteristic features of *Streptococcus mutans* which appear likely to account for its cariogenic potential include not only its ability to accumulate on tooth enamel, but also its ability to produce, via fermentative processes, large amounts of lactic acid.

In recent years, considerable success has been achieved in preventing and controlling certain bacterial infections by purposefully colonizing susceptible host tissues with nonvirulent analogs of disease-causing microorganisms. The basis of this phenomenon, termed bacterial interference, is in no single case completely understood, but in general terms appears to involve a competitive and/or antibiotic action of the nonvirulent strain, the so-called effector strain, on its pathogenic counterpart. Thus, for an organism to serve as an effector strain in the replacement therapy of a bacterial infection, it must be (a) nonvirulent itself, and (b) able to compete successfully with its pathogenic counterpart.

J.D. Hillman; U.S. Patent 4,324,860; April 13, 1982; assigned to Forsyth Dental Infirmary for Children has now isolated mutant strains from *Streptococcus mutans* strain BHT-2(str) following mutagenesis and plating on glucose tetrazolium medium. These strains are characterized by a single point mutation in the structural gene for the enzyme L(+) lactate dehydrogenase (LDH); gram positive, spheroidal cells occurring in pairs and chains; and by bright red colonies relatively larger in size than colonies of the parent strain on glucose tetrazolium medium.

These mutant strains when compared to the parent strain, in the case of resting cell suspensions, produce less total titratable acid when incubated in the presence of glucose; make no detectable lactic acid when incubated in the presence of glucose in the case of resting and growing cultures; adhere better to hydroxyapatite; and accumulate more plaque when grown in the presence of sucrose.

Quite advantageously, the characteristics of the isolated mutant strain provide a useful effector strain for the prevention and alleviation of dental caries in animals.

Two cultures, typical of the mutant strains which have been isolated, have been deposited with the American Type Culture Collection and are identified by deposit numbers ATCC 31,341 and ATCC 31,377. These cultures are further identified as *Streptococcus mutans* JH140; and *Streptococcus mutans* JH145. The mutagenesis was effected as follows: Exponentially growing cells of *Streptococcus mutans* strain BHT-2(str) cultured in Todd-Hewitt broth containing 0.5% glucose, were harvested by centrifugation and then washed twice in 0.1 M phosphate buffered saline (PBS, pH 7.0). These cells were then resuspended in PBS, pH 7.0 as before cultured, to a density of approximately 1.2 x 10^8 colony-forming units (cfu) per ml.

0.015 ml of ethylmethane sulfonate was then added per ml of cell suspension and vortexed into solution, according to usual techniques. After 60 minutes incubation in a 37°C water bath, mutagenesis was stopped by diluting the cell suspension with 9 volumes of PBS, pH 7.0. The cells were then harvested by centrifugation as before and resuspended in 10 ml of Todd-Hewitt broth containing 0.5% glucose. The culture was then grown to saturation overnight.

The useful biologically pure culture of a mutant of *Streptococcus mutans* strain BHT-2(str) has:

(a) bright red colonies, when grown on glucose tetrazolium medium "the colonies of relatively larger colony size than colonies of the parent strain grown on the same medium," spheroidal, gram-positive cells occurring in pairs and chains and having a mutation in the structural gene for the enzyme, lactate dehydrogenase;

(b) a lack of detectable lactic acid, when incubated in the presence of glucose;

(c) relatively better adhesion to untreated and saliva-treated hydroxyapatite than the parent strain;

(d) a capacity to produce more plaque on glass slides and tooth-enamel fragments than does the parent strain, when incubated in the presence of sucrose; and

(e) an ability to metabolize a member from the group consisting of amygdalin, cellobiose, dextran, esculin, fructose, galactose, glucose, inulin, lactose, maltose, mannitol, mannose, melibiose, raffinose, ribose, salicin, sorbitol, sucrose, trehalose and pyruvate, as evidenced by a higher pH of cultures of the mutant strain compared to that of the parent strain, when any of the members are incorporated in the culture medium of the parent strain and the mutant strain.

Recovering Hydrocarbons from Biomass

The objectives of *T.A. Weil, P.M. Dzadzic, C.-C.J. Shih and M.C. Price; U.S. Patent 4,338,399; July 6, 1982; assigned to Standard Oil Company (Indiana)* are to provide a process for production and recovery of hydrocarbons from hydrocarbon-containing plants which permits the effective utilization of whole plant biomass as a raw material source. Another object is to provide a process for production of liquid hydrocarbons in quantity from plant biomass. Another object is to provide a process for degradation of cellulose to provide glucose and to break the cellulosic walls of latex-containing cells, thereby releasing hydrocarbon-containing latex and increasing hydrocarbon yields.

The process particularly relates to the production of sugars and the recovering of hydrocarbons contained in laticifers of whole plant biomass which are rich in hydrocarbons. Examples of these plants are *Euphorbia heterophylla, Euphorbia lathyrus, Euphorbia marginata, Asclepias syriaca, Calotropis procera, Parthenium argentatum* (guayule) and *Apocynum sibiricum.* The process is not limited to these plants and can be applied to any hydrocarbon-containing plant. The plant material can be used fresh, frozen or dry. Enzymes which are effective include the cellulases, hemicellulases, and pectinases and also the combinations thereof. Either the isolated enzyme or the living enzyme producing culture can be used. A number of enzymes have been isolated from fungi and bacteria which hydrolyze cellulose and hemicellulose to simple sugars and degrade pectins. Commercially available enzymes produced by cultures of *Trichoderma viride* and *Aspergillus niger* are particularly useful.

The reaction can be run with or without mixing, preferably at a temperature between 30° and 60°C although temperatures of 10° to 90°C can be used. The reaction is run in an aqueous buffered solution preferably at pH 5 although a pH range from 0 to 10 is acceptable. Reaction times of from 1 to 2 hours up to several hundred hours can be used depending on enzyme concentration, temperature, mixing rate and pH. Enzyme to substrate ratios of less than 10^{-3} to 10 can be used. The process may be carried out in either a batch or continuous type operation. For a batch operation a reaction vessel is charged with the enzyme, plant material and solvent, and the reaction run at a particular temperature for a selected time. Following the reaction the contents of the reactor are distilled to recover hydrocarbons. The sugars produced from the cellulase and hemicellulase hydrolysis are easily recovered from the aqueous solution. For continuous operation either a plug flow or back mixed reactor is suitable.

The process comprises:

(a) hydrolyzing the whole plant biomass in the presence of enzymes selected from the group consisting of cellulase, hemicellulase and mixtures thereof under conditions which promote conversion of cellulose and hemicellulose to soluble sugars;

(b) removing a stream of hydrolysis products comprising a liquid sugar-containing phase, a liquid hydrocarbon-containing phase and a solid phase containing unhydrolyzed spent solids, all phases containing enzymes selected from the group consisting of cellulase and hemicellulase enzymes and mixtures thereof from the hydrolysis stage;

(c) separating liquid phases and solid phases of step (b);

(d) continuously adding a stream to step (a) of fresh whole plant biomass to replenish the whole plant biomass converted to hydrolysis products;

(e) separating the liquid phases from step (c) thereby recovering the enzymes present and the hydrocarbons present;

(f) recovering a stream of product sugar solution and enzyme solution from step (e);

(g) sterilizing a sugar product stream from step (f) to serve as sterile growth medium for microorganisms;

(h) adding the recovered solids from step (c) to the hydrolysis stage;

(i) providing a stream of makeup enzymes as in step (a) by the steps including:

    (1) in a first zone, growing in a suitable medium the enzyme-synthesizing organisms selected, the microorganism-containing medium comprising minor portions of the product solutions;

    (2) in a second zone contacting the microorganism-containing medium with an amount of cellulosic and hemicellulosic materials in an amount sufficient to induce formation of cellulase and hemicellulase under conditions substantially nonsupportive of growth of the cellulase and hemicellulase-secreting microorganisms.

Enzymatic Production of Peptides

The object of the process developed by *J.T. Johansen and F. Widmer; U.S. Patent 4,339,534; July 13, 1982; assigned to De Forenede Bryggerier A/S, Denmark* was to provide an enzymatic peptide synthesis of a general nature, not limited to specific amino acid components. They were also desirous of producing a process which does not involve risks of subsequent hydrolysis of internal peptide bonds.

They accomplished these objectives by a process for providing a peptide of the general formula A-B, in which A represents an N-terminal protected amino acid residue or an optionally N-terminal protected peptide residue and B represents an optionally C-terminal protected amino acid residue. This peptide is prepared by reacting a substrate component selected from the group consisting of:

(a) amino acid esters, peptide esters and depsipeptides of the formula $A{-}OR^1$ or $A{-}SR^1$ where A is as defined above and R^1 represents alkyl, aryl, aralkyl or an α-desamino fragment of an amino acid residue,

(b) optionally N-substituted amino acid amides and peptide amides of the formula $A{-}NHR^2$ wherein A is as defined above and R^2 represents hydrogen, alkyl, aryl or aralkyl, and

(c) optionally N-terminal protected peptides of the formula A–X wherein A is as defined above and X represents an amino acid

with an amine component selected from the group consisting of

(a) amino acids of the formula H–B–OH, and

(b) optionally N-substituted amino acid amides of the formula $H–B–NHR^3$ wherein B is an amino acid residue and R^3 represents hydrogen, hydroxy, amino or alkyl, aryl or aralkyl, and

(c) amino acid esters of the formula $H–B–OR^4$ or $H–B–SR^4$ wherein B is an amino acid residue and R^4 represents alkyl, aryl and aralkyl

in the presence of a carboxypeptidase enzyme in an aqueous solution or dispersion having a pH from 5 to 10.5, preferably at a temperature of from 20° to 50°C, to form a peptide, and subsequently cleaving a group R^3 or R^4 or an N-terminal protective group, if desired.

The aqueous reaction solution is preferably one containing up to 50% organic solvent, such as an alkanol, dimethyl sulfoxide, dimethyl formamide, dioxane, tetrahydrofuran, dimethoxy ethane, ethylene glycol and polyethylene glycols.

The amino acid or peptide ester used as substrate component is conveniently a benzyl ester or a straight chain or branched alkyl ester, selected from the group consisting of methyl, ethyl, propyl and butyl esters and optionally being substituted with inert substituents.

The N-substituted amino acid amide or peptide amide used as substrate is preferably p-nitroanilide; the amine component is an amide $H–B–NHR^3$ where R^3 is methyl, ethyl, propyl or isopropyl; and the amine component is an ester $H–B–OR^4$ where R^4 is methyl, ethyl, propyl or isopropyl. After the formation of the peptide containing a group R^3 or R^4 or an N-terminal protective group, that group is cleaved by means of the same L-specific serine or thiol carboxypeptidase enzyme as used in the formation of the peptide.

This process is based on a fundamental change in relation to the prior art, viz the use of exopeptidases instead of the enzymes employed till now which have all displayed predominant or at any rate significant endopeptidase activity.

The necessary enzyme characteristics are found in the group of exopeptidases called carboxypeptidases, which are capable of hydrolyzing peptides with free carboxyl groups. It has been found that a plurality of such carboxypeptidases exhibit different enzymatic activities which are very dependent on pH so that e.g. in a basic environment at a pH from 8 to 10.5 they display predominantly esterase or amidase activity and at a pH from 9 to 10.5 no or only insignificant peptidase activity. These properties can be advantageously used in the process because they contribute to the achievement of good yields.

Further, it has been found that the ability to form the acyl enzyme intermediates are particularly pronounced in serine or thiol proteases. A preferred group of carboxypeptidases is therefore serine or thiol carboxypeptidases. Such enzymes can be produced by yeast fungi, or they may be of animal, vegetable or microbial origin.

A particularly expedient enzyme is carboxypeptidase Y from yeast fungi (CPD-Y) which has been purified by affinity chromatography on an affinity resin comprising a polymeric resin matrix with a plurality of coupled benzylsuccinyl groups.

Other useful carboxypeptidase enzymes are penicillocarboxypeptidase S-1 and S-2 from *Penicillium janthinellum*, carboxypeptidases from *Aspergillus saitoi* or *Aspergillus oryzae*, carboxypeptidases C from orange leaves or orange peels, carboxypeptidase C_N from *Citrus natsudaidai Hayata*, phaseolain from bean leaves, carboxypeptidases from germinating barley, germinating cotton plants, tomatoes, watermelons and bromelain (pineapple) powder. An immobilized carboxypeptidase enzyme is preferable.

Production of Substance Which Stimulates Differentiation and Proliferation of Human Granulopoietic Stem Cells

A substance which acts directly on human granulopoietic stem cells and can be used as a diagnostic reagent and possibly a treatment for human granulocytopenia by stimulating the proliferation and differentiation of such cells (afterwards such a substance is referred to as CSF—colony stimulating factor) has been known since the early 1970s, but no method has until now been developed for the large-scale, low-cost production of a CSF product which has no side effects.

F. Takaku, K. Ogasa, M. Kuboyama, N. Yanai, M. Yamada and Y. Watanabe; U.S. Patent 4,342,828; August 3, 1982; assigned to Morinaga Milk Industry Co., Ltd., and The Green Cross Corp., both of Japan have provided a method for producing a substance capable of stimulating the proliferation and differentiation of human granulopoietic stem cells, which comprises cultivating monocytes and macrophages which have been isolated from human blood in a synthetic medium for tissue culture containing a glycoprotein isolated from human urine and capable of stimulating human or mouse bone marrow cells to form human granulocytes or mouse macrophages and granulocytes, thereby producing an active substance in the medium, and recovering the active substance from the medium.

The glycoprotein capable of stimulating the formation of human granulocytes [hereinafter referred to as glycoprotein (H)] which is isolated from human urine and used in this process, is fully described in Japanese patent application Laid-open No. 140,707/79, West German Patent Offenlegungschrift No. 2,910,745 and U.K. patent application Publication No. 2,016,477. A glycoprotein capable of stimulating the formation of mouse macrophages and granulocytes [herein referred to as glycoprotein (M)], which was isolated from human urine, was described as a known sialic acid-containing glycoprotein by Stanley and Metcalf, *Australian Journal of Experimental Biological Medical Science,* 47, 467-483 (1969); Lauke et al, *Journal of Cellular Physiology,* 94, 21-30 (1978) and others.

In preparing CSF for the pharmaceutical use, a medium with supplemented human serum or a serum-free medium is used in order to avoid side effects caused by foreign proteins. In preparing CSF for use as a diagnostic reagent, on the other hand, a medium added with bovine serum or fetal calf serum may be used.

Methods for isolating and cultivating monocytes and macrophages, isolation of the glycoprotein from human urine, purification processes, and recovery of the CSF from the conditioned medium are described in detail in the patent. Specific examples of each of these procedures are included.

Certain conclusions on the process of producing CSF were arrived at by experimental procedures. These include the findings that (a) a suitable incubation period for the procedure is 3 to 7 days; (b) CSF production is markedly increased by the presence of 10 to 100 μg of glycoprotein (H) or 1000 or more units of glycoprotein (M) per ml of medium; (c) a large quantity of CSF was produced when at least 10^5 monocytes and macrophages were inoculated into each ml of the medium; (d) four commercially available media for cell cultures were usable for production of CSF–McCoy's 5A and HAMF-10 (both Gibco Co.), RPMI-1640 and Eagle's MEM medium with amino acids (both Nissui Seiyaku Co.).

Production of Epoxides Using a Catalytic Bed of Resting Cells Exhibiting Oxygenase Activity

C.T. Hou; U.S. Patent 4,348,476; September 7, 1982; assigned to Exxon Research and Engineering Co. describes a process for advancing the oxidation state of a gaseous, oxidizable organic substrate through contact with oxygen and a solid state biocatalyst. The process comprises passing through a stationary catalytic bed comprising moist, resting cells exhibiting oxygenase activity, a gaseous oxidizable organic substrate and a gaseous source of oxygen, until the oxidative state of at least a portion of the substrate is increased, while maintaining the relative humidity in the bed at such a level that the cells remain moist and viable. Equipment for practicing this process is also described.

The process is particularly useful for carrying out oxidation reactions on gaseous hydrocarbon substrates containing up to and including 6 carbon atoms per molecule. Generally, the process is based upon the use of an oxygenase enzyme as a catalyst, for the incorporation of molecular oxygen directly into a specific organic molecule. It is of particular interest for the conversion of propylene to propylene oxide and similar reactions that are catalyzed by monooxygenase enzymes.

To practice this process, a reactor containing immobilized cells is prepared. Microbial cells in the resting stage, and known to have the desired biocatalytic activity, are formed into a thick paste with buffered solution. The paste is then coated on an inert carrier material such as porous glass beads, charcoal, dried silica gel, etc. Care is exercised that the cells remain moist. The immobilized cells prepared in this way are then packed in a suitable reactor, which may simply be a reactor tube. Generally, any suitable reactor may be used that will permit efficient contact between the substrate gases and the cells, while permitting the necessary temperature and humidity control.

Once the biocatalytic bed is prepared in the reactor, it is used by passing the gaseous substrate mixture through the reactor. Generally it is preferred to pass the substrate gas upwardly through the reactor bed, to avoid any settling and compaction of the bed. The reactor bed is maintained at a carefully controlled temperature slightly higher than the boiling point of the oxidized product, preferably about 5°C higher. In the case of epoxides of the kind produced by the oxidation process, generally the temperature within the reactor should be maintained in the range from about 5° to about 10°C above the boiling point of the desired product. In most cases, this means that the operating temperature of the reactor bed will fall in the range from about 15° to about 80°C.

The ozidizable substrate may be, for example, a C_{1-4} alkane; a C_{2-4} alkene or diene, selected from the group consisting of ethylene, propylene, butene-1, and butadiene; and generally other oxidizable organic substrates that vaporize at a relatively low temperature and that will remain in the gaseous stage until condensed from the effluent gas stream. These considerations generally limit the substrate to those molecules having at most 6 carbon atoms, and preferably, not more than 4 carbon atoms. Generally, naturally gaseous substrates are preferred, that is, those substrates that are gaseous at room temperature.

A gaseous source of oxygen is also an essential part of the substrate gas. It is preferably mixed with the oxidizable substrate gas prior to injection into the reactor. The source of oxygen may be air, oxygen itself, or a synthetically prepared mixture of oxygen and nitrogen, for example.

The immobilized cells, that are used in the biocatalyst, are each surrounded by a thin liquid phase. In order to maintain the catalytic activity of the biocatalyst reactor, it is essential that the relative humidity in the reactor be maintained at a level that avoids drying the cells and the liquid phase surrounding them. A preferred technique is simply to pass the substrate gases through a water bath, relying upon them to pick up water vapor and to entrain water droplets in doing so. Generally the relative humidity within the reactor should be maintained in the range from 50 to 100%, and preferably from 70 to 100%.

The catalytic bed in the reactor may be a dynamic bed or a stationary bed, the latter being preferred.

Product recovery can be accomplished by chilling the effluent stream to condense the product. The remaining gases from the effluent stream may be recycled. If air is used as the source of oxygen, and if recycling of the effluent gas remaining after condensation of the product is practiced, it may be desirable to inject supplemental oxygen directly into the gas supply to the reactor bed or into the recycling gases, in order to maintain the oxygen level at a sufficiently high value for good reactivity.

After a period of use, the biocatalytic activity of the reactor bed may drop off. If the cells remain viable, the biocatalytic activity can be restored at least in part by passing a suitable hydrocarbon, preferably a C_1 source, such as methanol vapor, for example, upwardly through the reactor bed for a long period of time.

MISCELLANEOUS PRODUCTS

Keratin Hydrolysate as Hair Waving Material

In general, cold waving of the hair is conducted by employing a waving lotion for first stage which contains a reducing agent such as thioglycollic acid or cysteine and is adjusted to pH about 9 to about 10 with an alkaline material, and a neutralizing lotion for second stage which contains an oxidizing agent such as sodium bromate or hydrogen peroxide.

In this cold waving process, it is necessary to employ an alkaline solution in the first stage to make the hair swell and, therefore the hair and skin are damaged by an alkaline material. Also, upon the oxidation in the second stage, a side reaction takes place partly to form keratin–S–S–CH_2COOH linkages, and as a result, the hair is greatly damaged.

The objective of *I. Yoshioka and Y. Kamimura; U.S. Patent 4,279,996; July 21, 1981; assigned to Seiwa Kasei Co., Ltd., Japan* is to provide a material useful as a hair fixative which can produce an excellent waving effect on the hair without damaging it.

This is accomplished by providing a water-soluble keratin hydrolyzate as a hair fixative, having at least two mercapto groups in one molecule and an average molecular weight of 2,000 to 20,000.

The process for preparing the water-soluble keratin hydrolyzate comprises reducing keratin in an aqueous solution of a reducing agent selected from the group consisting of mercaptans and sulfides under alkaline conditions and subjecting the resulting reduction product to enzymatic hydrolysis in an aqueous medium in the presence of an enzyme capable of hydrolyzing protein.

Any keratins such as wool, feathers, hair, horns, etc. may be usable as the starting material for preparing the keratin hydrolyzate. Wool is particularly preferred, since it is easy to obtain.

Examples of the mercaptan employed as the reducing agent are thioglycollic acid, cysteine, mercaptoethanol, thioglycerol and thiosalicylic acid. Examples of the sulfide employed as the reducing agent are sodium sulfide, potassium sulfide, calcium sulfide, triethanolamine sulfide, diethanolamine sulfide and monoethanolamine sulfide.

Proteases capable of being activated under acidic conditions such as pepsin, and proteases capable of being activated under neutral condition such as bromelin, thermolysin, trypsin or chymotrypsin are employed as the enzymes for hydrolyzing the reduction product of keratin.

Production of Iodine

There is an increasing demand for iodine and iodide salts. *S.L. Neidleman and J. Geigert; U.S. Patent 4,282,324; August 4, 1981; assigned to Cetus Corporation* have provided a process for producing iodine from such plentiful sources as natural brines which does not require the use of chlorine. The process uses enzymes to oxidize iodides to iodine.

The method comprises providing a reaction mixture of pH buffered water, a halogenating enzyme, an oxidizing agent and a source of ionic iodide. The reaction is run in the absence of iodine acceptor substrates which allows the recovery of iodine from the reaction mixture.

The starting material or source of ionic iodide may be any of a variety of brines or bitterns. The process may be used economically to recover iodine from natural brines, bitterns, salt lakes, and the like.

The halogenating enzyme used in the reaction mixture may be in pure form, free or immobilized, or may be in microbial cells which produce the enzyme. Among the useful halogenases are those derived from the microorganism *Caldariomyces fumago,* seaweed, milk (lactoperoxidase), thyroid (thyroid peroxidase), leukocytes (myeloperoxidase) and horseradish (horseradish peroxidase). Sufficient water is employed in the reaction mixture to wet the enzyme and, in fact, may be the major solvent in the reaction mixture.

The oxidizing agent employed in the reaction mixture is preferably hydrogen peroxide, which may be used in dilute form to reduce the cost of the process. The hydrogen peroxide may be added directly to the mixture in a single batch addition, or may be added in a continuous slow feed. Alternatively, the hydrogen peroxide may be generated as slow feed in situ by the use of a hydrogen peroxide-producing enzyme or chemical system. Such enzyme systems are well known in the art and include glucose oxidase in the presence of glucose; glucose-2-oxidase in the presence of glucose; methanol oxidase in the presence of methanol; D- and L-amino acid oxidases in the presence of D- and L-methionine; and diamine oxidases in the presence of histamine.

Any or all of the enzymes or cells producing the enzymes in this process may be used in either free or immobilized form. Reactions employing immobilized enzymes and cells may be run in columns or reaction tanks.

The iodine recovery reaction is conducted within the pH range of from about 2 to about 8, which is enzyme-dependent. The pH of the reaction is maintained within the desired range by use of a buffering agent, such as sodium or potassium phosphate, gluconate, citrate, formate, and acetate-based systems. The reaction may be conducted in an aqueous medium.

The production of iodine results during the course of the reactions which take place in the above reaction mixture. The iodine may be recovered by any convenient means, and may be produced either continuously or by batch processing. The reaction may be run in the presence of solvents which are immiscible in water and which can selectively extract the iodine as it is formed. Such solvents should, of course, have a nondetrimental effect on the enzymes used in the process. By extracting the iodine as it is formed, the use of such solvents will reduce the iodine toxicity to the enzyme and aid in the recovery of the free iodine. Among solvents that are suitable are benzene, toluene, xylene, mesitylene, ethyl acetate, ethyl ether, carbon tetrachloride and carbon disulfide.

Another simple technique for continuous removal of iodine from the reaction is to bubble an inert gas, such as nitrogen, through the reaction mixture. The iodine is thus carried in the gas phase, is removed and concentrated from the gas stream by sublimation or other procedures known to the art. The spent inert gas is recycled for reuse. The reaction is preferably conducted in the temperature range of 15° to about 50°C, preferably about 20° to about 30°C.

Bicyclodecadienes Which Induce the Production of Mold Spores

Although it is well known that in the life cycle of molds, yeasts and bacteria, the production of spores progresses differently with environmental changes, nothing has been published about any substance which could effect the sporulation of these microorganisms. *S. Marumo and M. Katayama; U.S. Patent 4,289,852; September 15, 1981; assigned to Noda Institute for Scientific Research, Japan* have found that two new bicyclodecadienes markedly induce the production of mold spores. These two substances were obtained by extraction of a cultivation of a

Sclerotinia S-1 species (FERM-P 4214; ATCC 20497) obtained from plums infected with Monilia disease.

These two compounds, which may be termed sporogenic substances, are 5-isopropyl-8-methyl bicyclo[5.3.0] deca-2,8-diene-2-carboxylic acid (hereinafter abbreviated to SF-1 substance) and 5-isopropyl-8-methyl bicyclo [5.3.0] deca-2,8-diene-2-carboaldehyde (hereinafter abbreviated to SF-2 substance).

The SF-1 and SF-2 compounds can be produced synthetically. In order to produce them by a fermentative process, however, one cultivates the Sclerotinia S-1 in a conventional medium containing nitrogen and carbon sources. The culture is carried out usually at a temperature of 20° to 32°C, at a pH value of 6.0 to 7.0, for a period of 5 to 10 days.

The SF-1 and SF-2 substances can be isolated from cultivated mixture and purified by extracting with an organic solvent by the usual method first of all. Examples of a solvent usable for the extraction include ethyl acetate, acetone, benzene, diethyl ether, chloroform and mixtures thereof. Subsequently, SF-1 substance is separated from SF-2 substance by an appropriate combination of various procedures such as fractional precipitation, distillation, adsorption, thin layer chromatography, column chromatography, countercurrent partition and the like, whereby SF-1 and SF-2 substances can be obtained in the purified forms.

Since these compounds markedly stimulate the sporulation of various microorganisms, particularly of molds, they are important from the viewpoint of industrial cultivation of molds. For example, the ability of molds to produce antibiotics such as cephalosporin or to produce enzymes such as protease and amylase would be remarkably improved by using these compounds.

Mass spectra proton nuclear magnetic resonance spectra and ultraviolet absorption spectra of SF-1 and SF-2 substances are illustrated in the patent.

Cacao Butter Substitute by Fermentation

T. Matsuo, M. Terashima, Y. Hasimoto and W. Hasida; U.S. Patent 4,308,350; December 29, 1981; assigned to Fuji Oil Company, Limited, Japan have developed a fermentation process which produces good yields of a cacao butter substitute having triglyceride components similar to those of natural cacao butter. (These components are 1,3-disaturated-2-unsaturated-triglycerides, referred to hereafter as SUS.) According to the process, the desired SUS-rich fats and oils can be produced by cultivating a microorganism being capable of assimilating a higher alkyl derivative and producing SUS-containing fats and oils in a specific medium containing a mixed carbon source comprising at least one stearyl compound and at least one of other carbon compounds selected from a palmityl compound, an oleyl compound and a saccharide under an aerobic condition, collecting the resulting cells of the microorganism and then recovering the fats and oils from the cells.

The microorganisms useful in the process include various microorganisms of the genus Candida, Torulopsis, Trichosporon, Pichia and Sporobolomyces, for instance, *Candida guilliermondii* IFO 0838 and IFO 0566; Candida sp. ATCC 20503 (FERM-P 3900) and ATCC 20504 (FERM-P 3901); Trichosporon sp. ATCC 20505 (FERM-P 3902) and ATCC 20506 (FERM-P 3903); Torulopsis sp. ATCC 20507 (FERM-P 3904); and the like. The microorganisms are cultivated in a specific medium containing a mixed carbon source under an aerobic condition to accumulate a large amount of fats and oils having similar components to those of cacao butter in the cells, from which the SUS-rich fats and oils having excellent properties suitable for cacao butter substitute are obtained, if desired, followed by fractionation thereof.

For instance, when Torulopsis sp ATCC 20507 is cultivated in a medium containing a mixed carbon source comprising methyl stearate, methyl palmitate and glucose, fats and oils having a total SUS content of 49.2% by weight, 12.8 wt % of other trisaturated compounds, 27.5 wt % of diunsaturated compounds and 6.9 wt % of 1- or 3-monounsaturated compounds are produced. These fats and oils, therefore, can be used as a cacao butter substitute just as they stand, and further, when they are fractionated once to remove the components having a lower melting point, an excellent cacao butter substitute can be obtained.

Maltase Made with a Mutant Yeast Strain

Maltase (α-glucosidase, EC 3.2.1.20) is an enzyme used in clinical assays for amylase. The maltase is utilized in an intermediate reaction wherein maltose, formed by the action of amylase on an oligosaccharide, is converted to glucose. The glucose is subsequently measured to determine the amylase activity. Maltase is produced from yeast strains such as *Saccharomyces cereviseae* or *Saccharomyces italicus.* The maltase produced from *S. italicus* is particularly useful in the assay for amylase in that it will not degrade higher oligosaccharides. Unfortunately, the carbon source in the growth medium for *S. italicus* is maltose which is very expensive as compared to possible alternative carbon sources and attempts to substitute relatively inexpensive sucrose or glucose as the carbon source have proven ineffective.

In addition, the use of maltose as a carbon source has presented production difficulties since the activity of the maltase produced peaks just before the maltose in the medium is depleted thereby requiring that the cells be harvested within a relatively narrow time span.

Accordingly, it would be desirable to provide a means for producing maltase useful in clinical assays for amylase from a yeast strain which does not require maltose as the carbon source and which would permit the use of inexpensive sucrose as the carbon source. Furthermore, it would be desirable to provide such a means wherein the activity of the maltase produced is not sharply reduced within a short time span so that the timing of cell harvesting is far less critical than in available processes.

C.L. Cooney and E.J. Schaefer; U.S. Patent 4,332,899; June 1, 1982; assigned to Massachusetts Institute of Technology have therefore provided a new strain of yeast comprising mutants of *S. italicus,* which, unlike the parent strain, will grow and produce maltase in a medium wherein the carbon source need not be maltose and can include sucrose as the carbon source. The maltase enzyme produced by the process is particularly useful for assay of amylase.

This strain of yeast is produced by mutation of *S. italicus* with a mutagenic agent such as ultraviolet light, nitrosoguanidine, ethyl methanesulfonate or other known agents.

The process for selecting the mutant capable of producing maltase while utilizing sucrose as the primary carbon source comprises exposing *S. italicus* to the mutagenic agent under conditions to effect mutation of *S. italicus,* dividing the cells into a plurality of samples, exposing the samples to a growth medium comprising sucrose as the carbon source, and selecting the samples capable of maltase production. The highest maltase-producing mutant of those tested was designated *S. italicus* ATCC 20601.

Surfactants from Various Microbe Cultures

J.E. Zajic, D.F. Gerson, R.K. Gerson and C.J. Panchal; U.S. Patent 4,355,109; October 19, 1982 discovered a certain class of microbes which can be grown by aerobic fermentation processes under controlled conditions to give materials of outstanding surfactant properties, in high yields. The microbes are certain cultures from the *Arthobacter-Corynebacterium-Nocardia* genera which are characterized by their ability to metabolize the protective waxy hydrocarbon material found naturally on the plant cuticles of water plants such as salvinia, to expose the plant to disease-causing infections. The cultures are represented by a culture referred to as *Corynebacterium salvinicum* strain SFC. The specific preferred microbial strain SFC, has been given the reference number ICPB 4312 (Cooper SFC).

It is preferred to use hydrocarbons as the carbon-containing substrate in the process, especially liquid aliphatic paraffinic hydrocarbons, straight or branched chain. Most preferably, the hydrocarbons have from 6 to 18 carbon atoms per molecule. Mixtures of hydrocarbons such as kerosene can be used also. The pH range for this process varies from 4 to 8.8 depending on the inorganic nitrogen source and buffer system used. The system is an aerobic process, thus either air or oxygen must be supplied in adequate amounts required for growth and product formation. Mixing is accomplished by shaking in small laboratory vessels, however in larger vessels the mixing required should be in the low to high turbulent range (Reynolds number of 1000 to 10,000). The optimal temperature is around 30°C, but high yields can be obtained over the range of 18° to 37°C.

It is preferred to use a concentration of hydrocarbon in the range from 0.5 to 9.0% (w/v). Inorganic salts beneficial to the surfactant production of the system include sodium nitrate, ammonium sulfate, ammonium chloride, ammonium nitrate, ammonium carbonate, and urea. Inorganic nitrogen compounds are beneficial to the surfactant production, the preferred range of concentrations of inorganic nitrogen being from 0.6 to 4.0% (w/v). It is also desirable to include vitamins and essential nutrients in the fermentation medium by adding yeast extract and nutrient broth. Most beneficial ranges of concentrations for these materials are 0.1 to 2% (w/v).

Isonicotinic acid hydrazide inhibits production of surfactant. However, if a small amount of surfactant of the synthetic type such as Tween 80 is added, the combination of both reagents, isonicotinic acid hydrazide and Tween 80, increases the production of surfactant by the *C. salvinicum* microbe several fold. The enzyme lyozyme can be used for treatment of the cells produced in the fermentation process, either after fermentation is completed or during the fermentation, to obtain a further significant increase in surfactant concentration. Preferred concentrations of lysozyme for this purpose are 0.005 to 0.01% (w/v). Diethoxymethane can be added to the fermentation broths to obtain a beneficial effect on surfactant production. Moreover, the total biomass produced in the presence of diethoxymethane is increased. Its effect on cell morphology is quite pronounced.

GENETIC ENGINEERING PROCESSES

RECOMBINANT DNA TECHNIQUES

Production of Biologically Functional Molecular Chimeras

Although transfer of plasmids among strains of *E. coli* and other Enterobacteriaceae has long been accomplished by conjugation and/or transduction, it has not been previously possible to selectively introduce particular species of plasmid DNA into these bacterial hosts or other microorganisms. Since microorganisms that have been transformed with plasmid DNA contain autonomously replicating extrachromosomal DNA species having the genetic and molecular characteristics of the parent plasmid, transformation has enabled the selective cloning and amplification of particular plasmid genes.

The ability of genes derived from totally different biological classes to replicate and be expressed in a particular microorganism permits the attainment of interspecies genetic recombination. Thus, it becomes practical to introduce into a particular microorganism genes specifying such metabolic or synthetic functions as nitrogen fixation, photosynthesis, antibiotic production, hormone synthesis, protein synthesis, e.g., enzymes or antibodies, or the like–functions which are indigenous to other classes of organisms–by linking the foreign genes to a particular plasmid or viral replicon.

In a historically important patent, *S.N. Cohen and H.W. Boyer; U.S. Patent 4,237,224; December 2, 1980; assigned to Board of Trustees of the Leland Stanford Jr. University* have provided methods and compositions for genetically transforming microorganisms, particularly bacteria, to provide diverse genotypical capability and producing recombinant plasmids. A plasmid or viral DNA is modified to form a linear segment having ligatable termini which is joined to DNA having at least one intact gene and complementary ligatable termini. The termini are then bound together to form a "hybrid" plasmid molecule which is used to transform susceptible and compatible microorganisms. After transformation, the cells are grown and the transformants harvested. The newly functionalized microorganisms may then be employed to carry out their new function; for example, by producing proteins which are the desired end product, or metabolites of enzymic conversion, or be lysed and the desired nucleic acids or proteins recovered.

In order to prepare the plasmid chimera, it is necessary to have a DNA vector, such as a plasmid or phage, which can be cleaved to provide an intact replicator locus and system (replicon), where the linear segment has ligatable termini or is capable of being modified to introduce ligatable termini. Of particular interest are those plasmids which have a phenotypical property which allow for ready separation of transformants from the parent microorganism. The plasmid will be capable of replicating in a microorganism, particularly a bacterium which is susceptible to transformation. Various unicellular microorganisms can be transformed, such as bacteria, fungi and algae—that is, those unicellular organisms which are capable of being grown in cultures of fermentation. Since bacteria are for the most part the most convenient organisms to work with, bacteria will be hereinafter referred to as exemplary of the other unicellular organisms. Bacteria which are susceptible to transformation include members of the Enterobacteriaceae, such as strains of *Escherichia coli;* Salmonella; Bacillaceae, such as *Bacillus subtilis;* Pneumococcus; Streptococcus, and *Haemophilus influenzae.*

A wide variety of plasmids may be employed of greatly varying molecular weight. The desired plasmid size is determined by a number of factors. First, the plasmid must be able to accommodate a replicator locus and one or more genes that are capable of allowing replication of the plasmid. Secondly, the plasmid should be of a size which provides for a reasonable probability of recircularization with the foreign gene(s) to form the recombinant plasmid chimera. Desirably, a restriction enzyme should be available, which will cleave the plasmid without inactivating the replicator locus and system associated with the replicator locus. Also, means must be provided for providing ligatable termini for the plasmid, which are complementary to the termini of the foreign gene(s) to allow fusion of the two DNA segments.

Another consideration for the recombinant plasmid is that it be compatible with the bacterium to be transformed. Therefore, the original plasmid will usually be derived from a member of the family to which the bacterium belongs.

The original plasmid should desirably have a phenotypical property which allows for the separation of transformant bacteria from parent bacteria. Particularly useful is a gene, which provides for survival selection. Survival selection can be achieved by providing resistance to a growth-inhibiting substance or providing a growth factor capability to a bacterium deficient in such capacity.

Conveniently, genes are available which provide for antibiotic or heavy metal resistance or polypeptide resistance, e.g., colicin. Therefore, by growing the bacteria on a medium containing a bacteriostatic or bacteriocidal substance, such as an antibiotic, only the transformants having the antibiotic resistance will survive. Illustrative antibiotics include tetracycline, streptomycin, sulfa drugs, such as sulfonamide, kanamycin, neomycin, penicillin, chloramphenicol, or the like.

Growth factors include the synthesis of amino acids, the isomerization of substrates to forms which can be metabolized or the like. By growing the bacteria on a medium which lacks the appropriate growth factor, only the bacteria which have been transformed and have the growth factor capability will clone.

One plasmid of interest derived from *E. coli* is referred to a pSC101. It has a molecular weight of about 5.8 x 10^6d and provides tetracycline resistance.

Another plasmid of interest is colicinogenic factor EI (ColE1), which has a molecular weight of 4.2 x 10^6d, and is also derived from *E. coli.* The plasmid has a single EcoRI substrate site and provides immunity to colicin E1.

In preparing the plasmid for joining with the exogenous gene, a wide variety of techniques can be provided, including the formation of or introduction of cohesive termini. Flush ends can be joined. Alternatively, the plasmid and gene may be cleaved in such a manner that the two chains are cleaved at different sites to leave extensions at each end which serve as cohesive termini. Cohesive termini may also be introduced by removing nucleic acids from the opposite ends of the two chains or alternatively, introducing nucleic acids at opposite ends of the two chains.

An alternative way to achieve a linear segment of the plasmid with cohesive termini is to employ an endonuclease such as Eco RI. The endonuclease cleaves the two strands at different adjacent sites providing cohesive termini directly.

The plasmid which has the replicator locus and serves as the vehicle for introduction of a foreign gene into the bacterial cell, will hereafter be referred to as "the plasmid vehicle."

It is not necessary to use plasmid, but any molecule capable of replication in bacteria can be employed. Therefore, instead of plasmid, viruses may be employed, which will be treated in substantially the same manner as the plasmid, to provide the ligatable termini for joining to the foreign gene.

The cleavage and noncovalent joining of the plasmid vehicle and the foreign DNA can be readily carried out with a restriction endonuclease, with the plasmid vehicle and foreign DNA in the same or differnt vessels. Depending on the number of fragments, which are obtained from the DNA endonuclease digestion, as well as the genetic properties of the various fragments, digestion of the foreign DNA may be carried out separately and the fragments separated by centrifugation in an appropriate gradient. Where the desired DNA fragment has a phenotypical property, which allows for the ready isolation of its transformant, a separation step can usually be avoided.

After preparation of the two double-stranded DNA sequences, the foreign gene and vector are combined for annealing and/or ligation to provide for a functional recombinant DNA structure. With plasmids, the annealing involves the hydrogen bonding together of the cohesive ends of the vector and the foreign gene to form a circular plasmid which has cleavage sites. The cleavage sites are then normally ligated to form the complete closed and circularized plasmid.

After the recombinant plasmid or plasmid chimera has been prepared, it may then be used for the transformation of bacteria.

Various techniques exist for transformation of a bacterial cell with plasmid DNA. One technique, which is particularly useful with *Escherichia coli*, is described in Cohen et al, *Proc. Nat. Acad. Sci:* 69, 2110 (1972). The bacterial cells are grown in an appropriate medium to a predetermined optical density. For example, with *E.*

coli strain C600, the optical density was 0.85 at 590 nm. The cells are concentrated by chilling, sedimentation and washing with a dilute salt solution. After centrifugation, the cells are resuspended in a calcium chloride solution at reduced temperatures (approximately 5° to 15°C), sedimented, resuspended in a smaller volume of a calcium chloride solution and the cells combined with the DNA in an appropriately buffered calcium chloride solution and incubated at reduced temperatures. The concentration of Ca^{2+} will generally be about 0.01 to 0.1 M. After a sufficient incubation period, generally from about 0.5 to 3.0 hours, the bacteria are subjected to a heat pulse generally in the range of 35° to 45°C for a short period of time; i.e., from about 0.5 to 5 minutes. The transformed cells are then chilled and may be transferred to a growth medium, whereby the transformed cells having the foreign genotype may be isolated.

In order to enhance the ability to separate the desired bacterial clones, the bacterial cells, which have been subjected to transformation, will first be grown in a solution medium, so as to amplify the absolute number of the desired cells. The bacterial cells may then be harvested and streaked on an appropriate agar medium. Where the recombinant plasmid has a phenotype, which allows for ready separation of the transformed cells from the parent cells, this will aid in the ready separation of the two types of cells. As previously indicated, where the genotype provides resistance to a growth-inhibiting material, such as an antibiotic or heavy metal, the cells can be grown on an agar medium containing the growth-inhibiting substance. Only available cells having the resistant genotype will survive.

Once the recombinant plasmid has been replicated in a cell and isolated, the cells may be grown and multiplied and the recombinant plasmid employed for transformation of the same or different bacterial strain.

By introducing one or more exogenous genes into a unicellular organism, the organism will be able to produce polypeptides and proteins (polyamino acids) which the organism could not previously produce. In some instances the polyamino acids will be useful in themselves, while in other situations, particularly with enzymes, the enzymatic product(s) will either be useful in itself or to produce a desirable product. One group of polyamino acids which are directly useful are hormones. Others are serum proteins, fibrinogin, prothrombin, thromboplastin, globulin, e.g., gamma globulins or antibodies, heparin, antihemophilia protein, oxytocin, albumins, actin, myosin, hemoglobin, ferritin, cytochrome, myoglobin, lactoglobulin, histones, avidin, thyroglobulin, interferon, kinins and transcortin.

Where the gene or genes produce one or more enzymes, the enzymes may be used for fulfilling a wide variety of functions. Included in these functions are nitrogen fixation, production of amino acids, e.g., polyiodothyronine, particularly thyroxine, vitamins, both water- and fat-soluble vitamins, antimicrobial drugs, chemotherapeutic agents, e.g., antitumor drugs, polypeptides and proteins, e.g., enzymes from apoenzymes and hormones from prohormones, diagnostic reagents, energy producing combinations, e.g., photosynthesis and hydrogen production, prostaglandins, steroids, cardiac glycosides, coenzymes, and the like.

DNA Joining Method

In recombinant DNA technology, small autonomously replicating DNA molecules in the form of closed loops, termed plasmids, are exploited. The DNA to be recombined with the plasmid may be obtained in a variety of ways, although

Federal safety requirements have made the in vitro formation of DNA complementary to isolated mRNA the method of choice. Such DNA is termed cDNA.

Recombinant plasmids are formed by mixing restriction endonuclease-treated plasmid DNA with cDNA containing end groups similarly treated. In order to minimize the chance that segments of cDNA will form combinations with each other, the plasmid DNA is added in molar excess over the cDNA. In prior art procedures this has resulted in the majority of plasmids circularizing without an inserted cDNA fragment. The subsequently transformed cells contained mainly plasmid and not cDNA recombinant plasmids. As a result, the selection process was very tedious and time-consuming. The prior art solution to this problem has been to attempt to devise DNA vectors having a restriction endonuclease site in the middle of a suitable marker gene such that the insertion of a recombinant divides the gene thereby causing loss of the function coded by the gene.

J. Shine; U.S. Patent 4,264,731; April 28, 1981; assigned to The Regents of the University of California has developed a method for treating DNA molecules that are to be combined in a subsequent joining reaction in which the frequency of joining an undesired combination is reduced and the frequency of joining the desired combination is enhanced. Specifically, the method involves the pretreatment of the reactant ends of the DNA to be joined to effect removal of certain 5' terminal phosphate groups. The ligase-catalyzed reaction is dependent upon the existence of a 5' phosphate at the end to be joined. Joining is therefore prevented between pairs of reactant end groups from which the 5' terminal phosphate has been removed.

The general method has been applied in two different circumstances. In the first circumstance, circular plasmid DNA is rendered linear by a double-strand scission and mixed with linear DNA in order to form a recombinant between the linear DNA and the plasmid. A major competing reaction, the head-to-tail joinder of the plasmid to reconstitute the closed loop without recombination with another DNA molecule, is prevented. In a second circumstance, linear DNA is subjected to chain scission to produce two subfragments which are to be purified separately, then rejoined. A major competing reaction, the rejoinder of the subfragments in opposite sequence from their original sequence, is prevented by application of this method.

Specific details of one method of using this process are presented in the following example.

Example: The formation of a recombinant plasmid and its characterization after replication is described. Plasmid pMB-9 DNA, prepared as described by Rodriguez et al, in *ICN-UCLA Symposium on Molecular and Cellular Biology*, (Academic Press, New York 1976) pp. 471-477, was cleaved at the Hind III restriction site with Hsu I endonuclease, then treated with alkaline phosphatase, type BAPF (Worthington Biochemical Corporation). The enzyme was present in the reaction mixture at the level of 0.1 U/μg DNA and the reaction mixture was incubated in 25 mM Tris-HCl, at pH 8 for 30 minutes at 65°C, followed by phenol extraction to remove the phosphatase. After ethanol precipitation, the phosphatase-treated plasmid DNA was added to cDNA containing Hind III cohesive termini at a molar ratio of 3 mols plasmid to 1 mol cDNA. The mixture was incubated in 66 mM Tris, pH 7.6, 6.6 mM $MgCl_2$, 10 mM dithiothreitol, and 1 mM ATP for one hour at 14°C in the presence of 50 U/ml of T4 DNA ligase.

The ligation mixture was added directly to a suspension of *E. coli* X-1776 cells prepared for transformation as follows: Cells were grown to a cell density of about 2 x 10^8 cells/ml in 50 ml of medium containing Tryptone 10 g/ℓ, yeast extract 5 g/ℓ, NaCl 10 g/ℓ, NaOH 2 mM, diaminopimelic acid 100 μg/ml and thymine 40 μg/ml, at 37°C. Cells were harvested by centrifugation for 5 minutes at 5,000 x *g* at 5°C, resuspended in 20 ml cold NaCl 10 mM, centrifuged as before and resuspended in 20 ml transformation buffer containing 75 mM $CaCl_2$, 140 mM NaCl and 10 mM Tris pH 7.5, and allowed to remain 5 minutes in ice. The cells were then centrifuged and resuspended in 0.5 ml transformation buffer. Transformation was carried out by mixing 100 μl of the cell suspension with 50 μl recombinant DNA (1 μg/ml). The mixture was incubated at 0°C for 15 minutes, then transferred to 25°C for 4 minutes, then at 0°C for 30 minutes. The cells were then transferred to agar plates for growth under selection conditions.

Screening for recombinant plasmids was carried out at 5 μg/ml tetracycline for transformation into the Hind III site. A selected recombinant, designated pAU-1, was isolated. Crude plasmid preparations of 2 to 5 μg DNA isolated from pAU-1 were digested with an excess of Hsu I endonuclease. EDTA-Na_2 10 mM, and sucrose 10% w/v (i.e., weight to volume) final concentration were then added and the mixture resolved on an 8% polyacrylamide gel. The DNA was found at a position corresponding to about 410 basepairs in length.

The relative frequency of transformation by nonrecombinant plasmids was measured, comparing alkaline-phosphatase-pretreated DNA with untreated DNA. Pretreatment reduced the relative frequency of transformation by nonrecombinant plasmids to less than 1 in 10^4 that of nontreated DNA.

Plasmid pUC6 from *Streptomyces espinosus*

The development of plasmid vectors useful for recombinant DNA genetics among microorganisms is well known. Similar DNA work is being done on industrially important microorganisms of the genus Streptomyces, but only one Streptomycete plasmid has been physically isolated and extensively characterized in the literature. The existence of other plasmids in the genus Streptomyces has been inferred from reported genetic data.

J.J. Manis; U.S. Patent 4,273,875; June 16, 1981; assigned to The Upjohn Company describes the plasmid pUC6 from the new microorganism *Streptomyces espinosus* biotype 23724a, NRRL 11439 obtained from an actinomycete culture from a soil sample isolated in the Upjohn screening laboratory. This plasmid can be obtained from NRRL 11439 by growing the culture on a suitable medium, fragmenting the mycelia, incubating the fragmented mycelia, harvesting the culture after a suitable time, and then lysing the mycelia. From this lysate, it is possible to isolate essentially pure pUC6. Plasmid pUC6 is advantageously small and present at many copies per cell. As such, pUC6 represents the only small high copy number plasmid described to date in Streptomyces. Further, its sensitivities to a variety of restriction endonucleases should allow its ready modification and adaptation to a number of host vector systems.

pUC6 is characterized by standard characterization tests which include its molecular weight, approximately 6.0 megadaltons, a restriction map as shown in Figure 10.1 (which follows) and presence at 20 to 40 copies per *S. espinosus* NRRL 11439 cell.

Figure 10.1: Restriction Endonuclease Cleavage Map

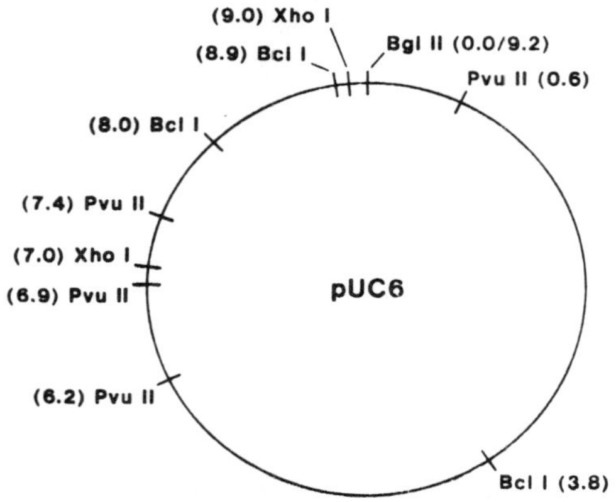

Source: U.S. Patent 4,273,875

The usefulness of plasmid pUC6 is derived from its capacity to function as a plasmid vector in industrially important microorganisms, e.g., Streptomyces. Hence, cloning of genetic information from Streptomyces into pUC6 provides a means of increasing the production of commercially important products from these organisms, e.g., antibiotics.

This approach is compared to the concept of cloning genes for antibiotic production into the well-characterized *Escherichia coli* K-12 host-vector system. The *E. coli* system has the disadvantage that it has been found that genes from some Gram-positive organisms, e.g., Bacillus, do not express well in the Gram-negative *E. coli* host. Likewise, plasmids from Gram-negative organisms are not maintained in Gram-positive hosts, and Gram-negative genetic information is either expressed poorly or not at all in Gram-positive hosts. This clearly argues for the advantage of a Gram-positive host-vector system and argues the usefulness of plasmid pUC6 in such a system.

Since pUC6 is a Streptomycete plasmid, it is ideally suited as a plasmid vector in the genus Streptomyces. Furthermore, since pUC6 is also a plasmid from a Gram-positive organism, it may serve as a vector in a number of other microorganisms, e.g., Bacillus, Arthrobacter, etc.

Streptomyces espinosus biotype 23724a, NRRL 11439, can be grown in a conventional aqueous nutrient medium under submerged aerobic conditions at a temperature of about 37°C for from 36 to 48 hours.

The process for isolating essentially pure pUC6 from *Streptomyces espinosus* biotype 23724a, NRRL 11439, comprises: (a) growing *S. espinosus* biotype 23724a, NRRL 11439, on a suitable *S. espinosus* growth medium until sufficient mycelial growth is obtained; (b) fragmenting the mycelia; (c) incubating the fragmented mycelia in a suitable growth medium containing sucrose and glycine; (d) harvesting the culture after a suitable time; (e) lysing the harvested mycelia; and (f) isolating essentially pure pUC6 from the lysate.

Method for Synthesizing DNA Sequentially

The rapidly developing art of recombinant DNA technology enables the introduction of genetic information into a foreign host and expression of this information in the new host. The genetic information may be DNA selected from or isolated from another organism or it may be DNA which is synthesized chemically by any of a variety of known methods. The chemical synthesis of DNA permits the production of genes which otherwise might be very difficult to retrieve from the cell in which they naturally originate.

Chemical synthesis of DNA molecules typically involves the synthesis of two different single strands of nucleotides which are complementary to each other. These are then joined together at the complementary base pairs to form the double-stranded DNA molecule. The single strand fragments may be formed by a number of known chemical schemes, and although known techniques have been successful, some difficulties arise in making individual DNA chains longer than about 10 to 15 nucleotide bases. This is because the yield from known chemical synthesizing techniques typically declines as the length of the synthesized nucleotide chain increases. The solution to this problem has been to synthesize fragments which are less than the full length of DNA desired to be synthesized, and to subsequently join the fragments together to form the desired relatively long DNA molecule. The two basic methods for accomplishing this have not been totally satisfactory, one being laborious and time-consuming and the other prone to error in the homology of the ligated fractions.

For this reason, *S.N. Cohen; U.S. Patent 4,293,652; October 6, 1981; assigned to Cetus Corporation* has developed a method for synthesizing double-stranded DNA which is highly efficient, provides for convenient purification as the synthesis proceeds, and provides amplification of the amount of new material for synthesis at each step.

The method builds the DNA molecules directly within the cloning vector. In a preferred form, complementary oligonucleotides are synthesized using a known procedure. These are annealed forming a double-stranded fragment. The vector is opened and the fragment of synthesized double-stranded DNA is inserted. After closing, the vector is cloned in an appropriate host, both purifying the synthetic DNA, by forming colonies derived from single instances of insertions, and amplifying the amount of the synthesized DNA by the replication process. The vector is then opened again, in one approach utilizing a unique class of restriction enzyme to cut exactly at the end of the first synthetic piece. Another synthesized double-stranded fragment of DNA can then be inserted. The vector is closed again and cloned. The process is repeated until the entire desired DNA molecule has been constructed, at which time it may be excised or cloned directly upon the vector with which it was made.

The cloning vehicles used in the method may be of any appropriate type. The plasmid or virus cloning vehicle selected should be stable in a suitable host, should have incorporated in it a replicator locus and genes capable of allowing replication of the vector, and should carry a phenotypic property allowing selection of transformed host. In addition, the cloning vehicle should have a single recognition sequence for a preselected restriction enzyme of the type that cuts at a site outside the recognition sequence. The single recognition sequence for the preselected restriction enzyme should not be such that the cut occurs in a needed informational region of the vector, as the exogenous or synthetic DNA will be inserted at that point. One vector satisfying the above requirements is the plasmid pBGP-120 modified so as not to be cut by the enzyme HphI. This plasmid contains a single-restriction site for the enzyme EcoRI.

The plasmid cloning vector which is selected is modified by inserting therein a DNA linker, i.e., a DNA segment having a particular nucleotide sequence. The DNA linker or insertion bridge, which may be synthesized by known techniques, is characterized by having a respective portion of a restriction enzyme recognition sequence at each of its ends corresponding to the separable portions of the single standard restriction enzyme recognition sequence in the cloning vehicle. Alternatively, the ends may be blunt with nucleotides suitable for blunt-end ligation.

In the foregoing example, pBGP120, the synthetic DNA linker would thus have a portion of an EcoRI enzyme recognition sequence at each end capable of linking with the open sticky ends of the severed EcoRI recognition sequence in the plasmid pBGP120. Alternatively, insertion of the linker in the vector can be accomplished by adding dAT or dGC tails. Between the two ends of the linker is built the recognition sequence for the restriction enzyme that cuts offset or downstream from the recognition sequence–in the example HphI. The enzyme HphI has the recognition sequence GGTGA. It cuts asymmetrically eight bases later on the plus strand and seven bases later on the minus strand. In addition to the recognition sequence in the linker between the ends, a spacer is included in the synthetic DNA linker which is at least the same number of nucleotides long as the cutting site of the offset cutting restriction enzyme. Thus, in the case of the enzyme HphI, the linker includes eight bases X on the plus strand and eight bases X' on the minus strand following the HphI recognition sequence.

In may also be preferred to incorporate, in the DNA linker, appropriate nucleotides for initiating transcription and translation of the synthesized DNA segments which are later appended to it. This is, in fact, necessary if the cloning vector itself does not have such nucleotides conveniently located for the correct reading frame. Thus, the DNA linker would be synthesized to include a promotor for transcription, a ribosome binding site, and translational start codon. Extra codons may also be employed in the linker to shift the reading frame as desired.

Once the cloning vector is selected and appropriately modified, if necessary, as described above, the method proceeds by preselecting a portion of the required or desired DNA strand and synthesizing it. The selected portion is the first block of nucleotides of the desired DNA molecule and is less than sixteen bases long for convenience in synthesizing by known techniques. This preselected DNA fragment is synthesized in any suitable manner, such as by synthesizing the individual complementary single strands and annealing them to form the desired double-stranded DNA fragment. The plasmid cloning vehicle is cut by the offset cutting restriction enzyme (HphI in the example). The cut cloning vehicle is then treated with a single strand specific exonuclease to remove the single-stranded tails, or is treated with DNA polymerase to fill in the missing nucleotides. In the illustrated case, one base is removed on each end of the cut cloning vehicle. In the illustrated embodiment, since the single-stranded portions are on 3' ends, the exonuclease of DNApolI can be used.

The open cloning vehicle, now with blunt ends, then has the first synthesized gene fragment inserted therein by blunt-end ligation. Two results are possible, depending upon the orientation of the inserted fragment. Only one of the possibilities is useful. The plasmids or cloning vehicles, including the inserted DNA fragments, are then placed in appropriate hosts and growth conditions are provided for the hosts to amplify and purify the inserted fragments. The cloning vehicles are then harvested from the colonies and are cut with the offset cutting restriction enzyme. The particular desired orientation is then verified by partial sequencing.

The foregoing procedure is then repeated by synthesizing the next fragment of the desired DNA sequence, and blunt-end ligating it into the cloning vehicle. The two fragment portions of the desired DNA are then amplified and purified by growing colonies.

After the second fragment is inserted, once again two different orientations are possible. The correctly placed or oriented pieces may be found after sequencing and cloning as before. Note that once a fragment has been placed accurately, it need not be sequenced again as it is carried on in the same orientation in future steps. The foregoing described procedure continues with cloning, cutting and blunt-end ligation of new synthetic pieces or fragments until the desired whole gene has been produced.

Process for Producing Hybrid Bacteria

J. Collins and B. Hohn; U.S. Patent 4,304,863; December 8, 1981; assigned to Gesellschaft für Biotechnologische Forschung mbH, Germany have developed a process for producing hybrid bacteria in which (a) a bacterial plasmid of not more than about 21 megadaltons and having one or several cos sites of phage lambda or a lambdoid phage and only one cut site per restriction endonuclease is cut in the presence of one or two restriction endonucleases, (b) the cut plasmid is coupled with a DNA fragment in the presence of a ligase to give a hybrid plasmid, (c) the resulting hybrid plasmid is packaged with the lysate of phage lambda or a lambdoid phage and (d) the packaging product is transduced into *Escherichia coli* to form hybrid bacteria.

It has been found that some plasmids cannot be cut by certain restriction enzymes at all. However, in order to be able also to use such restriction enzymes in this process, the following procedure is possible. It is possible to cut a plasmid with a restriction enzyme which cuts the plasmid at one site only. Parallel therewith a foreign DNA can be cut in such a manner that a DNA fragment results which can be cut at one site by a different restriction enzyme which does not cut the plasmid. Then the foreign DNA fragment can be coupled with the cut plasmid so that a hybrid plasma results which can now be cut by the different restriction enzyme at a certain site.

The foreign DNA fragment which is coupled with the plasmid at step (b) above is a fragment which gives the hybrid bacterium to be produced some desired feature.

Phage lambda or lambdoid phages like Phi 80 can be used for the production of a phage lysate. The production of the phage lysate, the packaging and the transduction can be carried out according to the process of Hohn & Murray, *Proc. Nat. Acad. Sci.*, 74 (1977) 3259-3263 or analogous methods.

The expert may use usual analytical methods to screen the produced hybrid bacteria and to calculate the efficiency of the process. In this connection the expert may select a feature of the hybrid bacteria which is suitable for a selection, for example, their resistivity against antibiotics, as, for example, rifampicin or ampicillin; preferably, this feature has been coded by the plasmid. It is possible to carry out the process in vitro.

The advantage of the process in comparison with other known transformation systems consists in the fact that a higher efficiency can be achieved and larger foreign DNA fragments can be transduced, for example, of a size of more than 2, especially 3 or 5 megadaltons and more, the upper limit being dependent on the selected packaging and transducing system. In this process hybrid plasmid DNA can be introduced into living bacteria at least one hundred times better than in the known transformation systems; in the case of larger hybrid plasmids the process can be carried out at least 1,000 to 100,000 times better. The better efficiency in the case of large hybrid plasmids results from the strong counterselection which occurs in transforming large molecules.

The particular advantage of this process over known packaging/transduction system with lambda vectors is the small size of the plasmid vectors which consequently allows packaging of large foreign DNA fragments. Preferably a plasmid is used having not more than about 18 megadaltons. Preferably a plasmid having only one cos site is used.

The patent contains detailed experimental procedures and data for the steps of the process.

Providing Maximum Frequencies of Transduction with *E. coli* Strains

Experiments with recombinant DNA are regulated by guidelines that have been established by the National Institute of Health. Many important experiments require effective containment to reduce the probability that experimental chimeras will escape from the laboratory.

Under the current NIH guidelines biological containment is to be used in combination with physical containment for safety purposes. Biological containment requires the use of a cripped strain of *Escherichia coli,* such as *E. coli* K12. *E. coli* K12 is a laboratory-adapted strain unable to survive more than 48 hours in the human intestinal tract. Several planned mutational defects in *E. coli* K12 have resulted in a self-destructing *E. coli* strain called *E. coli* K12 χ1776. This strain is acceptable for use in experiments requiring an EK2 host-vector system. *E. coli* K12 χ1776 is unable to synthesize its own cell wall or replicate its DNA outside a carefully controlled laboratory environment. In addition to the fact that its ability to survive in the human intestinal tract is further reduced, *E. coli* K12 χ1776 is also extremely sensitive to sunlight, moderately warm temperatures and detergents and chemicals frequently found in sewers and polluted waters. For this reason it has been approved as an EK2 host for biological containment.

In the combination of physical and biological containment outlined in the NIH guidelines, an EK2 host can be used in experiments at a reduced level of physical containment compared with similar experiments using an EK1 host.

The normal experimental protocol for transducing *E. coli* strains such as *E. coli* K12 involves accomplishing transduction during the exponential growth phase.

Because of the importance of *E. coli* K12 χ1776, the ability to transduce genetic markers into this strain is extremely important. Derivatives of χ1776 for this purpose, however, could not be isolated using the procedure for transducing *E. coli* K12. Previously, transduction of *E. coli* during the stationary growth phase was unknown.

C.L. Hershberger; U.S. Patent 4,322,497; March 30, 1982; assigned to Eli Lilly and Company has provided a method for facilitating transduction of genetic markers into any strain of *E. coli* that is transduced poorly in the exponential growth phase. This method comprises growing the recipient *E. coli* strain to the stationary phase, during which time the culture is susceptible to transduction, and carrying out transduction by standard procedures during the stationary phase, thereby introducing genetic markers into the *E. coli* strain.

Strains of *E. coli* K12 χ1776 modified by transduction using this method are useful tools for biological containment and are contemplated as a part of this process.

In carrying out the method, standard procedures for transduction may be used. The standard procedure for transduction involves exposing a recipient bacterial culture to a bacteriophage-lysate prepared on a suitable donor. The term lysate as used herein refers to the bacteriophage-lysate prepared by infecting a suitable bacterial donor as described in the art. Any donor which is susceptible to a phage, the lysate of which will infect the *E. coli* strain, can be used.

In carrying out the transduction, the multiplicity of exposure can vary with different strains and is normally within a range from about 0.01 to 10. To facilitate transduction, the most effective multiplicity of exposure (moe) can be determined by standard laboratory procedures. With χ1776, for example, to obtain the maximum frequency of tranductants per plaque-forming units (pfu), an moe of 0.4 is preferable; to obtain a maximum frequency of transductants per cell, an moe of about 4 is preferable.

The bacteriophage-lysates most frequently used for tranductions of this type are prepared from bacteriophage P1 or a derivative of P1. Examples of commonly used bacteriophages which are derivatives of P1 include P1 vir a, P1 Harris and P1 kc.

Variations in several experimental parameters for transduction, such as concentration of calcium or magnesium, time for absorbing bacteriophage and concentration of cells, do not significantly alter the frequency of transduction when using this method.

This method is applicable for the introduction of any genetic marker. Since markers with selectable phenotypes constitute the group most useful for the genetic selection of chimeras, this procedure will facilitate construction of strains for isolating recombinant DNA.

Preparation of a Recombinant DNA Phage

Genetic manipulation is a new research field initiated as the result of timely union between the rapid progress in the research on genetic and chemical properties of replicons such as plasmids and bacteriophages and progress in the research on enzymes associated with DNA, especially endonucleases which recognize the specific sequences of nucleotides in DNA and cause cleavage of polynucleotide chains (restriction enzymes) and DNA ligases.

Various procedures have previously been proposed for the gene recombination. As an example, there is a known procedure for the in vitro recombination of *Drosophila melanogaster* DNA with λ (lambda) phage DNA. However, since the site of recombination is the DNA region carrying the genetic information necessary for lysogenization of the phage, it is impossible to integrate the resulting hybrid DNA into the host DNA.

Consequently, practically it is necessary to always preserve the host cell and the phage for ready use, but the preservation of the phage presents a problem. In addition, when a plasmid having recombined genetic information for a specific enzyme protein infects the host, production of the specific enzyme takes place continuously (owing to the gene dosage effect) even when the host cells are in a preserved state, thus causing considerable disturbance in host metabolism and inducing various secondary variations to compensate the disturbance. For instance, some of the plasmids undergo variation to decrease the number of copies of the plasmid, others acquire through variation the behavior to decrease the function or synthetic activity of the enzyme or to affect other metabolic systems so as to correct the distorted character. In actual cases, because of such secondary variations which profoundly affect the results, genetic recombination procedures have been far from satisfactory.

The phage particle generally consists of protein and nucleic acid (DNA or RNA) forming a structure in which the nucleic acid is surrounded by the protein (called coat protein). The nucleic acid bears in memory all of the genetic information necessary for the bacteriophage to multiply on infecting a host cell. In the case of the bacteriophage λ, for example, one half of the DNA chain carries genes which are necessary for self-replication and the other half carries the genetic information for the synthesis of coat proteins.

If a phage DNA segment carrying the genetic information coding for the synthesis of coat proteins were replaced by another DNA segment carrying the genetic information coding, for example, for the synthesis of a useful enzyme by the aid of endonuclease and DNA ligase, it would seem possible to induce the phage, on infecting a host cell, to synthesize a large amount of the useful enzyme in place of the coat protein.

However, since the bacteriophage generally has endonuclease-sensitivity even in its DNA region participating in self-replication, it undergoes scission (cleavage) in the region by the action of endonuclease, and the self-replication becomes impossible.

The objective of *E. Nakano, N. Saito and D. Fukushima; U.S. Patent 4,332,897; June 1, 1982; assigned to Noda Institute for Scientific Research, Japan* was, therefore, to provide a method for preparing a useful new bacteriophage composed of the DNA molecule sensitive to an endonuclease only in the region where the genetic information for the production of coat proteins is located and, therefore, capable of replacing the genetic information of that region by other intended genetic information to produce a self-reproducible hybrid DNA molecule.

Thus, this bacteriophage is composed of a DNA molecule made endonuclease-sensitive only in the DNA region carrying genetic information for the production of coat proteins. Such a bacteriophage can be obtained by making a bacteriophage of the lambdoid phage species endonuclease-resistant and mating the resulting bacteriophage with a lambdoid phage having endonuclease-sensitivity in the DNA region responsible for the production of coat proteins.

Although bacteriophage generally multiplies depending more or less on the function of its host, it is a replicon capable of multiplying independently outside the host chromosome, viz, it is in the autonomous state. The bacteriophage used in this process is a temperate phage having such property that when the phage infects a host cell, its phage DNA can be integrated in the host cell DNA under appropriate conditions (lysogeny). Preferable temperate phages are lambdoid phages including λ (IFO 20016), 434 (IFO 20018), 82 (IFO 20019), φ80 (IFO 20020), φ170 (IFO 20021), etc. Also usable are those phages lysogenized in appropriate host cell, such as, for example, φ80 lysogenized in *E. coli* W3110 (ATCC 31277).

As to the preparation of an endonuclease-resistant phage, a preferred endonuclease is a restriction enzyme such as Eco RI, Bam HI, or Hind III.

The endonuclease-resistant phage such as, for example, a mutant phage absolutely resistant to cleavage by a restriction enzyme can be obtained, for example, in the following manner. Alternate cultivation of a lambdoid phage in a host containing a restriction enzyme and in another host containing no restriction enzyme results in extinction of a phage susceptible to the action of the restriction enzyme and continuous increase in the population of a mutant difficultly susceptible to this action. By such microbial concentration, it is possible to obtain finally a phage (restriction enzyme-resistant phage) containing DNA absolutely unsusceptible to the action of the restriction enzyme (i.e., DNA whose chain is perfectly uncleavable by the action of the restriction enzyme).

In order to insert the sites cleavable by the restriction enzyme into an intended region of the resulting phage DNA, which is absolutely free from the sites cleavable by the restriction enzyme, the phage is mated with a lambdoid phage having endonuclease-sensitivity in the DNA region carrying genetic information for the production of coat proteins. The mating is carried out by addition of two phage suspensions, one resistant and the other sensitive to the restriction enzyme (10^9 to 10^{10}/ml, each), successively (or simultaneously) after mixing them to a suspension of *E. coli* (10^8 to 10^9/ml) sensitive to both phages.

The mating is also possible by inducing multiplication of an endonuclease-resistant or sensitive phage lysogenized in *E. coli* strain K12 and then infecting the induced lysogen with another endonuclease-sensitive or -resistant phage. The *E. coli* cells infected with these phages as above are shaken at 37°C for 1 to 2 hours in a medium such as, for example, Tryptone medium (any medium can be freely used so long as growth of the host cells is possible).

For the sensitive *E. coli* any of the K12 strains may be used including, for example, W 3110 (ATCC 27325) and W 3350 (ATCC 27020). The *E. coli* may be used in the form of culture fluid or in the form of suspension prepared by centrifuging the culture to remove the supernatant, suspending the sediment in a 10 mM

$MgCl_2$ solution and shaking for 1 hour at 37°C. The latter form is better for the adsorption and infection of the phage.

In a manner as described above, it is possible to obtain about 10^5/ml of a phage having endonuclease-sensitivity only in the DNA region carrying genetic information for the production of coat proteins.

The prepared phage is isolated by growing it in the *E. coli* strain which cannot adsorb the endonuclease-resistant bacteriophage and in which the endonuclease-sensitive phage is lysogenized.

A recombinant DNA is formed by replacing the DNA sequences carrying genetic information for the production of coat proteins of the new phage DNA obtained above with the intended DNA sequences. The resulting recombinant DNA is allowed to infect the host so as to be integrated into the host DNA. The host cell thus treated (lysogen) can be stored for future use. Thus, it is unnecessary to preserve always the host cell and the recombinant phage, preservation of the latter being rather difficult. When needed, multiplication of the recombinant phage is induced to "amplify" the genetic information and the induced cells are cultured in a medium such as, for example, Tryptone medium. In such a manner, a great amount of specific protein (as much as 50% of the soluble protein of the host cell) can be produced owing to the amplified genetic information, which enables the purification of the specific protein to take place very easily and efficiently. For these reasons the new bacteriophage obtained above will contribute greatly to the industry.

It has been found that when *E. coli trp*$^-$ host lysogenized with the recombinant phage constructed by introducing *E. coli trp* gene into the new phage obtained as above is cultured after induction of phage multiplication, a large amount of enzyme for tryptophan biosynthesis is produced.

In further work by *E. Nakano, T. Masuda, N. Saito and D. Fukushima; U.S. Patent 4,348,477; September 7, 1982; assigned to Noda Institute for Scientific Research, Japan*, there is described a method for preparing a recombinant DNA which comprises (1) cleaving with an endonuclease a phage DNA having an endonuclease-sensitive region not in the DNA segment participating in temperate phage DNA replication and integration of DNA into a host chromosome but in other DNA segments, a temperate phage DNA having an endonuclease-sensitive region in the DNA segment carrying genetic information for the production of coat protein, and a DNA carrying the intended genetic information; (2) mixing together all fragments produced by this cleavage; (3) adding DNA ligase to the mixture; and (4) recovering from the resulting mixture a phage DNA having its coat protein producing ability deleted by the replacement of the DNA segment carrying genetic information for coat protein production with a DNA fragment carrying the intended genetic information.

The recombinant DNA obtained in this way can be incorporated into a host chromosome and the recombinant DNA integrated in the host can be reproduced in large amounts by induction. Since the recombinant DNA of this process has no coat protein producing ability, it cannot produce an active form of phage particles and, hence, presents no pollution problem. Moreover, a DNA having a molecular weight higher than that of conventional one can be used for the DNA carrying intended genetic information. The bacteriophage used in this process is a temperate phage having the property such that when the phage infects a host cell, its phage DNA can be integrated in the host cell DNA (lysogeny). Preferable temperate phages are lamboid phages such as those described in the previous patent. Preferred endonucleases are also those used in the previous process–Eco RI, Bam I or Hind III.

The endonuclease-resistant phage and endonuclease-sensitive phages are obtained as explained in the last patent and mated in the same way. The donor DNA carrying the intended genetic information can originate in microorganisms (bacteria, molds, yeasts), higher organisms, transducing phages, or the like. Examples of the genetic information to be incorporated in the recombinant DNA include cystine synthetase, suppressor gene, DNA ligase, tryptophan synthetase, gene participating in the synthesis of silkworm fibroin, gene participating in the hormone synthesis, etc.

Further, the recombinant DNA can be efficiently prepared by use of a transducing phage DNA prepared by integrating the intended genetic information into a phage DNA having an endonuclease-sensitive region in the DNA segment participating in the synthesis of coat protein and a phage DNA having an endonuclease-sensitive region not in the DNA segment participating in the DNA replication and integration into a host chromosome but in other segments.

In cleaving a phage DNA or a DNA carrying the intended genetic information by use of endonuclease, it is suitable to allow the enzyme to act at a DNA concentration of 20 to 200 μg/ml, an enzyme concentration of 100 to 2,000 U/ml and at a temperature of 26° to 42°C, preferably 37°C for 10 minutes to 2 hours. The cleaving can be effected in a mixture of a phage DNA and a DNA carrying the intended genetic information.

DNA ligase is then added to a mixture of generally equal amounts (in terms of DNA) of each suspension which has been subjected to the action of endonuclease. The DNA ligase used can be *E. coli* DNA ligase, T4 phage DNA ligase, or the like. Of these, T4 phage DNA ligase is most easily available. The DNA ligase is allowed to act at a DNA concentration of 10 to 80 μg/ml, DNA ligase concentration of 1 to 10 U/ml and at a temperature of 0° to 10°C for 1 to 14 days.

The recovery of the intended recombinant DNA from the obtained mixture of various recombinant DNA's and other substances is performed as follows.

At first *E. coli* is lysogenized with a temperate phage having the same cohesive ends and immunity as those of the phage used in preparing the recombinant DNA but having a different attachment site (the region integrated into a host chromosome on lysogenization). The resulting *E. coli* is infected with a large amount of a temperate phage having the same cohesive ends and immunity as those of the abovementioned phage but having no attachment site. The infected bacterium is mixed with the recombinant DNA and kept at 20° to 40°C to allow the latter to be incorporated into the bacterium cell. Since the phage having the same immunity as that of the recombinant DNA has been integrated into the host, the recombinant DNA entering the cell cannot multiply and becomes readily integrated into the host chromosome.

When the donor DNA used for the recombination is originated from yeasts, bacteria, transducing phages, or the like, the separation of cells containing the intended DNA from the cells lysogenized with the recombinant DNA can be effected by collecting the cells producing the products of the intended gene. For instance, if the intended gene is a gene of tryptophan synthetase, the recombinant DNA is allowed to be integrated into *E. coli* cells incapable of synthesizing tryptophan and collecting the cells which restored the tryptophan synthesizing ability, i.e., the cells capable of growing in a medium lacking tryptophan.

When the donor DNA is originated from molds or higher organisms, it is used after having been combined, by means of DNA ligase, with a gene capable of expression within *E. coli*, such as, for example, a fragment of plasmid DNA having a drug-resistant gene. The separation of the intended recombinant DNA can be achieved in this case by collecting the cells manifesting the genetic information (drug resistance).

The recombinant DNA thus prepared by deleting the phage DNA segment carrying genetic information for the coat protein production and recombining with the DNA fragment carrying the intended genetic information can be preserved by infecting a host cell and integrating into the host DNA. When required, the recombinant DNA preserved in host cells can be induced to "amplify" the genetic information. When the host cells are cultured in, for example, Tryptone medium, a large amount of specific protein can be produced in accordance with the "amplified" genetic information. Thus, the industrial usefulness of this process is believed to be very great.

Yet another method of preparing a new recombinant DNA is provided by *E. Nakano, T. Masuda, N. Saito and D. Fukushima; U.S. Patent 4,348,478; Sept. 7, 1982; assigned to Noda Institute for Scientific Research, Japan.*

This method comprises (1) cleaving with an endonuclease a temperate phage DNA having an endonuclease-sensitive region not in the DNA segment participating in the replication of phage DNA and the integration of phage DNA into a host chromosome but at least in the DNA segment carrying genetic information for the coat protein production and another DNA carrying intended genetic information, (2) adding DNA-ligase to the mixture of both cleft DNAs, and (3) recovering from the mixture a phage DNA having its coat protein producing ability deleted by the replacement of the DNA segments carrying genetic information for the coat protein production with a DNA fragment carrying the intended information.

The experimental procedures used in this patent are those already described in the two previous patents.

Construction of a Replicable Cloning Vehicle Having Quasi-Synthetic Genes

The DNA (deoxyribonucleic acid) of which genes are made comprises both protein-encoding or "structural" genes and control regions that mediate the expression of their information through provision of sites for RNA polymerase binding, information for ribosomal binding sites, etc. Encoded protein is "expressed" from its corresponding DNA by a multistep process within an organism by which: (1) the enzyme RNA polymerase is activated in the control region (hereafter the "promoter") and travels along the structural gene, transcribing its encoded information into messenger ribonucleic acid (mRNA) until transcription of translatable mRNA is ended at one or more "stop" codons; and (2) the mRNA message is translated at the ribosomes into a protein for whose amino acid sequence the gene encodes, beginning at a translation "start" signal, most comonly ATG (which is transcribed "AUG" and translated "f-methionine").

In accordance with the genetic code, DNA specifies each amino acid by a triplet or "codon" of three adjacent nucleotides individually chosen from adenosine, thymidine, cytidine and guanine or, as used herein, A,T,C, or G. These appear in the coding strand or coding sequence of double-stranded ("duplex") DNA, whose remaining or "complementary" strand is formed of nucleotides ("bases") which hydrogen bond to their complements in the coding strand. A complements T, and C complements G.

A variety of techniques are available for DNA recombination, according to which adjoining ends of separate DNA fragments are tailored in one way or another to facilitate ligation. The latter term refers to the formation of phosphodiester bonds between adjoining nucleotides, most often through the agency of the enzyme T4 DNA ligase. Thus, blunt ends may be directly ligated.

Alternatively, fragments containing complementary single strands at their adjoining ends are advantaged by hydrogen bonding which positions the respective ends for the subsequent ligation. Such single strands, referred to as cohesive termini, may be formed by the addition of nucleotides to blunt ends using terminal transferase, and sometimes simply by chewing back one strand of a blunt end with an enzyme such as λ-exonuclease.

Again, and most commonly, resort may be had to restriction endonucleases (hereafter, "restriction enzymes"), which cleave phosphodiester bonds in and around unique sequences of nucleotides of about 4 to 6 base pairs in length ("restriction sites"). Many restriction enzymes and their recognition sites are known. Many make staggered cuts that generate short complementary single-stranded sequences at the ends of the duplex fragments. As complementary sequences, the protruding or "cohesive" ends can recombine by base pairing. When two different molecules are cleaved with this enzyme, crosswise pairing of the complementary single strands generates a new DNA molecule, which can be given covalent integrity by using ligase to seal the single strand breaks that remain at the point of annealing. Restriction enzymes which leave coterminal or "blunt" ends on duplex DNA that has been cleaved permit recombination via, e.g., T4 ligase with other blunt-ended sequences.

A "cloning vehicle" is an extrachromosomal length of duplex DNA comprising an intact replicon such that the vehicle can be replicated when placed within a unicellular organism by transformation. An organism so transformed is called a "transformant." The cloning vehicles commonly in use are derived from viruses and bacteria and most commonly are loops of bacteria DNA called "plasmids."

Advances in biochemistry in recent years have led to the construction of "recombinant" cloning vehicles in which, for example, plasmids are made to contain exogenous DNA. In particular instances the recombinant may include "heterologous" DNA, by which is meant DNA that codes for polypeptides ordinarily not produced by the organism susceptible to transformation by the recombinant vehicle.

Thus, plasmids are cleaved with restriction enzymes to provide linear DNA having ligatable termini. These are bound to an exogenous gene having ligatable termini to provide a biologically functional moiety with an intact replicon and a phenotypical property useful in selecting transformants. The recombinant moiety is inserted into a microorganism by transformation and the transformant is isolated and cloned, to obtain large populations that include copies of the exogenous gene and also, in particular cases, to express the protein for which the gene codes.

Aside from the use of cloning vehicles to increase the supply of genes by replication, there have been attempts, some successful, to actually express proteins for which the genes code. In the first such instance a gene for the brain hormone somatostatin under the influence of the lac promoter was expressed in *E. coli* bacteria. [K. Itakura et al, *Science* 198, 1056 (1977).] More recently, the A and B chains of human insulin were expressed in the same fashion and combined to form the hormone. [D.V. Goeddel et al, *Proc. Natl. Acad. Sci., U.S.A.* 76, 106 (1979).] In each case the genes were constructed in their entirety by synthesis. In each case, proteolytic enzymes within the cell would apparently degrade the desired product, necessitating its production in conjugated form, i.e., in tandem with another protein which protected it by compartmentalization and which could be extracellularly cleaved away to yield the product intended.

While the synthetic gene approach has proven useful in the several cases thus far discussed, real difficulties arise in the case of far larger protein products, e.g., growth hormone, interferon, etc., whose genes are correspondingly more complex and less susceptible to facile synthesis. At the same time, it would be desirable to express such products unaccompanied by conjugate protein, the necessity of whose expression requires diversion of resources within the organism better committed to construction of the intended product.

D.V. Goeddel and H.L. Heyneker; U.S. Patent 4,342,832; August 3, 1982; assigned to Genentech, Inc. have provided methods and means for expressing quasi-synthetic genes. In this method, enzymatic reverse transcription provides a substantial portion, preferably a majority, of the coding sequence without laborious resort to entirely synthetic construction, while organic synthesis of the remainder of the coding sequence affords a completed gene capable of expressing the desired polypeptide unaccompanied by bioinactivating leader sequences or other extraneous protein.

Alternatively, the synthetic remainder may yield a proteolysis-resistant conjugate so engineered as to permit extracellular cleavage of extraneous protein, yielding the bioactive form. There are, accordingly, made available methods and means for microbial production of numerous materials previously produced only in limited quantity by costly extraction from tissue, and still others previously incapable of being industrially manufactured.

In its most preferred embodiment, the process represents the first occasion in which a medically significant polypeptide hormone (human growth hormone) has been bacterially expressed while avoiding both intracellular proteolysis and the necessity of compartmentalizing the bioactive form in extraneous protein pending extracellular cleavage. Microbial sources for human growth hormone made available by this process offer, for the first time, ample supplies of the hormone for treatment of hypopituitary dwarfism, together with other applications previously beyond the capacity of tissue-derived hormone sources, including diffuse gastric bleeding, pseudarthrosis, burn therapy, wound healing, dystrophy and bone knitting.

In this method of constucting a replicable cloning vehicle capable, in a microbial organism, of expressing a particular polypeptide of a known amino acid sequence in which a gene coding for the polypeptide is inserted into the cloning vehicle and placed under the control of an expression promoter, changes from the usual procedures are as follows:

(a) A first gene fragment for an expression product other than the polypeptide is obtained by reverse transcription from an RNA. This fragment preferably comprises a majority of the coding sequence for the polypeptide.

(b) In cases where the first fragment comprises protein-encoding codons for amino acid sequences other than those of the polypeptide, these codons are eliminated while at least a substantial part of the coding sequence is retained although the resulting fragment codes for an expression product other than the polypeptide.

This product of step (a) or (b) is a fragment which encodes less than the entire amino acid sequence of the polypeptide.

(c) One or more synthetic nonreverse transcript-gene fragments encoding the rest of the amino acid sequence of the polypeptide is provided by organic synthesis, and at least one of the fragments codes for the amino-terminal section of the polypeptide.

(d) The synthetic gene fragments of step (c) and those produced by (a) or (b) are ligated together and then deployed in a replicable cloning vehicle in proper reading phase relative to each other and under the control of an expression promoter. This cloning vehicle is conveniently a bacterial plasmid.

The synthetic fragment encoding the amino-terminal portion of the polypeptide additionally codes for expression of a specifically cleavable amino acid sequence, and the fragments are deployed downstream from and in reading phase with expressed protein-encoding codons, so that the conjugated plasmid expression product may be specifically cleaved to yield the polypeptide.

In a preferred embodiment, the polypeptide is a human growth hormone. The first fragment, produced in step (a) above, comprises protein-encoding codons for amino acid sequences other than those in the human growth hormone so that elimination step (b) is needed and yields the Hae III restriction enzyme fragment of the first fragment. Step (b) includes digestion of the Hae III fragment with the restriction enzyme Xma I, which cleaves away the codons for untranslated mRNA and simultaneously provides a single-stranded terminus at one end of the resulting fragment.

Included in the patent are examples of the construction and expression of the human growth hormone. Figures are included depicting the codons for amino acids 1 to 24 of the human growth hormone, and the construction of the cloning vehicle for a gene fragment coding for the amino acids of the human growth hormone, etc.

Alteration of a Nucleotide at a Predetermined Position in a Gene

A major difficulty in the use of recombinant DNA techniques for the production of desired protein is in isolating or cloning the gene which encodes production of the desired protein. This is particularly true where the gene desired is a human gene, e.g., one encoding production of a hormone such as insulin. It is possible where the amino acid sequence of a particular protein is known to synthesize a gene for its production. However, synthesis of a gene can be a lengthy and laborious technique. Moreover, many laboratories do not have the facility nor the trained personnel to take advantage of gene synthesis techniques.

In some cases, a gene which has already been cloned may differ structurally from an uncloned or difficult to clone gene by only one or a few bases or nucleotides. For example, animal insulin differs from human insulin in most cases by only one to several amino acids. Although animal insulin is taken by most diabetics, some patients do suffer immunological responses as a result of the amino acid differences, making it desirable to be able to supply human insulin to those individuals. If an animal insulin gene, such as rat insulin, could be altered at the specific nucleotides in the gene which differ from the corresponding nucleotides in human insulin so as to change the animal gene nucleotides to those present in the human gene, the resulting changed gene could be used to produce the human type of insulin. Other examples do exist where an already isolated or cloned gene may differ from an uncloned gene by only one or a few nucleotides.

It is an object of the process developed by *C.P. Bahl; U.S. Patent 4,351,901; September 28, 1982; assigned to Cetus Corporation* to provide a method which makes it possible to alter a gene which has already been cloned, and which is similar but not identical to a desired but uncloned gene, so as to create the desired gene or, more specifically to provide a method for altering a gene at an individual nucleotide to change that nucleotide to a desired nucleotide.

Very generally, in this method a nucleotide at a predetermined position in an available DNA sequence is altered. The alteration is such as to change the sequence of the available DNA to produce a desired DNA sequence such as a gene, initiation site, stop signal or restriction site. The alteration is accomplished by isolating single strand fragments of the gene, each having a substantial number of bases on the 5' side of the predetermined position and terminating with the base immediately prior to the predetermined position. A ribonucleotide or a protected deoxyribonucleotide corresponding to the desired altered nucleotide is attached to each of the isolated fragments at the predetermined position. Single strand templates of DNA are provided, each corresponding to at least a portion of the strand of the gene complementary to the strand having the isolated fragment.

The templates each have a first part complementary to the isolated fragments and a second part extending at least a substantial number of bases past the predetermined position. The isolated fragments are annealed to the first parts of the templates and are extended beyond the predetermined position complementary to the second parts of the templates. The resulting partial mismatched double strands may then be used to produe pure DNA containing altered deoxyribonucleotides at the predetermined positions.

In the above general description the following details may be added. The single strand fragments are isolated by first isolating double-stranded DNA gene fragments having a substantial number of bases on each side of the predetermined position, separating the strands of these DNA gene fragments and isolating the single strands which contain the nucleotide to be altered, fragmenting the isolated single strands and selecting those having all the bases preceding but not including the predetermined position. The selection is accomplished by electrophoresis on a polyacrylamide gel, and the single strands are fragmented by treatment with dimethyl sulfate or hydrazine followed by sodium hydroxide.

The templates are DNA fragments having 5' ends corresponding to the 5' ends of the isolated fragments.

The desired single nucleotide is added (1) by initially attaching between one and six ribonucleotides with the enzyme terminal transferase and then removing all but one by treatment with alkali or (2) by attaching a single protected deoxyribonucleotide with the enzyme terminal transferase and then converting the protecting group to a 3'OH.

The patent explains this nucleotide alteration process in detail, using drawings. Three examples of nucleotides are also given. In the first, four different codons in a cloned gene for rat insulin are altered to produce the gene coding for human insulin. In the second example, bovine ACTH undergoes a single nucleotide conversion to produce human-type ACTH. In the third example, the possibility of altering bovine growth hormone to make the more useful human growth hormone is examined.

Recombinant Microbial Cloning Vehicles

Advances in biochemistry in recent years have led to the construction of "recombinant" cloning vehicles in which, e.g., plasmids are made to certain exogenous DNA. In particular instances the recombinant may include "heterologous" DNA, by which is meant DNA that codes for polypeptides ordinarily not produced by the organism susceptible to transformation by the recombinant vehicle.

Thus, plasmids are cleaved to provide linear DNA having ligatable termini. These are bound to an exogenous gene having ligatable termini to provide a biologically functional moiety with an intact replicon and a desired phenotypical property. The recombinant moiety is inserted into a microorganism by transformation and transformants are isolated and cloned, with the object of obtaining large populations capable of expressing the new genetic information. Methods and means of forming recombinant cloning vehicles and transforming organisms with them have been widely reported in the literature.

Despite wide-ranging work in recent years in recombinant DNA research, few results susceptible to immediate and practical application have emerged. This has proven especially so in the case of failed attempts to express polypeptides and the like coded for by "synthetic DNA," whether constructed nucleotide by nucleotide in the conventional fashion or obtained by reverse transcription from isolated mRNA (complementary or "cDNA").

K. Itakura; U.S. Patent 4,356,270; October 26, 1982; assigned to Genentech, Inc. describes what appears to represent the first expression of a functional polypeptide product from a synthetic gene, together with related developments which promise widespread application. The product referred to is somatostatin (U.S. Patent 3,904,594), an inhibitor of the secretion of growth hormone, insulin and glucagon whose effects suggest its application in the treatment of acromegaly acute pancreatitis and insulin-dependent diabetes.

The process provides a method of preparing a structural gene coding for the microbial expression of a polypeptide wherein a series of oligodeoxyribonucleotide fragments are prepared and assembled by:

(a) Preparing a first series of oligodeoxyribonucleotide fragments which, when joined in the proper sequence, yield a DNA coding strand for the amino acid sequence of the polypeptide;

(b) Preparing a second series of oligodeoxyribonucleotide fragments which, when joined in proper sequence, yield a DNA strand complementary to the coding strand;

(c) Effecting hydrogen bonding between mutually complementary portions of the first and second series fragments to form a double-stranded structure; and

(d) Completing the respective strands by ligation, wherein:

    (1) The resulting gene codes for the expression of a mammalian polypeptide;

    (2) At least a majority of the codons in the coding are those preferred for the expression of microbial genomes; and

    (3) The fragments joined in step (c) lack complementarity, one with another, save for fragments adjacent one another in the structural gene.

There is also provided a double-stranded polydeoxyribonucleotide having cohesive termini each comprising one strand of a double-strand restriction endonuclease recognition site and, between the termini, a structural gene coding for the expression of a mammalian polypeptide, at least a majority of the codons in the coding strand of the gene being codons preferred for the expression of microbial genomes.

Further there is provided a recombinant microbial cloning vehicle comprising a first restriction endonuclease recognition site, a structural gene coding for the expression of a polypeptide and a second restriction endonuclease site, in which the gene codes for the expression of a mammalian polypeptide or intermediate therefor, at least a majority of the codons of the structural gene being codons preferred for the expresssion of microbial genomes.

The following figures illustrate one context in which preferred embodiments of the process find application, i.e., expression of the hormone somatostatin by bacterial transformants containing recombinant plasmids.

Figure 10.2(a) is a schematic outline of the process: the gene for somatostatin, made by chemical DNA synthesis, is fused to the *E. coli* β-galactosidase gene on the plasmid pBR322.

Figure 10.2: Recombinant DNA Cloning Vehicle

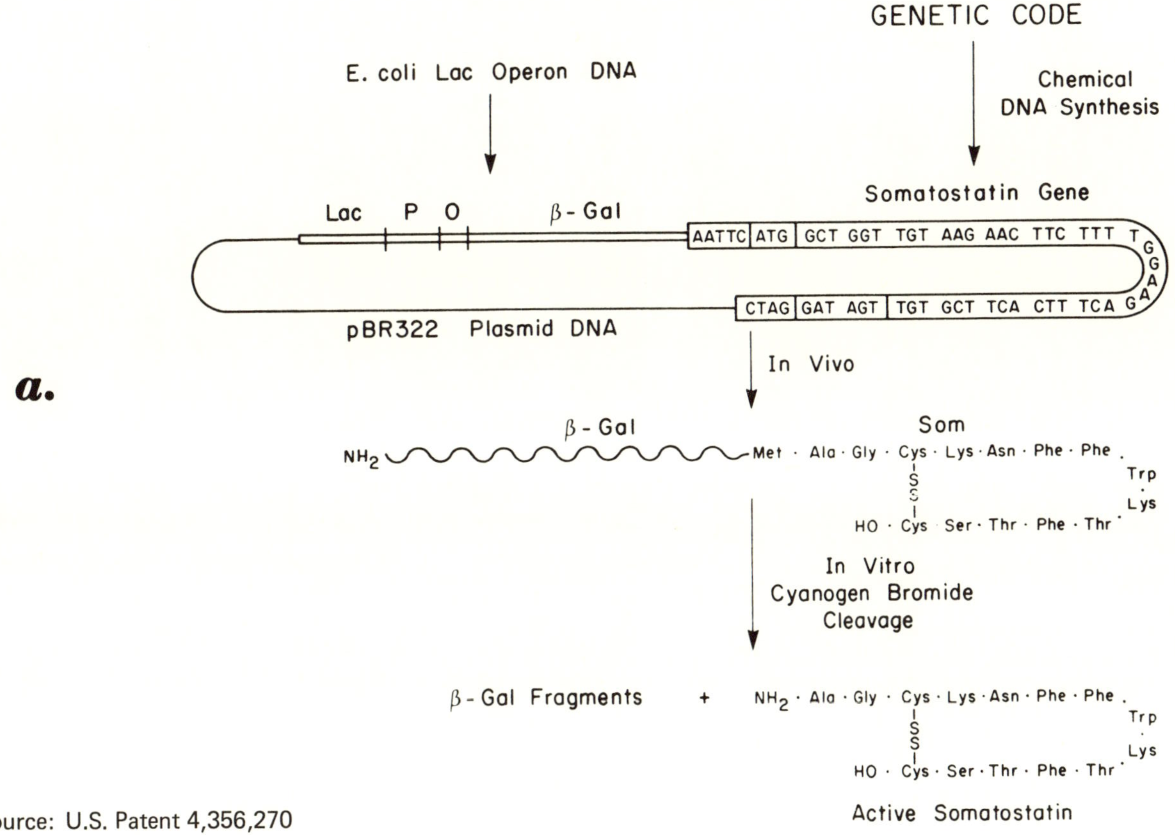

Source: U.S. Patent 4,356,270

Figure 10.2: (continued)

b.

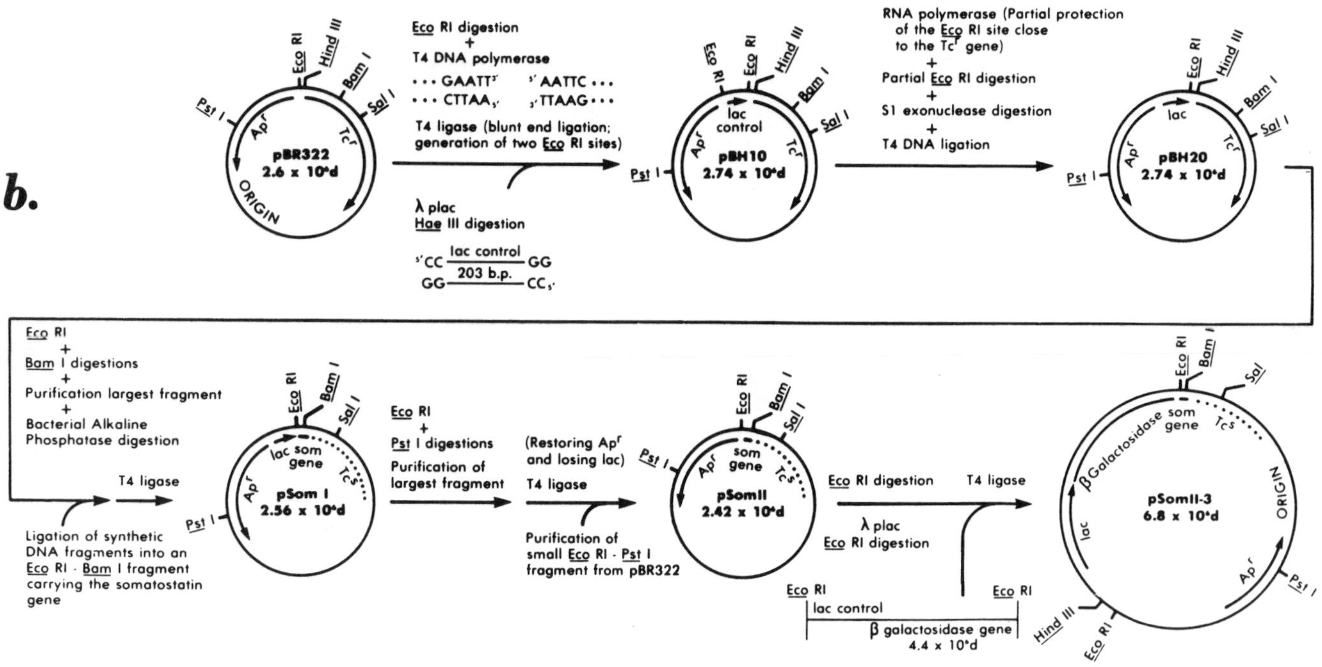

(a) Schematic outline of the process.
(b) Flow chart for construction of recombinant plasmid.

Source: U.S. Patent 4,356,270

After the transformation into *E. coli*, the recombinant plasmid directs the synthesis of a precursor protein which can be specifically cleaved in vitro at methionine residues by cyanogen bromide to yield active mammalian polypeptide hormone. A, T, C and G denote the characteristic bases (respectively, adenine, thymine, cytosine and guanine) of the deoxyribonucleotides in the coding strand of the somatostatin gene.

Figure 10.2(b) is a flow chart for the construction of a recombinant plasmid (e.g., pSOM11-3) capable of expressing a somatostatin ("SOM")-containing protein, beginning with the parental plasmid pBR322. In Figure 10.2(b) the approximate molecular weight of each plasmid is stated in daltons ("d"). Ap^r and Tc^r respectively denote genes for ampicillin and tetracycline resistance, while Tc^s denotes tetracycline susceptibility resulting from excision of a portion of the Tc^r gene. The relative positions of various restriction endonuclease specific cleavage sites on the plasmids are depicted (e.g., Eco RI, Bam I, etc.).

PRODUCTION AND/OR USE OF PLASMID VECTORS

Hybrid Plasmid pUC1021

The development of plasmid vectors useful for recombinant DNA genetics among microorganisms has become a practical technique. *F. Reusser; U.S. Patent 4,332,898; June 1, 1982; assigned to The Upjohn Company* has constructed a new plasmid pUC1021 (*E. coli* HB101; NRRL B-12167) from *Bacillus megaterium* (NRRL B-12165) chromosomal DNA, and plasmid pBR322, a well-known plasmid which can be obtained from *E. coli* RR1 (NRRL B-12014), whose restriction endonuclease map has been published by J.G. Sutcliff in *Nucleic Acids Research* 5 (1978), 2721-2728. Thus, hybrid plasmid pUC1021 contains a functional tet gene promoter region composed of *B. megaterium* DNA and pBR322 DNA.

The inserted *B. megaterium* DNA was mapped for restriction enzyme sites, and the DNA base sequence was determined for the first 50 to 60 bases from each end of the insert. The sequencing results show that the promoter region is purine-rich and contains three A-clusters.

Figure 10.3 shows the restriction enzyme map for the *B. megaterium* pBR322 promoter region. The inserted *B. megaterium* segment is located between the two Hind III sites.

The DNA insert is further characterized by (1) the following partial nucleotide sequence which is located between the two Hind III sites of pUC1021:

CCGGGAAGTGAAGTCAGAGAAAAGGAAAAGTGCGAGAGGGAAGGAAAAAGAGGGGA

(2) being 350 base pairs in length; and (3) having a single Mbo II restriction site.

The process for preparing hybrid plasmid pUC1021 comprises: (a) linearizing plasmid pBR322 with Hind III to obtain linear plasmid DNA; (b) obtaining chromosomal DNA from *Bacillus megaterium* NRRL B-12165; (c) digesting this chromosomal DNA with Hind III; and (d) ligating the digested linear plasmid DNA from pBR322 and the digested chromosomal DNA from *Bacillus megaterium,* NRRL B-12165, to obtain hybrid plasmid pUC1021.

Figure 10.3: Restriction Map for *Bacillus megaterium* pBR322 Promoter Region

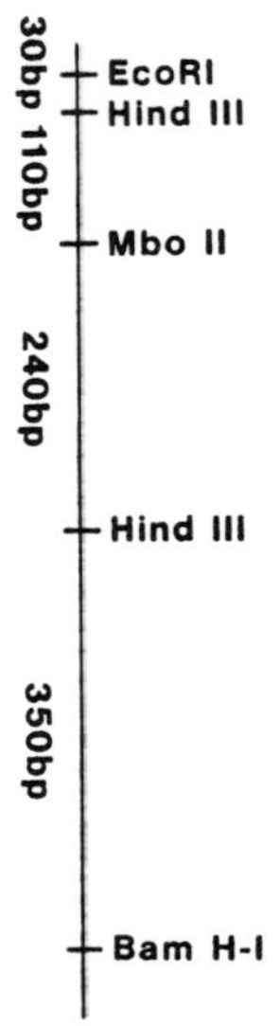

Source: U.S. Patent 4,332,898

pUC1021 can be used to create recombinant plasmids which can be introduced into host bacteria by transformation. The process of creating recombinant plasmids is well known in the art.

Such a process comprises cleaving the isolated vector plasmid, e.g., pUC1021, at specific sites by means of restriction endonucleases, e.g., Mbo II, Hpa II or Hae III. The plasmid, which is a circular DNA molecule, is thus converted into linear DNA molecules by the enzymes which cut the two DNA strands at a specific site. Other nonvector DNA is similarly cleaved with the same enzyme(s). Upon mixing the linear vector or portions thereof and nonvector DNA's, their single-stranded or blunt ends can pair with each other, and in the presence of a second enzyme known as polynucleotide ligase can be covalently joined to form a single circle of DNA.

The above procedure also can be used to insert a length of DNA from any plant or animal into pUC1021. For example, the DNA which codes for ribosomal RNA in the frog can be mixed with pUC1021 DNA that has been cleaved. The resulting circular DNA molecules consist of plasmid pUC1021 with an inserted length of frog rDNA.

The recombinant plasmids containing a desired genetic element, prepared by using pUC1021, can be introduced into a host organism for expression. Examples of valuable genes which can be inserted into host organisms by the abovedescribed process are genes coding for somatostatin, rat proinsulin, and proteases.

Cointegrate Plasmids from Plasmids of Streptomyces and Escherichia

J.J. Manis and S.K. Highlander; U.S. Patent 4,332,900; June 1, 1982; assigned to The Upjohn Company have obtained the new cointegrate plasmids–pUC 1019 (*E. coli* CSH50; NRRL B-12252) and pUC1020 (*E. coli* CSH50; NRRL B-12253) by the in vitro linkage of the approximately 4.2 kilobase (kb) BclI restriction endonuclease fragment of the *Streptomyces espinosus* plasmid pUC6 (from *S. espinosus* biotype 23724a; NRRL B-11439) into the BamHI endonuclease site of the plasmid pBR322 obtained from *E. coli* RR1 (NRRL B-12014). They have also constructed the plasmid pUC1024 (*E. coli* RR1, NRRL B-12254).

Plasmids pUC1019 and pUC1020 constitute the insertion of this BclI fragment in the two possible orientations in pBR322. In a like manner, the 4.1 and 0.9 kb Bc1I restriction fragments of pUC6 can be recombined with pBR322. Plasmid pUC1024 is derived from pUC1019 by in vitro deletion of DNA sequences between endonuclease PvuII sites. Plasmid pUC1024 is ca 3.5 kb smaller than pUC1019. The plasmids, advantageously, are transformed into a suitable host, e.g., *E. coli*.

Plasmids pUC1019 and pUC1020 are recombinant DNA molecules consisting of the entire genome of the small (ca 2.6 x 10^6 daltons) high copy number (ca 30/chromosome) *E. coli* plasmid pBR322 and ca 46% (4.2 kb) of the genome of the small (ca 6.0 x 10^6 daltons) high copy number (ca 30/chromosome) *S. espinosus* plasmid pUC6. Plasmids pUC1019 and pUC1020 contain single sites for the restriction endonucleases EcoRI, PstI, HindIII and XhoI. The XhoI site will also allow the cloning of SalI restriction fragments. Hence plasmids pUC1019 and pUC1020 represent DNA molecules which may function as vectors in both *E. coli* and Streptomyces and represent valuable intermediates for the development of better host-vector systems.

Plasmids pUC1019, pUC1020 and pUC1024 are characterized by the restriction maps shown in Figure 10.4(a), (b) and (c).

Figure 10.4: Restriction Endonuclease Cleavage Maps

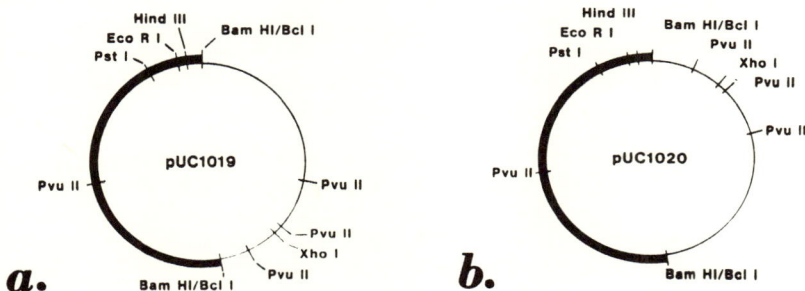

(continued)

Figure 10.4: (continued)

c.

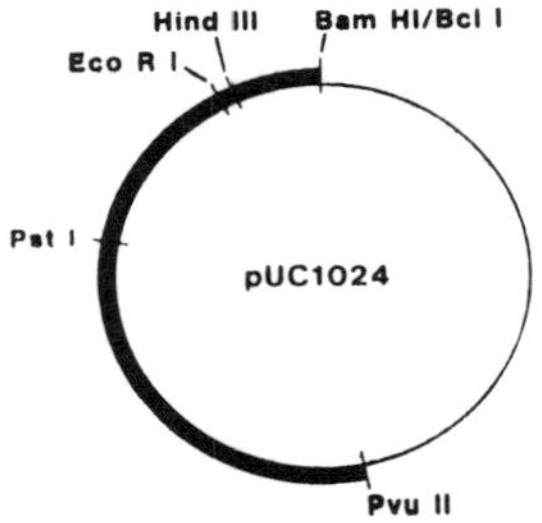

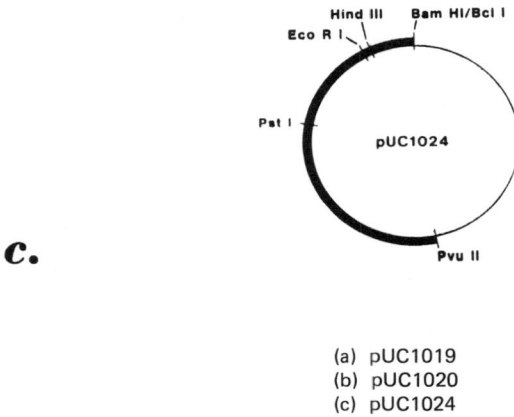

(a) pUC1019
(b) pUC1020
(c) pUC1024

Source: U.S. Patent 4,332,900

Plasmid pUC6 can be obtained from NRRL B-11439 by growing the culture on a suitable medium, fragmenting the mycelia, incubating the fragmented mycelia, harvesting the culture after a suitable time, and then lysing the mycelia. From this lysate it is possible to isolate essentially pure pUC6.

The restriction endonuclease cleavage maps in Figure 10.4 are constructed on the basis of plasmids pUC1019 and pUC1020 having a molecular weight of ca 5.7 megadaltons or a molecular weight of ca 8.6 kb. Plasmid pUC1024 has a molecular length of ca 5.1 kb. It retains the locus conferring genetic instability to plasmid pUC1019 in *E. coli* hosts. Plasmid pUC1024 has single restriction sites for the endonucleases EcoRI, HindIII, PstI and PvuII. The PvuII site will allow the cloning of blunt-ended DNA fragments from a wide range of restriction enzyme digests or from other properly prepared DNA fragments. The restriction endonuclease abbreviations are as follows: (1) BglII is an enzyme from *Bacillus globigii*; (2) BclII is an enzyme from *Bacillus caldolyticus*; (3) PvuII is an enzyme from *Proteus vulgaris*; and (4) XhoI is an enzyme from *Xanthomonas holicola.*

pUC1019, pUC1020 and pUC1024 can be used to create recombinant plasmids which can be introduced into host microbes by transformation as described in the previous patent.

The usefulness of plasmids pUC1019, pUC1020 and pUC1024 is derived from their capacity to function as plasmid vectors in industrially important microorganisms, e.g., Streptomyces. Also, pUC1019, pUC1020 and pUC1024 are especially useful because of their single restriction sites. Hence, cloning of genetic information from Streptomyces into pUC1019, pUC1020 or pUC1024 provides a means of increasing the production of commercially important products from these organisms, e.g., antibiotics.

J.J. Manis and S.K. Highlander; U.S. Patent 4,338,400; July 6, 1982; assigned to

The Upjohn Company have also obtained two new plasmids—pUC1026 (*E. coli* RR1; NRRL B-12329) and pUC1027 (*E. coli* RR1; NRRL B-12330) from the same parents as the plasmids described in the previous patent—the *E. coli* plasmid pBR322 to portions of the *S. espinosus* plasmid pUC6. These plasmids lack a locus contained in the pUC6 genome which causes the genetic instability of other dual vector pUC6:pBR322 recombinant plasmids in *E. coli* hosts.

The process for preparing recombinant plasmid pUC1026 comprises: (a) digesting plasmid pBR322 with BamHI to obtain linear plasmid DNA; (b) purifying the linear plasmid DNA; (c) digesting plasmid pUC6 with BglII and XhoI and purifying the resulting largest DNA fragment; (d) precipitating the purified DNAs from steps (b) and (c) and ligating to obtain ligated DNA; (e) deproteinizing the ligated sample and then digesting with BclI to obtain linear DNA; and (f) deproteinizing the linear DNA and ligating to obtain recombinant plasmid pUC1026.

The process for preparing recombinant plasmid pUC1027 comprises: (a) transferring plasmid pUC1013 to *E. coli* (GM 119), NRRL B-12328; (b) isolating pUC-1013 from *E. coli* (GM 119), NRRL B-12328; (c) digesting pUC1013 with BclI and purifying the resulting largest DNA fragment; and (d) ligating the purified DNA fragment to obtain recombinant plasmid pUC1027.

Figure 10.5(a) shows the restriction endonuclease cleavage map of pUC1026 and Figure 10.5(b) shows the map for pUC1027.

Figure 10.5: Restriction Endonuclease Cleavage Maps

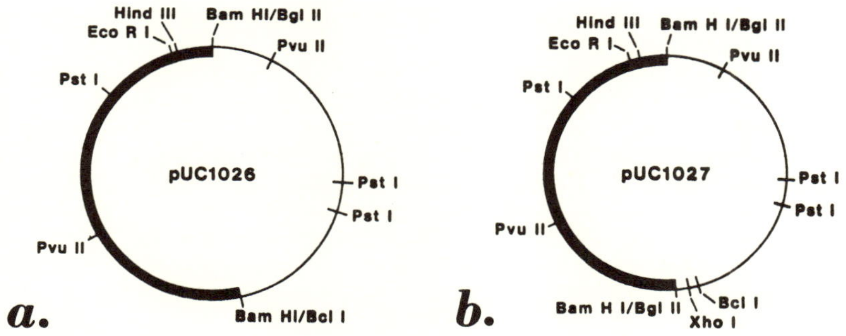

(a) pUC1026
(b) pUC1027

Source: U.S. Patent 4,338,400

The plasmids pUC1026 and pUC1027 are unique in that they do not contain the pUC6 sequences mapping between 3.8 and 6.2 kb on the pUC6 map. This region of the pUC6 genome contains a genetic locus which causes pUC6:pBR322 recombinant plasmids to be unstable in *E. coli* hosts. The genomic composition of several pBR322:pUC6 recombinant plasmids, showing what pUC6 genome they contain was ascertained. The stability of these plasmids in *E. coli* hosts was tabulated. It was found that only those plasmids containing the 3.8 to 6.2 kb region of pUC6 were unstable. It was therefore realized that the deletion or omission of this region of pUC6 from recombinant plasmids makes them more stable in hosts which are naive (i.e., have never contained) the plasmid vector.

The identification of such a plasmid instability locus is an important factor in the development of vector systems using Streptomycete plasmids alone or in combination with plasmids and/or viruses from other organisms. Though the vector system exemplified herein is specifically to a dual vector, i.e., pUC6:pUC322, it is probable that single plasmid vectors exhibiting instability in a host can likewise be stabilized by elimination of an instability locus. Such elimination can be by deletion of the particular locus, mutation, and the like.

More cointegrate plasmids have been constructed by *J.J. Manis and S.K. Highlander; U.S. Patent 4,340,674; July 20, 1982; assigned to The Upjohn Company* from the in vitro covalent linkage of the *E. coli* plasmid pBR322 and the *S. espinosus* plasmid pUC6. The procedure for constructing these plasmids comprises cutting pBR-322 and pUC6 to give linear plasmid DNA, and ligating the T_4 DNA ligase to give the cointegrates. Specifically, plasmid pUC6 is inserted via its BglII restriction site into the BamHI site of the *E. coli* vector plasmid pBR322. Because of the two possible orientations of insertion of pUC6 into pBR322, two different recombinant plasmids, i.e., pUC1012 and pUC1013, are constructed and isolated. The plasmid cointegrates, advantageously, are transferred into a suitable host, e.g., *E. coli*.

Plasmids pUC1015 and pUC1016 are obtained by restructuring plasmids of pUC-1012 and pUC1013, respectively. This restructuring is achieved by digestion of the parent plasmids with restriction endonuclease XhoI to obtain a digest of plasmid DNA, and ligating the largest DNA fragment after purification by agarose gel. These plasmids are transformed into *E. coli* CSH50 by use of well-known procedures.

pUC1015 is derived from plasmid pUC1012 by in vitro deletion of ca 2.0 kb of DNA. pUC1016 is derived from plasmid pUC1013 by deletion of ca 2.0 kb of DNA. The smaller plasmids pUC1015 and pUC1016 contain single restriction sites for the restriction enzymes PstI, EcoRI, HindIII, XhoI and BclI. Plasmid pUC-1022 is derived from plasmid pUC1012 by the in vitro deletion of approximately 9.1 kb of the pUC1012 DNA sequences. The deleted sequences are between PvuII restriction endonuclease sites in pUC1012, thus giving pUC1022 a single PvuII site. pUC1022 confers ampicillin resistance upon its host and has a molecular weight of ca 2.8 x 10^6 daltons. Plasmid pUC1023 is similarly derived from plasmid pUC1013 by in vitro deletion between PvuII restriction sites. Plasmid pUC-1023 has a single PvuII cleavage site, confers ampicillin resistance upon its host, and has a molecular weight of ca 2.2 x 10^6 daltons.

Restriction maps characterizing these plasmids are given in the patent. Northern Regional Research Laboratory accession numbers for the plasmids from this process are as follows: NRRL B-12106, *E. coli* CSH50 (pUC1012); NRRL B-12107, *E. coli* CSH50 (pUC1013); NRRL B-12150, *E. coli* CSH50 (pUC1015); NRRL B-12151, *E. coli* CSH50 (pUC1016), NRRL B-12183, *E. coli* RR1 (pUC-1022); and NRRL B-12184, *E. coli* RR1 (pUC1023).

These plasmids are useful as cloning vehicles in recombinant DNA work. For example, using DNA methodology, a desired gene, e.g., the insulin gene, can be inserted into the plasmids and the resulting plasmids can then be transformed into a suitable host microbe which, upon culturing, produces the desired insulin.

Plasmid Having an Insertion Site for a Eukaryotic DNA Fragment

N.H. Carey, J.S. Emtage, W.C.A. Tacon and R.A. Hallewell; U.S. Patent 4,349,629; September 14, 1982; assigned to G.D. Searle & Company have developed a method for preparing plasmid vectors based upon the provision of an insertion site, particularly a Hind III site, for a chosen eukaryotic DNA fragment. This insertion site is adjacent to a bacterial promoter, e.g., a trp promoter, so that the transcription and translation of the DNA fragment are controlled by the promoter. These plasmids are generally also provided in the region immediately following the insertion site, with the gene for tetracycline resistance, so that after insertion of the chosen DNA at the insertion site, transcription and translation of the inserted DNA controlled by the promoter may conveniently be confirmed, since, when it has occurred, tetracycline resistance is preserved, and, in most cases, when it has not, tetracycline resistance is destroyed.

One embodiment relates to a plasmid having an insertion site for a eukaryotic DNA fragment adjacent to a bacterial promoter and downstream from a prokaryotic ribosome binding site and initiator codon such that the bacterial promoter controls transciption and translation of an inserted DNA fragment. Such a plasmid may be prepared by a process which comprises cloning a bacterial promoter into an isolated plasmid, determining the position of an intended insertion site and providing the insertion site, if it is not already present. It may also be produced using such a plasmid by digesting cellular DNA with a specific restriction enzyme, isolating a promoter-containing fragment and cloning the fragment into a suitable bacterial plasmid.

Another embodiment relates to such a plasmid having inserted therein at the insertion site a eukaryotic DNA fragment, which may be inserted into the prepared plasmid by known techniques.

For purposes of exemplification, one of the plasmids is that having the following characteristics: a molecular length of 6,750 base pairs (bp), a Hind III site, a Bam HI site at 346 bp from the Hind III site, a Sal I site at 275 bp from the Bam HI site, a Hpa I site at 4230 bp from the Sal I site and 1950 bp from the Hind III site, the plasmid having the trp promoter, the E gene and part of the D gene in the portion of 1950 bp extending between the Hpa I site and the Hind III site and the gene for ampicillin resistance in the portion of 4800 bp between the Hind III site and the Hpa I site. This plasmid, referred to herein as pEH5, may be produced as follows:

(a) Isolation of the Hind III trpE Fragment – This fragment may be obtained either by Hind III digestion of DNA from most strains of *E. coli,* e.g., *E. coli* strain W3110, or by Hind III digestion of the plasmid, pRH1/trpE. pRH1 was itself obtained by Eco RI digestion of *E. coli* plasmids pDS1118 and pML2, ligation of the thus-obtained fragments and selection for ampicillin and kanamycin resistance. Hind III digestion of pRH1 and ligation with Hind III digested DNA from strain *E. coli* W3110, followed by transformation into strain W3100 trp oE∇1 and culture in the absence of tryptophan, produced pRH1/trpE.

(b) Cloning the Hind III trpE Fragment – The Hind III trpE fragment, obtained as described above, was covalently ligated to a fragment obtained by Hind III restriction of the plasmid pBR322. The ligated DNA was used to transform *E. coli* W3110 trp oE∇1 and colonies selected for ampicillin resistance and tryptophan complementation. The plasmids obtained were found to be of two types,

depending on the orientation of the fragments during ligation; these two plasmids are referred to herein as pEH3 and pEH4. The plasmid pEH3 was selected on the basis of its tetracycline resistance and used for the next stage of the production of the present plasmids. The plasmid pEH3 has the following characteristics.

A molecular length of 9866 bp; a Hind III site; a Bam I site 346 bp from the Hind III; a Sal I site 275 bp from the Bam I; and Eco RI site 3709 bp from the Sal 1; a second Hind III site 31 bp from the Eco RI; a Hpa I site 305 bp from the second Hind III; a second Hpa I site 3250 bp from the first Hpa I and 1950 bp from the first Hind III, the plasmid having the trp promoter, the E gene and part of the D gene in the portion of 1950 bp between the second Hpa I and the first Hind III sites; the gene for tetracycline resistance immediately following the first Hind III site and the gene for ampicillin resistance in the portion 3709 bp between the Sal I and the Eco RI sites.

(c) Deletion of the Hind III Site Proximal to the Eco RI Site in pEH3 – pEH3 was digested with Eco RI to produce linear molecules which were digested with exonuclease III (EC 3.1.4.27) and then with S1 nuclease (EC 3.1.4-) to remove the 5'-protruding tails and to produce blunt ends. Then, the linear molecules were ligated with T4-induced DNA ligase (EC 6.5.1.1) and the thus-obtained plasmid was used to transform the trp oE∇1 strain and selected for trp complementation and ampicillin resistance to obtain the plasmid referred to as pEH5.

As mentioned above, a particular utility of these plasmids is that they have a clearly defined site convenient for the introduction of a desired eukaryotic DNA fragment and that the structure of the plasmids is such that the transcription of the introduced DNA is controlled by the promoter, particularly, a trp promoter, associated with the defined introduction site, for example, the Hind III site of the plasmid pEH5. This plasmid therefore represents a valuable final intermediate from which may be produced plasmids having inserted at the insertion site, for example the Hind III site, genetic information in the form of suitably modified DNA designed to produce a desired polypeptide product.

USE OF OTHER GENETIC ENGINEERING PROCEDURES

Production of Specific Proteins Unencumbered by Extraneous Peptides

M. Ptashne, G.D. Lauer, T.M. Roberts and K.C. Backman; U.S. Patent 4,332,892; June 1, 1982; assigned to President and Fellows of Harvard College have provided a method of producing native, unfused prokaryotic or eukaryotic protein in bacteria which comprises inserting into a bacterial plasmid a gene for a prokaryotic or eukaryotic protein and a portable promoter consisting of a DNA fragment containing a transcription initiation site recognized by RNA polymerase and containing no protein translational start site. This promoter is inserted ahead of a protein translational start site of the gene to form a fused gene having a hybrid ribosome binding site. The plasmid is then inserted into the bacteria to transform it with the plasmid-containing fused gene, and the transformed bacteria is then cultured to produce the prokaryotic or eukaryotic protein.

The process utilizes nucleases, restriction enzymes, and DNA ligase to position a portable promoter consisting of a DNA fragment containing a transcription site but no translation initiation site near the beginning of the gene which codes for the desired protein to form a hybrid ribosomal binding site. The protein produced by the bacterium from this hybrid is the native derivative of the implanted gene. It has been found that the endonuclease digestion product of the *E. coli* lac operon, a fragment of DNA which contains a transcription initiation site but no translational start site, has the required properties to function here as a portable promoter, being transcribed at high efficiency by bacterial RNA polymerase. The mRNA produced contains a Shine-Dalgarno (S-D) sequence but it does not include the AUG or GUG required for translation initiation. However, in accordance with the process, a hybrid ribosomal binding site is formed consisting of the S-D sequence and initiator from the lac operon and the ATG sequence of the gene, and such a fused gene is translated and transcribed efficiently. Using the enzymes exonuclease III and S1, the promoter may be put at any desired position in front of the translational start site of the gene in order to obtain optimum production of protein.

Since the promoter can be inserted at a restriction site ahead of the translational start site of the gene, the gene can first be cut at the restriction site, the desired number of base pairs and any single stranded tails can be removed by treating with nucleases for the appropriate time period, and religating. The following specific example will illustrate more fully the nature of the method.

Example: A rabbit β-globin gene was first cloned into the Hin III site of pBR322, a plasmid of the *E. coli* bacteria, via restriction enzyme cuts of the initial DNA, reconstitution of the gene by T4 ligase, insertion of the reconstituted gene into the Hin III site using chemically synthesized Hin III linkers, and religating with DNA ligase.

The Hin III cut at the carboxyl end of the cloned gene was removed by partially digesting with Hin III, filling in the resulting Hin III "sticky ends" with *E. coli* DNA polymerase I, and religating with T4 ligase. This left in the resulting plasmid a single Hin III cut 25 base pairs ahead of the amino terminus of the globin gene.

Differing number of the 25 base pairs between the Hin III cut and the ATG signalling the start point of translation were removed from different samples of the cloned gene as follows: the plasmid was cut with Hin III, resected for various times from 0.5 to 10 minutes with Exo III, then treated with SI to remove single-stranded tails.

The portable promoter of the lac operon, an R1-Alu restriction fragment of *E. coli* DNA, was then inserted by treating each sample of the plasmid with R1 which cuts at a unique site some 30 base pairs upstream from the Hin III site, and the portable promoter was inserted into the plasmid backbone at this site. This requires one "sticky end" and one "flush" end, both of which are ligated by the same treatment with ligase.

Colonies of *E. coli* each containing one of these resulting plasmids were then screened for β-globin production using RIA-screening techniques to identify the one or more producing β-globin.

The globin gene in the above construction can be any gene coding for prokaryotic or eukaryotic proteins, and any other unique restriction site can be employed in place of the Hin III site. If the restriction site is located inconveniently far from the beginning of the gene, it may be moved (for example, a Hin III site may be moved by opening the plasmid with Hin III, digesting with Exo III and S1, then religating the resulting plasmid in the presence of excess Hin III linkers). Any suitable restriction site can be employed for insertion of the portable promoter in place of the R1 site (e.g., Pst, BAM, or Sal I). Finally, it should be emphasized that the most difficult step, the cloning of the gene into the plasmid, is done once and left unchanged. The promoter fragment will confer its constitutive expression on the cell so it is easy to screen for the intact promoters.

Procedure for Making and Using a Fused Gene

T.J. Silhavy, H.A. Shuman, J. Beckwith and M. Schwartz; U.S. Patent 4,336,336; June 22, 1982; assigned to President and Fellows of Harvard College have devised a method for making a fused gene. In this procedure a bacterial gene which codes for a bacterial cytoplasmic protein is fused to the noncytoplasmic directing portion of a gene which codes for a noncytoplasmic protein, that is, a protein which is normally transported to the bacterial outer membrane, the periplasmic space between the cytoplasmic and outer membranes, or outside the bacterial cell, so that a hybrid protein is produced. The cytoplasmic protein fused to the noncytoplasmic carrier protein is thereby transported to, or beyond the cell surface, or to the periplasmic space, thus facilitating its collection and purification, or enabling the bacteria to be used as an immunogen. After the gene fusion described herein is accomplished, the fused gene can be isolated and a gene coding for any protein can be inserted into it using conventional techniques. This fused gene can then be used to infect, transform, or transduce other bacterial cultures, so that they will produce hybrid proteins which are carried to the cell surface.

This process can be applied to any bacteria including Gram-positive bacteria, whose excreted proteins can act as carriers, and, even more usefully, to Gram-negative bacteria, whose cell surface and periplasmic space proteins can act as carriers.

The genes which produce the noncytoplasmic carrier proteins can include any genes which code for proteins which travel to the periplasmic space or the cell surface or which are excreted by the cell. The genes which code for the cytoplasmic proteins include any genes for cytoplasmic proteins which can be fused to genes for carrier proteins, and which themselves produce useful proteins, or into which can be inserted, by conventional methods, genes coding for useful proteins. These useful proteins include eukaryotic cell proteins such as insulin, peptide hormones such as somatostatin, and specific viral and bacterial antigens.

When ordinarily produced in *E. coli,* for example, the cytoplasmic and useful proteins previously mentioned remain in the bacterial cytoplasm, making purification from the bulk of *E. coli* proteins a difficult problem. Also, the useful foreign proteins can be degraded by proteases in the cytoplasm, so that they may not be recovered at all.

However, if a gene coding for a useful protein in inserted into the gene for a cytoplasmic protein, after the gene for the cytoplasmic protein has been fused to a gene coding for a carrier protein according to the method of this process, collection and purification can be greatly facilitated, and the degradation problem can be diminished as well.

The procedure has additional advantages when the useful protein is a bacterial or viral antigen normally produced by a pathogenic organism. For example, if the host bacterium is *E. coli,* when the gene for the antigen of a pathogen is inserted into the gene for a cytoplasmic protein which has been fused to the gene for a carrier protein according to this method, there is produced, a hybrid protein, part of which is the pathogenic antigen, which can be transported to the cell surface. The whole killed *E. coli* can thus be used as an immunogen; the need for purification is eliminated. Also, the immunological effectiveness of these cells is likely to be greater than that of purified antigen because insoluble antigens are generally more effective than soluble ones.

The pathogenic antigens on the *E. coli* cell surface can be particularly useful in localized immunization against diarrhetic diseases of the lower gut. The only protective immunity to these diseases is mediated by secretory antibodies (IgA) produced in the vicinity of the lower gut. Administration of soluble antigens is ineffective in eliciting the appropriate secretory antibody response, but oral administration of killed *E. coli* with the antigen on the cell surface can be effective. *E. coli* is a natural inhabitant of the gut and is itself not harmful; neither is the surface antigen from the pathogen itself pathogenic. The *E. coli* with the surface antigen would, however, elicit the appropriate protective localized secretory antibody response.

It can be seen that the fused gene, even without another gene inserted into it, is itself useful in three ways, implicit in the aforementioned description: First, the bacterial cytoplasmic protein coded for may itself be a useful protein and therefore advantageously exported to the bacterial cell surface. Second, the fused gene is the vehicle into which a gene for any other useful protein which is advantageously exported is inserted. Third, the isolated fusion, which is sometimes incorporated into a phage, can be used to infect, transform, or transduce other cultures. The patent gives a detailed example of the method of preparing a fused gene.

Synthesizing a Mature Protein Within a Bacterial Host

W. Gilbert and K. Talmadge; U.S. Patent 4,338,397; July 6, 1982; assigned to President and Fellows of Harvard College have devised a method of synthesizing within a bacterial host, and secreting through the membrane of the host, a selected protein or polypeptide, such as a eukaryotic cell protein, e.g., proinsulin, serum albumin, human growth hormone, parathyroid hormone and interferon. This process particularly relates to a method of obtaining from a bacterial host a mature protein or polypeptide and thereby avoiding the need to treat the protein or polypeptide further to remove the signal sequence of other chemical substituents, such as an f-met (i.e., the formaldehyde group on its first methionine group), which are present on its precursor, as synthesized by the bacterial host. The method comprises:

(a) cleaving a cloning vehicle to form a cleavage site after a promoter of either (1) a bacterial or phage gene within the cloning vehicle or (2) a DNA fragment of the bacterial or phage gene;

(b) forming a hybrid gene by inserting into the cleavage site a nonbacterial DNA fragment which codes for a precursor of the selected protein or polypeptide, including the signal sequence of the selected protein or polypeptide;

(c) transforming the host with the cloning vehicle; and then

(d) culturing the transformed host to secrete the selected protein or polypeptide.

By this method, mature proteins or polypeptides can be produced, free of signal sequences or other chemical substituents. In this regard, the proteins or polypeptides can be recovered either from the periplasmic space of the bacterial host cell or from the medium in which the host is cultured, depending on the size of the proteins or polypeptides.

Any cloning vehicle that contains a bacterial or phage gene for a protein, a polypeptide or an RNA molecule or a DNA fragment of such a bacterial or phage gene, including a promoter of the bacterial or phage gene, and that can be cleaved to form a cleavage site after the promoter, can be utilized. Preferably, the cloning vehicle contains a bacterial or phage gene or a DNA fragment thereof which codes for an extracellular or periplasmic protein or polypeptide (i.e., a protein or polypeptide that is normally secreted from a host cell). Examples of such a gene include the gene for antibiotic resistance, e.g., the gene for penicillin resistance (i.e., penicillinase), the gene for chloramphenicol resistance, and the gene for tetracycline resistance, the gene for alkaline phosphatase, and the gene for bacterial ribonuclease.

Also, any DNA fragment of a nonbacterial and nonphase gene ("nonbacterial DNA fragment") can be inserted into the cleavage site in the cloning vehicle to form a hybrid gene, provided the fragment:

(1) contains its gene's complete structural DNA sequence for the selected protein or polypeptide and its gene's signal DNA sequence, so that the fragment codes for a precursor of the selected protein or polypeptide, including the signal sequence of the precursor; and

(2) contains its gene's translational start signal if the cleavage site in the cloning vehicle is before or within the translational start signal of the bacterial or phase gene or the DNA fragment.

Preferably, the nonbacterial DNA fragment utilized, contains its gene's complete DNA sequence which codes for a precursor of a protein or polypeptide that is normally secreted from a host cell. Among the nonbacterial DNA fragments which preferably are utilized are those which code for a precursor of a eukaryotic cell protein, such as preproinsulin, preserum albumin, prehuman growth hormone, preparathyroid hormone and preinterferon.

The specific location of the cleavage site in the cloning vehicle into which the nonbacterial DNA fragment is inserted is not critical for making the selected mature protein or polypeptide. However, it is particularly preferred that the cleavage site be within or no more than approximately 40 nucleotides after the translational start signal of the bacterial or phage gene or the DNA fragment thereof—regardless of whether the bacterial or phage gene or the DNA fragment thereof codes for a normally secreted or normally nonsecreted protein or polypeptide.

The nonbacterial DNA fragment is inserted into the cleavage site, after the promoter of either the bacterial or phage gene or the DNA fragment thereof, so that there is, in order, in the resulting hybrid gene: the promoter of the bacterial or phage gene or the DNA fragment thereof; a translational start signal (either from the bacterial or phage gene or the DNA fragment thereof or from the nonbacterial DNA fragment); the signal DNA sequence of the nonbacterial DNA fragment; and the structural DNA sequence of the nonbacterial DNA fragment.

Preferably, for most efficient expression of the selected protein or polypeptide, a ribosome binding site (either from the bacterial or phage gene or the DNA fragment thereof or from the nonbacterial DNA fragment) is provided in the hybrid gene between the promoter and the translational start signal. Moreover, if the nonbacterial DNA fragment is inserted after the translational start signal of a bacterial or phase gene or a DNA fragment thereof, the reading frame of the nonbacterial DNA fragment should be located in the reading frame defined by the translational start signal of the bacterial or phage gene or the DNA fragment thereof.

A selected protein or polypeptide can be secreted in high yields from a bacterial host, transformed with a cloning vehicle of this process, without there being a hydrophobic leader sequence or any other chemical substituents, such as a f-met, on the secreted protein or polypeptide. The hydrophobic leader sequence and any other chemical substituents which were present on the precursor of the selected protein or polypeptide as synthesized within the host are cleaved from the selected protein or polypeptide. It is believed that this happens during secretion from the host.

Detailed examples of the synthesis are given in the patent, using starting materials as prepared by the following papers (references to be made by number): (1) Bolivar et al, *Gene* 2, 95-113 (1977); (2) Villa-Komaroff et al, *P.N.A.S.* 75, 3727-3731 (1978); (3) Johnsrud, *P.N.A.S.* 75, 5314-5318 (1978); (4) Boyer et al, *J. Mol. Biol.* 41, 459-472 (1969); (5) Bedbrook et al, *Cell* 9, 707-716 (1976); (6) Reiner, *J. Bact.* 97, 1522-1523 (1969).

(a) The cloning vehicles were derived from a starting cloning vehicle which was the small plasmid pBR322 as constructed and described by (1). The cloning vehicles could all be cleaved at a Pst I restriction site within the gene for *Escherichia coli (E. coli)* penicillinase.

(b) The nonbacterial DNA fragments were: a Pst-ended, poly-G tailed, cDNA copy of the DNA fragment of the gene for rat preproinsulin that was described by (2) and is hereinafter called "DNA fragment 19" [called "PI 19" by (2)]; and a derivative of DNA fragment 19. The nonbacterial DNA fragments encode all but the translational start signal (ATG) and the first two codons of the gene for rat preproinsulin.

Four well-known strains of *E. coli* K-12 were used as bacterial hosts: MM294, described by (3); HB101, described by (4); FMA 10 described by (5); and PR13 described by (6). The hosts were selected for the presence within them of a plasmid which codes for resistance to the antibiotic tetracycline, the gene for which resistance is also encoded on pBR322.

Monoclonal Antibody Specific for High MW Carcinoembryonic Antigen

The substance now known as carcinoembryonic antigen (CEA) was first described by Gold and Freeman in 1965. CEA was, at first, believed to be diagnostic for the presence of colorectal carcinoma but subsequently it was shown to be present in patients with a variety of other conditions.

H. Koprowski, K.F. Mitchell and Z. Steplewski; U.S. Patent 4,349,528; September 14, 1982; assigned to The Wistar Institute have produced a monoclonal hybrid cell (hybridoma) which produces an antibody with reactivity directed specifically to CEA having a molecular weight of about 180,000 daltons. The antibody produced by the hybrid cell and the method of testing for the CEA are also described. The hybridoma has been assigned the ATCC accession number CRL-8019.

The authors of this patent have determined that 180,000 molecular weight CEA has an antigen site not shared with other molecular weight components of CEA and have discovered the antibody which is specific for that antigen site. Moreover, the antibody of this process, while reactive for all colorectal carcinoma cells tested, has not been reactive with normal cells or cells of other types of tumor. The ability of this hybridoma antibody to react specifically with the 180,000 MW form of CEA provides a basis for an assay with exquisite specificity.

The general method used for production of antibody-secreting somatic cell hybrids has been previously described in *Proc. Natl. Acad. Sci. U.S.A.* 75, 3405-3409, (1978) and in Koprowski U.S. Patent 4,172,124.

Briefly, a BALB/c mouse was immunized with CEA excreting colorectal carcinoma cells from cell line SW1116 (obtained from the Scott and White Clinic, Temple, Texas—cells from that source are designated SW) and spleen cells from this mouse were fused with nonsecretor myeloma cells of cell line P3X63 Ag8, variant 653, [Kearney et al, *J. Immunol.* 123, 1548-1550 (1979)]. More particularly, BALB/c mice were immunized for a secondary response as described in *J. Exp. Med.* 121, 439-462, (1965). Three days prior to their sacrifice the mice were given a second injection intravenously of 1×10^6 immunizing cells (SW 1116). After sacrifice, a spleen cell suspension was prepared in the manner described in *Proc. Natl. Acad. Sci. U.S.A.* 74, 2985-2988 (1977). Immune splenocytes were fused with the variant 653 of the myeloma cell line P3X63 Ag8 in the presence of polyethylene glycol as also described in that reference, with one variation. Prior to fusion, cells were washed in Ca^{2+} and Mg^{2+} free medium, and the fusion was performed in the absence of Ca and Mg ions.

Hybrids were selected in a medium which contained hypoxanthine/aminopterin/thymidine (HAT selective medium) and fused cells were seeded in wells of 24 well tissue culture plates (Limbro FB-16-24TC). Approximately 20 days after fusion, single colonies were picked from each well and processed as described in *Proc. Natl. Acad. Sci. U.S.A.* 76, 1438-1442 (1979).

The hybridoma cells can be grown in vitro to produce these antibodies which are secreted by the cells into the medium. Any of a variety of known media can be used, and the choice of a medium is easily within the skill of the art. One particularly satisfactory medium is Dulbecco's minimal essential medium (MEM) containing about 20% fetal calf serum. Desirably, the medium chosen will be substantially free of antibodies so that purification steps to remove unwanted antibodies can be avoided.

In a typical procedure MEM+20% fetal calf serum is inoculated with the hybridoma cells and incubated at about 37°C. After a suitable period, for example, 4 or 5 days, the liquid containing secreted antibodies is separated from the cells by centrifugation or the like. If desired, the cells can be added to fresh media to produce additional antibodies. The separated liquid can be used as such for analytical tests. If desired, however, the liquid can be subjected to any of the variety of purification steps known to the art. Indeed, procedures are well known to isolate the antibody if such is desired.

As noted above, CEA is secreted by colorectal carcinoma cells. For diagnostic purposes, this antibody is added to blood serum of serum extract and the reactivity of the antibody with the antigen is measured. Any of a variety of known antigen assay techniques may be used to measure the reactivity.

The diagnostic method for detecting the presence of colorectal carcinoma is performed as follows:

(a) an aliquot of 180,000 MW ^{125}I-carcinoembryonic antigen is contacted with the antibody and the antibody is agglutinated;

(b) a second aliquot of 180,000 MW ^{125}I-carcinoembryonic antigen is mixed with blood serum and the mixture is contacted with the antibody and the antibody is agglutinated; and

(c) the radioactivity of the bound material of step (b) is compared with the radioactivity of the bound material step (a), a decrease in the radioactivity of bound material in step (b) indicating the presence of 180,000 carcinoembryonic antigen in the blood serum.

This antibody detected CEA from the colorectal carcinoma cells. Immune complexes were formed between the hybridoma antibody of this process and radioiodinated, solubilized, cleared, SW948 colorectal carcinoma cells (obtained from Scot and White Clinic). Complexes, recovered by binding to rabbit antimouse immunoglobulin coated SaCI were solubilized in SDS sample-buffer and subjected to SDS-PAGE analysis. A dense radioactive band was formed in the upper region of the gel that had a MW of 180,000 daltons. Similar bands of varying intensities were seen in preparations from other colorectal carcinoma cell lines. Control antibody, P3, did not precipitate the 180,000 MW antigen.

The antibody was tested for reactivity with carcinoma and other cell lines. Importantly, it bound to all eight colorectal cell lines tested but did not bind to any of 23 other cell lines tested. More particularly, it did not bind to any of 3 lung carcinoma cell lines, 1 breast carcinoma cell line, 10 melanoma cell lines, 2 astrocytoma cell lines, 2 sarcoma cell lines or to 5 fibroblasts.

Production of a Highly Active Bacteriolytic Protein

Recent developments in biochemistry have made available recombinant bacteria that synthesize enzymes and other nonbacterial proteins. These genetically engineered bacteria differ from those that hitherto occurred naturally by containing, along with their own genes, at least portions of genes inserted from other organisms which have instructions encoded in their DNA for synthesizing proteins having important biomedical applications. When the animal gene is properly integrated into the bacterial genome, the resulting recombinant bacteria produces the protein specified by the animal gene. Cultures of the recombinant bacteria are easily grown at low cost and hold out the promise of efficiently producing important proteins.

The useful proteins produced by such recombinant bacteria are typically trapped within the bacterial protoplasm and it is necessary to remove the cell wall surrounding the bacteria in order to free the useful protein. This has been done in the past by techniques such as sonification, freeze thawing or grinding techniques that physically destroy the bacterial cell wall. These techniques are both time consuming and nonspecific and they are likely to interact with and cause denaturation and/or inactivation of the useful proteins.

D. Hultmark, H. Steiner, T. Rasmuson and H.G. Boman; U.S. Patent 4,355,104; October 19, 1982; assigned to Kabigen AB, Sweden provide processes and substances to significantly increase the yields from genetically engineered bacteria by enabling an efficient lysis of the bacterial cell wall. These new immune proteins constitute a group herein termed P9. The P9 proteins also appear to be useful as pharmacological substances to control certain bacterial infections. The amino acid compositions of two such distinguishable proteins, termed P9 A and P9 B, are very similar. The molecular weight of the monomer of P9 A is believed to be 3564 and that of the dimer of P9 B is believed to be within 20% of about 7,000. Both P9 A and P9 B are heat stable, but differ in their amino acid contents of glutamic acid and methionine. A preferred procedure for producing protein P9 B has the following steps:

(1) treating an insect hemolymph to induce an immunity against *E. coli;*

(2) applying the immunized hemolymph to a first acid chromatographic column equilibrated with 0.1 M ammonium formate buffer at a pH of about 7.6 containing a detergent to reduce the adsorption of lipid material to the column;

(3) washing the column a first time with the formate buffer containing the detergent;

(4) washing the column a second time with 0.4 M ammonium acetate buffer at a pH about 5.1;

(5) eluting the first acid chromatographic column with a linear gradient of ammonium acetate buffer at a pH about 5.1;

(6) collecting and pooling a linear portion of the buffer gradient from about 0.60 to about 0.75 M;

(7) applying the pooled linear portion to a second acid chromatographic column equilibrated with 0.1 M ammonium formate;

(8) eluting the second acid chromatographic column with a linear gradient of ammonium formate at pH about 6.6 with a 0.2 to 0.6 M gradient;

(9) collecting and pooling a linear portion of the formate buffer gradient from about 0.4 to about 0.45 M; and

(10) freeze-drying the collected and pooled linear portion having a buffer gradient from about 0.4 to about 0.45 M.

Protein P9 A is made in a similar manner, with collection of a different linear portion of the buffer gradient.

A preferred utility of P9 proteins is in connection with the yield of useful proteins from cultures of recombinant *E. coli* bacteria. The use of P9 proteins to lyse a broth of *E. coli* that produce human growth hormone yields on the order of 150 mg per liter of broth per 5-hour interval. This represents a 50 to 85% gain in yield over prior methods. This increased efficiency is possibly tied to the fact that the P9 proteins nondestructively disassemble both the outer and the inner membranes of Gram-negative bacteria.

Another intended use of P9 proteins is as a pharmacological antibiotic for those strains for which it has specific potent effect. In particular, P9 protein is observed to be potent against bacterial strains which are streptomycin and penicillin resistant. In addition, there are no known antibodies for the P9 proteins indicating a wide acceptability for human and veterinary applications. One apparently useful application would be for surface infections because of the high activity against Pseudomonas.

SPECIFIC PRODUCTS MADE BY GENETIC ENGINEERING TECHNIQUES

Interferon

Interferon is a glycoprotein whose synthesis is induced in cultured cells principally by viruses or by natural or synthetic double stranded RNAs. This induction requires de novo macromolecular synthesis as indicated by its sensitivity to inhibitors of both RNA and protein synthesis. Interferon is believed to be secreted by the induced cell, whereby the secreted interferon interacts with other cells resulting in the establishment, maintenance and expression of an intracellular antiviral state.

The amount of interferon which is produced is exquisitely small, so that its isolation has been extremely elusive. The primary source of human interferon is from Helsinki, Finland, where partially purified human leukocyte interferon is obtained from blood given by blood donors. Because of the limited source of interferon, and the difficulties in purification and concentration, the cost of interferon is prohibitively high. In view of its antiviral nature and its acceptability by a mammalian host, the production of interferon in useful amounts holds great promise for its use in treatment of a wide variety of viral induced disorders or malignancies.

C. Colby, Jr. and D.W. Denney, Jr.; U.S. Patent 4,262,090; April 14, 1981; assigned to Cetus Corporation have developed new methods and compositions for producing significant amounts of mRNA which codes for mammalian interferon (IF mRNA). By employing recombinant DNA technology, the mRNA can be used to produce interferon, the interferon gene, and IF dsDNA (double stranded DNA) containing recombinant DNA. Prior to this, in view of the extremely small amounts of interferon that are produced by virus induced cells and as a concomitant, even smaller amounts of mRNA, it has not been feasible to employ recombinant DNA technology to produce interferon. By virtue of this process, cell cultures can be produced which provide useful amounts of mRNA, so as to allow for the first time the production of the interferon gene for incorporation into recombinant DNA and transformation of microorganisms.

The process is predicated upon the ability to mutate and isolate a mammalian cell which is semiconstitutive in its production of interferon. What this intends is that the cell continuously produces interferon at low level. In effect, the regulatory system which suppresses the production of interferon except when induced by viruses or dsRNA, has been at least partially inactivated so that the interferon gene is continuously transcribed to produce mRNA, which is then translated to interferon.

The mutant cell is hybridized with an inter- or intraspecific cell having a wild type IF operon for IF synthesis and regulation, and desirably one or more phenotypic markers to allow for selection of hybrid cells. Upon induction of interferon production by employing viruses or double-stranded RNA, it is found that the hybrid cell produces substantially greater amounts of IF mRNA and interferon than is normally produced by either the original mutant cell or the wild type cell. By harvesting mRNA which is enriched for IF mRNA, complementary DNA (cDNA) may be produced by conventional means, followed by the production of double-stranded DNA (dsDNA) from the cDNA. The dsDNA may then be used to form a recombinant DNA with a replicon compatible with a microorganism host to provide for production of the interferon gene, interferon, and recombinant DNA containing the IF gene. The following is a list of the involved steps:

(1) mutagenesis, selection and screening of a mammalian cell semiconstitutive for IF production (IF^{sc});

(2) hybridization of the mutant cell with a mammalian cell having a wild type IF gene and optionally one or more phenotypic properties for selection;

(3) inducing IF mRNA production with dsRNA, or with certain viruses, and isolation of mRNA;

(4) preparation of IF cDNA from IF mRNA and of IF dsDNA from IF cDNA;

(5) joining IF dsDNA with a replicon recognized by a microorganism host to form a functional recombinant DNA and transforming the host with the recombinant DNA; and

(6) cloning the transformed host to replicate the IF dsDNA and produce interferon and isolating the interferon as a concentrate.

Selection of appropriate mutated mammalian cells is perhaps the most difficult and time consuming part of the process. Cells having semiconstitutive interferon production are selected by their viral resistance.

By plating the cells on an appropriate nutrient medium and infecting the cells with viruses, surviving clones may then be screened for their interferon production ability.

A test has been developed to establish the semiconstitutive production of interferon and is described in detail in the patent, as is all the experimental detail of the production.

Amino Acid-Producing Microorganism Strains

It is an object of the process by *V.G. Debabov, J.I. Kozlov, N.I. Zhdanova, E.M. Khurges, N.K. Yankovsky, M.N. Rozinov, R.S. Shakulov, B.A. Rebentish, V.A. Livshits, M.M. Gusyatiner, S.V. Mashko, V.N. Moshentseva, L.F. Kozyreva, and R.A. Arsatiants; U.S. Patent 4,278,765; July 14, 1981* to use genetic engineering techniques to prepare strains which produce amino acids possessing enhanced capability of producing amino acids without additional growth factors.

This object is accomplished by a process in which a chromosome DNA fragment of a donor microorganism containing genes controlling the synthesis of a selected amino acid and having a mutation breaking the negative regulation of the synthesis of this amino acid is combined with a vector molecule of DNA with the formation of a hybrid DNA molecule. In so doing, use is made of a vector molecule of DNA capable of ensuring amplification of the hybrid DNA molecule. The resulting hybrid molecule of DNA is used for transforming cells of a recipient strain having a mutation blocking the synthesis of the selected amino acid in this strain and a mutation partly blocking the related step of metabolism of this amino acid to give a strain possessing increased productivity with regard to the selected amino acid.

To remove the ballast genetic material and to increase the stability of the hybrid plasmid, as well as to increase the number of its copies in a cell, it is advisable that the resulting hybrid molecule of DNA be treated, prior to transformation of cells of the recipient strain, with specific endonucleases to ensure its cleavage at definite sites, followed by joining the required DNA fragments with polynucleotide ligase.

In accordance with this process, the method for preparing a strain which produces an amino acid such as L-threonine starts with a fragment of DNA chromosome of a donor strain *E. coli* VNIIGenetika MG442 containing genes of threonine operon wherein enzyme products of the gene thrA are stable to inhibition with threonine as a result of mutation. The fragment is produced by means of endonuclease Hind III combined with a vector molecule of DNA (such as plasmid pBR322) to form a hybrid plasmid having a molecular weight of 11.4 megadaltons and consisting of two copies of the plasmid pBR322 and the chromosome DNA fragment of the donor strain. This ensures resistance of cells to penicillin and tetracycline and may be contained in cells in the stage of logarithmic growth in an amount of about 10 copies.

The resulting hybrid plasmid is used to transform cells of the recipient strain *E. coli* VL334 having mutations blocking the synthesis of L-threonine and L-isoleucine. This blocking is partial with respect to L-isoleucine and may be compensated by an increased content of threonine in a cell. These mutations ensure a selective advantage to the cells containing the hybrid plasmid over the cells which lost the plasmid during culturing. The strain produced is *E. coli* VNIIGenetika VL334 (pYN6) which produces L-threonine and is deposited in the Central Museum of Industrial Microorganisms of the All-Union Research Institute of Industrial Microorganisms identified by the registration number CMIM B-1649. The parent strains of VNIIGenetika MG442 and VNIIGenetika VL334 are also deposited in the aforesaid Central Museum and are identified by the registration numbers CMIM B-1628 and CMIM B-1641, respectively.

Another embodiment of the method for preparing a strain which produces the amino acid L-threonone resides in that a chromosome DNA fragment of the donor strain *E. coli* VNIIGenetika MG442 prepared by means of the endonuclease Hind III which contains genes of threonine operon with enzyme products of the gene thrA, becomes resistant, as a result of mutation, to inhibition by threonine, and is combined with a vector molecule of DNA. As the vector molecule, use is made of the plasmid pBR322 with the formation of a hybrid plasmid having MW of 11.4 megadaltons and containing 2 copies of the plasmid pBR322 and the chromosome DNA fragment of the donor strain.

The resulting hybrid plasmid is treated with specific endonucleases Hind III and Bam HI and the thus-produced fragments are joined by polynucleotide ligase. The resulting hybrid plasmid has a MW of 5.7 megadaltons and consists of one molecule of plasmid pBR322 and the chromosome DNA fragment of the donor strain. It ensures resistance of cells against penicillin and may be contained in cells in the stage of logarithmic growth in an amount of about 20 copies. This resulting hybrid plasmid is used to transform cells of the recipient strain *E. coli* VL 334 having mutations blocking the synthesis of L-threonine and L-isoleucine. The blocking of L-isoleucine is partial and may be compensated by an increased content of threonine in a cell. These mutations ensure a selective advantage to cells containing the hybrid plasmid during the process of cultivating, and the strain

E. coli VNIIGenetika VL334 (pYN7) is prepared which produces L-threonine. This strain is deposited in the Central Museum of Industrial Microorganisms of the All-Union Research Institute of Genetics and Selection of industrial microorganisms under the registration number CMIM B-1684.

A series of experiments on cloning genes of threonine operon in cells of *E. coli* K 12 have been carried out; the results of these experiments justify use of the process. The experiments have been performed in the following manner. The donor strain used is *E. coli* W3350 bearing threonine operon of the wild type; the plasmid pBR322 is used as the vector molecule of DNA. Using restrictional endonuclease Ecc RI for the preparation of DNA fragments, the hybrid plasmid pEKI is obtained. The subsequent treatment of the plasmid pEKI with endonuclease Hind III and polynucleotide ligase results in the creation of the plasmid pYNI which differs from the plasmid pEKI in that it does not carry the fragment of chromosome between two sites of endonuclease Hind III splitting. The plasmids pEKI and pYNI contain only two of the three genes of the threonine operon, i.e., genes thrA and thrB.

Transformation of the strain *E. coli* C600 with the resulting hybrid plasmid has not given any noticeable increase in the synthesis of threonine by the cells.

Using the same donor strain and plasmid pBR322 as the vector, but employing the specific endonuclease Hind III, a hybrid plasmid pYNII is obtained which contains all three genes of threonine operon.

Transformation of the recipient strain *E. coli* VL334 with the resulting hybrid plasmid caused an increased output of threonine: during fermentation in the culture liquid there accumulated up to 4 g/ℓ of the amino acid. In this manner it has been found that in order to increase productivity with respect to threonine, it is necessary to amplify all three genes of threonine operon.

This method for preparing strains producing amino acids makes it possible, by using genetic engineering techniques directed to increasing the dose of genes required for biosynthesis of the needed amino acid, to produce strains necessitating no additional growth factors and possessing an increased capability of producing the required amino acid.

Antibiotic-Producer Micromonospora Strains

It is widely known that antibiotic production is a genetically determined property. Consequently any change induced in the genetic material of a producer strain may alter its antibiotic-producing property, too. These stable genetic changes (mutations) occur also spontaneously, though very rarely. The incidence of mutations may be enhanced by known methods, i.e., UV, near-UV, gamma and x-ray irradiations as well as treatment with different mutagenic agents. These methods have been successfully applied at a variety of strains to enhance their antibiotic-producing capacity but the mutation induced is a random process, cannot be controlled, and not all survivors of the mutagenic treatment become mutants. So a large number of survivors has to be screened in order to obtain finally a better producer mutant.

It is known that several enzyme levels are raised by enhancing the gene dose, i.e., if the gene is doubled within a given chromosome, the level of the protein coded by it is doubled, too. Similarly, the synthesis of antibiotics may be increased by raising the level of enzymes responsible for antibiotic synthesis. It can be concluded that any process able to increase significantly the number of gene duplications may be of major importance in augmenting the productivity of antibiotic-producer strains.

For industrial application, a new species can be prepared from two Micromonospora strains producing antibiotics of related structure which has more advantageous properties both as regards cultivation or yield or which can produce a valuable antibiotic having a similar structure to the antibiotics formed by the parent strains. The process of protoplast fusion may be the method of choice to solve the above problems.

L. Alföldi, K. Bálint, C. Kari, I. Török, G. Szvoboda, T. Láng, I. Gádo and G. Ambrus; U.S. Patent 4,294,927; October 13, 1981; assigned to Gyogyszerkutato Intezet, Hungary have provided a process for the preparation of antibiotic-producer Micromonospora strains having modified genetic material. After cultivation in a glycine medium, a protoplast suspension is prepared with lysozyme, under osmotically buffered conditions ensured by sucrose, from each culture of the two genetically labelled mutants of the antibiotic-producer Micromonospora. The two suspensions obtained are combined in the presence of polyethylene glycol and incubated at room temperature. The resulting fused protoplasts are suspended in soft agar, then plated on an agar plate containing proline and inorganic salts, incubated for 20 to 30 days, and finally all mutants are selected from the regenerated colonies which are sure to have genetic material different from that of the parent strains.

To promote the lysis of the cell wall by means of lysozyme, glycine (0.2 to 0.5%) is preferably added to the culture medium. Due to the inhibitory effect exerted by glycine on cell wall synthesis, it is a common feature of these glycine-sensitized cultures that mycelial hyphae are both shorter and thicker compared to those grown in the absence of glycine, and in addition bulges appear on them. Only cultures showing similar microscopic pictures can yield protoplast suspensions with satisfactory efficacy.

These sensitized cultures are treated for 30 to 90 minutes with lysozyme (5 to 10 mg/ml) under osmotically buffered conditions (in a medium containing preferably 0.2 to 0.35 M sucrose).

Combining the individual protoplast suspensions prepared from various mutants at a ratio of 1:1, and treating the mixture with polyethylene glycol (MW 6000) the protoplasts undergo fusion. The suspension containing the fused protoplasts is incubated in a regenerating agar under osmotically buffered conditions ensured by sucrose, and proline as well as inorganic salts, preferably $CaCl_2$, $MgCl_2$ and KH_2PO_4, promoting regeneration, in a temperature range of 30° to 37°C. The protoplasts are regenerated within 20 to 30 days.

The genetic properties of each parent pair chosen should ensure the successful selection of recombinants formed in the course of fusion. For this purpose antibiotic resistance or various auxotrophs may be applied which have already been used with success in conventional microbial genetics.

In accordance with this procedure the case of the sisomicin-producer *Micromonospora inyoensis* ATCC 27600 a casamino acid dependent but rifampicin resistant, and a casamino acid nondependent but rifampicin sensitive (sensitive already to 0.5 μg/ml of rifampicin) parent pair was utilized. Following completed fusion mutants were looked for which are nondependent on casamino acid for growth but are rifampicin resistant, i.e., are forming colonies in minimal medium containing 10 μg/ml of rifampicin. Regeneration is carried out in nonselective medium ensuring optimal conditions for regeneration. In order to select recombinant phenotypes, the regenerated colonies are washed off, and the suspension obtained is plated onto a minimal medium containing 10 μg/ml of rifampicin. The colonies formed from individual cells having recombinant phenotypes are obtained as a result of fusion.

A selection process is used whereby the protoplasts of one of the parents are inactivated by heat treatment. The heat treatment should be effective enough to ensure loss of viability, at the same time should be mild enough to preserve the genetic material intact. In this case the complete genetic material of one partner is transferred into the heat-inactivated cytoplasm of the other partner. This procedure provides a means for selection when only one of the parents is marked genetically.

This principle was used for the gentamicin-producer *Micromonospora purpurea* var. nigrescens (MNG 00122) strain by applying a streptomycin-resistant-streptomycin-sensitive parent pair. For the selection of recombinants the protoplasts of the resistant parent were cautiously heat-inactivated (55°C) prior to fusion. As streptomycin sensitivity is a dominant feature over streptomycin resistance, 8 to 10 days have to elapse prior to the selection of recombinants (phenotype lag). After 8 to 10 days 10,000 μg/ml of streptomycin is layered on the regeneration plates. The genetic material of the colonies formed after 20 to 30 days is different from that of the parent pairs.

Applying heat inactivating techniques and utilizing the fact that *Micromonospora purpurea* var. nigrescens is more resistant to streptomycin by two orders of magnitude than *Micromonospora inyoensis,* the interspecific hybrid of the two strains is prepared by protoplast fusion. According to this procedure the protoplast suspension of streptomycin-resistant *Micromonospora purpurea* var. nigrescens is heat-inactivated and fused with the protoplast suspension of streptomycin-sensitive *Micromonospora inyoensis.* After 8 to 10 days agar containing 1,000 μg/ml of streptomycin is layered on the fused protoplasts to be regenerated. The colonies formed after 20 to 30 days are interspecific hybrids.

The above principle was also applied for parent pairs bearing other genetic marking. One parent is casamino-acid dependent while the other is nondependent. For the selection of recombinants the protoplasts of the casamino-acid dependent parent are cautiously heat-inactivated. The fused protoplasts are layered on minimal agar. The colonies formed after 20 to 30 days have recombinant properties.

It was found that the protoplast fusion process is suitable for the transfer of genetic information into other aminoglycoside-producing Micromonospora, too, at a fairly high frequency. In addition studies were carried out with the strains *Micromonospora olivoasterospora, Micromonospora echinospora* and *Micromonospora purpurea* (holotype).

Anucleated, Live *E. coli* Vaccine

G.G. Khachatourians; U.S. Patent 4,311,797; January 19, 1982; assigned to University of Saskatchewan, Canada describes the production of an *E. coli* strain which produces metabolically active, nonreproductive anucleated live K99+ *E. coli* cells and a vaccine made from this strain. This process further relates to a vaccine that is effective against enteropathogenic colibacillosis. Inability of these anucleated live K99+ *E. coli* cells (ALEC) to multiply in vitro and to infect animals is shown.

Strains of microorganisms such as *E. coli,* which produce metabolically active but nonreproductive, anucleated live cells (minicells) are known in the art. During the reproduction of such an *E. coli* strain the parental cells will generate two different sets of offspring. Instead of dividing in the middle of the cell to form equally-sized daughter cells which grow to form full-size *E. coli,* this mutant of *E. coli* divides very close to the pole of the cell, thereby giving one very small cell, called a minicell, and one very large cell. The minicells are live and metabolically active yet are unable to reproduce, because they do not have a nucleus. The large cells, on the other hand, contain a nucleus and are reproductive. A strain of *E. coli* which reproduces by dividing into a minicell and a regular cell will be called "minicell-producing *E. coli.*" For every 10^9 viable cells in a culture, at least 10^9 anucleated cells (minicells) are produced. These live anucleated and nonreproductive cells accumulate in the culture.

The method for making an *E. coli* strain, preferably *E. coli* GK 500 (ATCC 31563) which produces metabolically active, nonreproductive ALEC cells comprises the steps of:

(a) mixing an *E. coli* strain which produces metabolically active, nonreproductive, anucleated live cells (minicells) with a K99+ enteropathogenic *E. coli* strain;

(b) incubating the mixed cells in a culture medium, preferably at a cell density of approximately 2-3 x 10^8 cells/ml culture medium and preferably at 37°C for about 18 hours;

(c) separating minicell-producing *E. coli* cells which contain the K99 plasmid from other cells in the culture medium;

(d) growing the separated minicell-producing K99+ *E. coli* in a second culture medium, preferably on agar plates containing streptomycin and more preferably on agar plates containing 200 μg/ml streptomycin.

The process of making a vaccine which contains metabolically active, nonreproductive anucleated live K99+ *E. coli* cells and which is useful for preventing neonatal diarrhea in cattle comprises the steps of:

(a) transferring the K99 plasmid from an enteropathogenic *E. coli* strain into a minicell-producing *E. coli* strain;

(b) growing the resulting minicell-producing K99+ *E. coli* strain in broth culture medium under aerobic conditions with agitation, whereby it is preferred to select colonies of minicell-producing K99+ *E. coli* GK 500 and to grow them in a minimal salts based synthetic medium for 15 to 18 hours at 36° to 37.5°C, most preferably at 37°C;

(c) separating the bacterial cells from the broth culture medium by centrifugation and suspending them, preferably in a buffered salt solution;

(d) separating minicells from complete reproductive cells by sucrose gradient centrifugation.

The preferred sucrose gradient centrifugation method is the following: layering the suspension containing the bacterial cells on a first sucrose gradient, which preferably comprises a 5% sucrose solution layered over a 20% sucrose solution, and centrifuging; collecting and concentrating the top fractions of the first sucrose gradient; layering the concentrated fractions on a second sucrose gradient, which preferably comprises a linear sucrose gradient of 10 to 5% layered over a 10% sucrose solution, and centrifuging; and collecting, combining and concentrating the top fractions of the second sucrose gradient.

The combined concentrated top fractions of the second sucrose gradient may be further purified by freezing the concentrated fractions at -20°C, preferably for at least 2 to 3 days, more preferably for 9 to 10 days, followed by thawing slowly at 20° to 22°C and by centrifuging the thawed suspension and resuspending the pellet, preferably in a buffered salt solution.

The bacterial vaccine for inducing immunity by promoting the formation of antibodies against enteropathogenic K99+ *E. coli* in cattle is made up of an aqueous suspension containing metabolically active but nonreproductive anucleated live cells (minicells) of the ALEC strain preferably of *E. coli* of serotype K99+:K12+:0 101:H, and most preferably of *E. coli* GK500.

In a preferred embodiment the vaccine comprises an aqueous suspension which is stable for at least 3 months and which contains between 10^{12} and 10^{13} minicells/ml and less than one reproductive cell per 10^6 anucleated cells. Preferably the antigenicity of the vaccine is dependent on the presence of K99 antigen on the surface of the minicells.

Neonatal diarrhea in calves is prevented by administering the vaccine to the pregnant cow prior to parturition, preferably at least one week prior to parturition and more preferably about 6 weeks and again about 3 weeks prior to parturition.

L-Threonine Prepared by Using an *E. coli* Strain

L-threonine is known to be an essential amino acid extensively applicable as the component of diverse nutritive mixtures of medical use. In addition, L-threonine can be used as an additive to man's food and animal fodder, as well as a reagent for pharmaceutical and chemical industries. Various methods of preparing L-threonine by means of microbiological synthesis are known to the art. However, many of these processes require expensive additions to the culture medium and none of them has a very good yield.

V.G. Debabov, N.I. Zhdanova, A.K. Sokolov, V.A. Livshits, J.I. Kozlov, E.M. Khurges, N.K. Yankovsky, M.M. Gusyatiner, A.F. Sholin, V.P. Antipov and T.M. Pozdnyakova; U.S. Patent 4,321,325; March 23, 1982, therefore, have as an objective an increase in the rate of L-threonine accumulation in the culture medium in the course of fermentation of the L-threonine producer.

This and other objects are accomplished by a process for producing L-threonine by submerged cultivation of a microorganism on a nutrient medium containing carbon and nitrogen sources and mineral salts in the presence of antibiotics, followed by separating the biomass from the culture fluid and isolating the end product. The microorganism used is the *E. coli* strain VNIIGenetika M-1 deposited in the Central Museum of Commercial Microorganisms as No. IIMIIB.

This *E. coli* strain was prepared by introducing a hybrid plasmid into a mutant recipient strain, the cultivation being carried out in the presence of penicillin, resistance to which is accounted for by the above hybrid plasmid. It is expedient to introduce penicillin into the original nutrient medium in an amount of 0.1 to 0.5 g/ℓ, which makes it possible to use the culture medium enriched with some nutrients for the fermentation process, whereby the duration of the incipient (nonproductive) stage of fermentation is drastically reduced due to a large increase in the specific rate of growth of the culture within that period without the loss of any principal producer properties. In order to enrich the original nutrient medium use is made of an acid or enzymatic yeast hydrolyzate, or else yeast autolyzate, taken in an amount of 1 to 5 g/ℓ in terms of yeast dry weight; while in order to maintain optimum concentrations of carbon and nitrogen and the pH value in the fermentation medium throughout the fermentation period, a balance mixture is periodically added to the nutrient medium, containing an ammoniac solution and a carbon source.

After termination of the fermentation, the resultant biomass is separated from the culture medium and the end product, L-threonine, is isolated. The yield is excellent, varying in the examples given from 80 to 90% by weight.

K. Miwa, T. Tsuchida, O. Kurahashi, S. Nakamori, K. Sano and H. Momose; U.S. Patent 4,347,318; August 31, 1982; assigned to Ajinomoto Company, Incorporated, Japan have developed a process for producing L-threonine by fermentation, using genetic engineering techniques. An auxotrophic mutant of the genus Escherichia resistant to alpha-amino-beta-hydroxyvaleric acid (referred to hereafter as AHV) was chosen as the deoxyribonucleic acid (DNA) donor, then there was obtained from the mutant a DNA fragment possessing genetic information relating to L-threonine synthesis, and finally the DNA fragment was inserted into a plasmid obtained from *E. coli*. The recombinant plasmid was successfully introduced into a microorganism of Escherichia. Further, it was found that a strain having higher L-threonine productivity can be obtained when a microorganism which does not require L-threonine is used as the recipient of the recombinant plasmid. This microorganism produces L-threonine in a yield much higher than hitherto known L-threonine producing mutants.

The mutant of the genus Escherichia resistant to AHV is obtained by usual artificial mutation techniques. Chromosomal DNA is extracted from the mutant by the usual manner and treated with restriction endonuclease by the usual methods. Hind III has been used as the restriction endonuclease to obtain the DNA fragment possessing the genetic information relating to the L-threonine synthesis.

Plasmid extracted from *E. coli* is used as the vector DNA. Ordinary methods can be applied for inserting the DNA fragment, which is obtained from an AHV-resistant strain of Escherichia and possesses the genetic information related to L-threonine biosynthesis, into the vector. The recombinant plasmid thus obtained can be incorporated into the Escherichia microorganism by conventional transformation techniques.

As the recipient of the recombinant plasmid, an L-threonine requiring mutant of Escherichia is usually used, since such a mutant is convenient for selection and isolation of the transformant. When a mutant which does not require L-threonine for growth, especially that resistant to AHV, is used as the recipient, higher productivity is obtained. Transformants thus obtained can be selected and isolated by conventional methods based on the characteristics possessed by the vector DNA and/or the recipient.

The methods of culturing the L-threonine-producing strain thus obtained are conventional, except that the addition of 50 mg/dl of L-aspartic acid to the medium has been found to improve the yield of L-threonine.

Pseudomonas Strains Containing at Least Two Compatible Plasmids

The biodegradation of aromatic hydrocarbons such as phenol, cresols and salicylates has been studied rather extensively with emphasis on the biochemistry of these processes, notably enzyme characterization, nature of intermediates involved and the regulatory aspects of the enzymic actions. The genetic basis of such biodegradation, on the other hand, has not been as thoroughly studied because of the lack of suitable transducing phages and other genetic tools.

The work of Chakrabarty and Gunsalus [*Genetics,* 68, No. 1, page S10, (1971)] has showed that the genes governing the synthesis of the enzymes responsible for the degradation of camphor constitute a plasmid. Similarly, this work has shown the plasmid nature of the octane-degradative pathway. However, attempts by the authors to provide a microorganism with both CAM and OCT plasmids were unsuccessful, these plasmids being incompatible.

If the development of microorganisms containing multiple containing energy-generating plasmids specifying preselected degradative pathways could be made possible, the economic and environmental impact of such a development would be vast. For example, there would be immediate application for such versatile microbes in the production of proteins from hydrocarbons, in cleaning up oil spills, and in the disposal of used automotive lubricating oils.

A.M. Chakrabarty; U.S. Patent 4,259,444; March 31, 1981; assigned to General Electric Company describes unique microorganisms that have been developed by the application of genetic engineering techniques. These microorganisms contain at least two stable (compatible) energy-generating plasmids, these plasmids specifying separate degradative pathways. The techniques for preparing such multiplasmid strains from bacteria of the genus Pseudomonas are described in the patent.

Living cultures of two strains of Pseudomonas [*P. aeruginosa* (NRRL B-5472) and *P. putida* (NRRL B-5473)] are prepared. The *P. aeruginosa* NRRL B-5472 was derived from *Pseudomonas aeruginosa* strain 1c by the genetic transfer thereto, and containment therein, of camphor (CAM), octane (OCT), salicylate (SAL), and naphthalene (NPL) degradative pathways in the form of plasmids. The *P. putida* NRRL B-5473 was derived from *Pseudomonas putida* strain P_pG1 by genetic transfer thereto, and containment therein, of camphor, salicylate and naphthalene degradative pathways and drug resistance factor RP-1, all in the form of plasmids.

A transmissible plasmid has been found that specifies a degradative pathway for SAL, an aromatic hydrocarbon. In addition, a plasmid has been identified that specifies a degradative pathway for NPL, a polynuclear aromatic hydrocarbon. The NPL plasmid is also transmissible.

Having established the existence of (and transmissibility of) plasmid-borne capabilities for specifying separate degradative pathways for salicylate and naphthalene, unique single-cell microbes have been developed by the application of genetic engineering techniques containing various stable combinations of the CAM, OCT, SAL, and NPL plasmids. In addition, stable combinations in a single cell of these plasmids together with a nonenergy-generating plasmid (drug resistance factor RP-1) have been achieved. The versatility of these microorganisms has been demonstrated by the substantial extent to which degradation of such complex hydrocarbons as crude oil and Bunker C oil has been achieved thereby.

In brief, the process for preparing microbes containing multiple compatible energy-generating plasmids specifying separate degradative pathways is as follows:

(1) selecting the complex or mixture to be degraded;

(2) identifying the plurality of degradative pathways required in a single cell to degrade the several components of the complex or mixture therewith;

(3) isolating a strain of some given microorganism on one particular selective substrate identical or similar to one of the several components (the selection of the microorganism is generally on the basis of a demonstrated superior growth capability);

(4) determining whether the capability of the given strain to degrade the selective substrate is plasmid-borne;

(5) attempting to transfer this first degradative pathway by conjugation to other strains of the same organism (or to the same strain which has been cured of the pathway) and then verifying the transmissible nature of the plasmid;

(6) purifying the conjugatants (recipients of the plasmids by conjugation) and checking for distinctive characteristics of the recipient to insure that the recipient did, in fact, receive the degradative pathway;

(7) repeating the process so as to introduce a second plasmid to the conjugatants;

(8) rendering the first and second plasmids compatible, if necessary, by fusion of the plasmids; and

(9) repeating the process as outlined above until the full complement of degradative pathways desired in a single cell has been accomplished by plasmid transfer (and fusion, when required).

In the first reported instance (Chakrabarty et al article mentioned above) in which the attempt was made to locate more than one energy-generating degradative pathway in the same cell, it was found that CAM and OCT plasmids cannot exist stably under these conditions. In spite of the implication from these results that multiple energy-generating plasmid content in a single cell could be achieved but not maintained, it was decided to attempt to discover some way in which to overcome this problem of plasmid incompatibility. The problem of plasmid instability was solved by bringing about fusion of the plasmids in the recipient cell.

The development of single cell capability for the degradation and conversion of complex hydrocarbons was selected as the immediate beneficial application with particular emphasis on the genetic control of oil spills by the way of a single strain of Pseudomonas. In order to be able to cope with crude oil and Bunker C oil spills it was decided that the single cells of Pseudomonas derivative produced by this method should possess degradative pathways for linear aliphatic, cyclic aliphatic, aromatic and polynuclear aromatic hydrocarbons. *Pseudomonas aeruginosa* (NRRL B-5472) strain, which displays these degradative capabilities was thereupon eventually developed.

By establishing that SAL and NPL degradative pathways are specified by genes borne by transmissible plasmids in Pseudomonas and by the discovery that plasmids can be rendered stable (e.g., CAM and OCT) by fusion of the plasmids, it has been made possible, for the first time, to genetically engineer a strain of Pseudomonas having the single cell capability for multiple separate degradative pathways. Such a strain of microbes equipped to simultaneously degrade several components of crude oil can degrade an oil spill much more quickly (days) than a mixed culture meanwhile bringing about coalescence of the remaining portion into large drops. This action quickly removes the opportunity for spreading of the oil thereby enhancing recovery of the coalesced residue.

L-Lysine

Hitherto, in order to render a wild strain capable of producing L-lysine from carbohydrates, it has been necessary to induce artificial mutants from the wild strain. There are many known lysine-producing artificial mutants. Most of the known lysine-producing mutants are resistant to lysine-analogs such as S-(2-aminoethyl)-cysteine (AEC), and/or require homoserine for growth, and belong to the genus Brevibacterium or Corynebacterium. These microorganisms produce L-lysine in a yield of from 40 to 50%.

It has however, become difficult to increase the yields of L-lysine using the artificial mutation techniques. A need therefore continues to exist for the development of new microorganisms capable of producing L-lysine in high yields.

K. Sano and T. Tsuchida; U.S. Patent 4,346,170; August 24, 1982; assigned to Ajinomoto Company, Incorporated, Japan have succeeded in obtaining an L-lysine-producing microorganism of the genus Escherichia, which produces L-lysine in a yield higher than artificially induced mutants of Escherichia.

This microorganism may be used in a method for producing L-lysine by fermentation, which comprises: culturing in a culture medium an L-lysine-producing microorganism constructed by incorporating a hybrid plasmid in a recipient of the genus Escherichia and recovering the L-lysine accumulated in the culture medium, the hybrid plasmid containing a deoxyribonucleic acid fragment possessing genetic information related to L-lysine production and obtained from a microorganism of the genus Escherichia resistant to a lysine-analog.

The DNA-donor strain used to construct this L-lysine producer is a microorganism of the genus Escherichia possessing genetic information related to L-lysine production. Strains having higher productivity of L-lysine are preferably used as the DNA-donor. The mutant resistant to the lysine-analog used as the DNA-donor can be obtained by conventional mutation techniques.

The lysine-analogs are those which inhibit the growth of Escherichia strains, but the inhibition is suppressed partially or completely when L-lysine coexists in the medium. Examples of lysine-analogs are oxo-lysine, lysine-hydroxamate, AEC, γ-methyl-lysine, and β-chlorocaprolactam.

Chromosomal DNA is extracted from the DNA-donor in a well-known manner and treated with a restriction endonuclease by a well-known method. The plasmid or phage DNA used as the vector in the synthesis procedure is also treated with a restriction endonuclease in an analogous manner. Various kinds of restriction endonucleases can be used, if the digestion of the chromosomal DNA is done partially. Thereafter, the digested chromosomal DNA and vector DNA are subjected to a ligation reaction.

Recombination of DNA to prepare the recombinant plasmid can be carried out by incorporating with terminal transferase deoxyadenylic acid and thymidylic acid, or deoxyguanylic acid and deoxycytidylic acid into the chromosomal DNA fragment and cleaved vector DNA, and by subjecting the modified chromosomal DNA fragment and cleaved DNA to an annealing reaction. As a suitable vector DNA, a conventional vector can be employed such as Co1 E1, pSC 101, pBR 322, pACYC 177, pCR 1, R6K, or λ-phage, or their derivatives.

The hybrid DNA thus obtained can be incorporated into a microorganism of the genus Escherichia by conventional transformation techniques. The desired transformant is screened using a medium on which only a clone, having one or both of the characteristics of L-lysine productivity possessed by the chromosomal DNA fragment and those possessed by vector DNA, can grow.

As the recipient microorganism for the hybrid DNA, an L-lysine-auxotroph is usually used, since it is conventional to distinguish the lysine-producing transformant from the recipient. Desirably, a mutant already having higher productivity of L-lysine is used as the recipient, to obtain better results.

The methods of culturing the L-lysine producing strains thus obtained are conventional, as is the recovery of the accumulated L-lysine. By this method, L-lysine can be produced in higher yields than has been achieved in previously known methods using artificial mutants of Escherichia.

β-Endorphin

The process of *J.D. Baxter, I. Fettes and J. Shine; U.S. Patent 4,350,764; September 21, 1982; assigned to The Regents of the University of California* is believed to provide the first instance of the synthesis of a mammalian hormone by a microorganism transformed by a coding sequence comprising naturally occurring mammalian codons, wherein the biological activity of the product was demonstrated. This process is exemplified by the bacterial synthesis of mouse β-endorphin. The mouse endorphin differs from human endorphin merely by two amino acids and the mouse endorphin is thought to be biologically active in humans.

DNA comprising the naturally occurring nucleotide sequence coding for amino acids 44-90 of β-lipotropin and including the entire coding region for β-endorphin with the exception of the C-terminal glutamine was modified, transferred to an expression transfer vector, and expressed as a fusion protein. The fusion protein was further modified in vitro to yield mature β-endorphin. β-endorphin was purified from a bacterial lysate. The structure and biological activity of the resulting product was proven by immunological assay, and by two independent assays designed to demonstrate biological activity.

COMPANY INDEX

INVENTOR INDEX

U.S. PATENT NUMBER INDEX

NOTICE